ELECTRICAL ENGINEERING DEVELOPMENTS

THE CONTROL OF THE PUMPING COMPLEX ELECTRIC DRIVE IN NON-STEADY OPERATION STATES

Electrical Engineering Developments

Additional books and e-books in this series can be found on Nova's website under the Series tab.

ELECTRICAL ENGINEERING DEVELOPMENTS

THE CONTROL OF THE PUMPING COMPLEX ELECTRIC DRIVE IN NON-STEADY OPERATION STATES

MYKHAYLO V. ZAGIRNYAK
TETYANA V. KORENKOVA
OLEKSANDR O. SERDIUK
OLEKSII M. KRAVETS
AND
VIKTORIYA G. KOVALCHUK

nova
science publishers
New York

Library of Congress Cataloging-in-Publication Data

ISBN: 978-1-53615-017-9

Published by Nova Science Publishers, Inc. † New York

CONTENTS

PREFACE

Modern pumping complexes (PC) represent complicated power-consuming technological systems including one or several pumping units (PU) operating in parallel for the common manifold, a network of pipelines with the installed stop-control and safety valves. During PC function, it is necessary to change PU operating mode in accordance with the current water consumption. The most efficient method of PC parameter regulation consists in the variation of the rotation frequency of the pump impeller by means of a variable-frequency electric drive (ED). It provides the most accurate coincidence with the required schedule of water consumption with minimal power losses. This approach makes it possible to save up to 30-40% of the energy consumed by PC and up to 10-20% of the pumped liquid.

When PC operation modes change, complex hydrodynamic processes occur in the pipeline; they are accompanied by increased dynamic loads in the form of surges, oscillations and pressure difference, by the vibration of the technological equipment, etc. Such processes result in the time-dependent change of the pressure, flow velocity, hydraulic resistance and power. The above said is caused by the following: the emergency shutdown of the PU power supply; the actuation of the reverse valves, fast shutting/opening the safe or stop-control valves; periodic stops, repeated starts as well as the commutation switching the pumps, hydraulic consumers, etc. PC operation in such modes results in the failure of the pumping equipment, reduction of the operation resource of the pipeline valves, breakage of the pipeline network.

The power efficiency of PC functioning can be improved due to the use of the systems of the pump electric drive control or stop-control valves in the non-steady operation states.

To decrease the dynamic loads in PC pipeline network it is proposed to generate an irregular rate of the pipeline valves control based on of the variable-frequency electric drive of the stop-control stopcock. The optimal law of the stopcock variable-frequency electric drive control is obtained. It enables the alteration of the pressure in the hydrosystem within the admissible limits and operative control of the pipeline valves under the emergency operation modes.

The principles of the creation of the pipeline valves variable-frequency ED systems via the use of the reverse relations of the pressure and discharge in the hydronetwork, the degree of the stopcock opening, ED power backup from a regular supply or an active flow energy damper are further developed. It allowed, unlike the conventional systems, eliminating the inadmissible increase of the pressure in the pipeline and the efficient use of the energy of the reverse flow of the liquid.

A method for the determination of the limits of the non-cavitation operation of PC with a variable-frequency electric drive is proposed. It is based on the analysis of the mutual operation modes of the pipeline network pumping unit and the time-variable water consumption. It makes it possible to determine the range of the admissible values of the variation of the pump electric drive rotation frequency, which correspond to the non-cavitation mode of PC operation when PU discharge is regulated within the required limits.

To minimize the power losses, caused by the cavitation in the pipeline, an electromechanical device is designed. It allows the takeoff of the gas-vapor mixture discharged from the liquid with its simultaneous recuperation into electricity.

A PC power model in the form of the equations of the balance of the hydraulic power harmonic components of the power supply and the hydrosystem elements at the periodic change of the process variables is proposed. Such an approach is the basis for the solution to the problems of the determination of the parameters of PC whose flow sheets take into account the group operation of PU with a variable-frequency and an unregulated electric drive, occurrence of periodic processes both in turbomechanisms and at the sections of the pipeline network. The described method makes it possible to single out the power variable component caused by the nonlinear periodic processes in the hydronetwork, to obtain the frequency characteristic of the power processes in the hydrosystem.

The presented monograph contains the theoretical and experimental research enabling the solution to the topical scientific problem of the improvement of the energy efficiency and reliability of PC functioning. We used the fundamental postulates of the classical theory of electrical engineering, hydrodynamics, electric drive, the methods of the finite elements, approximation and optimization, the theory of Fourier series, the mathematical and experimental research.

The monograph has been written in Kremenchuk Mykhailo Ostrohradskyi National University. The authors are grateful to Professors A. V. Maliar, N. Ya. Ostroverkhov and V. P. Rozen for reviewing and support of the monograph, master R.M. Manko for assisting in the editing of some chapters of the monograph. In addition, the authors are grateful to a senior teacher of Kremenchuk Mykhailo Ostrohradskyi National University K. V. Kovalenko for the assistance in the preparation of the monograph in the English language.

ABBREVIATIONS

ACS	automatic control system
ADE	active dissipator of energy
AT	air turbine
C	consumer
CD	coupling device
CoD	control device
CS	control system
CV	check valve
DABA	device for automation bleeding of the air
ED	electric drive
EDCCP	electromechanical device for the control of cavitation processes
EG	electric generator
EM	electric machine
ERW	emergency recovery work
ESDDL	electromechanical system of the decrease of dynamic loads
FC	frequency converter
IM	induction motor
LC	logic controller
P	pump
PC	pumping complex
PP	pumping plant
PS	pressure sensor
PS	pumping station
PU	pumping unit
R	receiver
S	stopcock
SW	software

SYMBOLS

A_{eq}	the averaged norm of the depreciation for the electrical engineering equipment
b	the width of the stopcock sealing ring
c	the velocity of sound in the liquid at rest
C_{ef}	the economic effect of the introduction of ACS of PC ED
C_{ef1}	the economic effect of the reduction of the non-productive power losses caused by the presence of cavitation processes
C_{ef2}	the economic effect of the recuperation of the steam-gas mixture energy
C_{ef3}	the economic effect of the reduction of the equipment wear and the decrease of PC breakdown rate
C_{ERW}	ERW cost
C_{PL}	labor cost
C_M	material cost
C_{TM}	transport and mechanisms cost
C_{LW}	the cost of the water loss
C_{LC}	the cost of the liquidation of the possible breakdowns of the pipeline network
C_{1a}	the cost of the liquidation of one breakdown of the pipeline
d	pipe diameter
d_{in}	the internal diameter of the stopcock sealing ring
d_{out}	the external diameter of the stopcock sealing ring

d_{sp}	the diameter of the stopcock spindle
E	the reduced modulus of elasticity of the pipeline
E'	Young's modulus for the water-air mixture
E''	the modulus of elasticity of the pipe material
f_b	the coefficient of friction in the stopcock bearings
f_c	the coefficient of friction on the stopcock sealing surface
f_{sb}	the coefficient of friction in the stopcock gaskets
f_{kav}	the frequency of cavitation oscillations
f_0	the frequency of the voltage supplied to IM stator winding
F_1	the friction in the stopcock body sealing surface on the side of the medium input
F_2	the friction in the stopcock body sealing surface on the side of the medium output
G	the gravitation acting on the stopcock wedge
g	acceleration of gravity
H	head
H_{st}	static head
H_{kr}	the critical head at the pipeline section, corresponding to beginning of the development of cavitation processes
H_{cn}	the rated head in the consumer network
H_{pl}	the head in the pipeline
H_p	the head at the pump output
H_0	the pump head at zero discharge
H_{v0}	the head at the stopcock before the start of shutting
H_c	the head in the consumer network
H_i	the head at the pipeline section output
H_{i-1}	the head at the pipeline section input
$H_{zad}(t)$	the set value of the head
$H_{tek}(t)$	the current value of the head
ΔH	head losses
ΔH_{dir}	the increase of the head at the direct surge

Δh cavitation reserve

$\Delta h(t)$ head discrepancy

Δh_{kav} the head losses caused by the presence of cavitation processes in the hydronetwork

Δh_{fr} the friction head losses in the pipeline

h_{sb} the height of the stopcock gasket

I quality criterion

$\vec{i_1}$ the vector of IM stator winding current

$\vec{i_2}$ the vector of IM rotor winding current

I_m the values of the quality criterion at the current step

I_{m-1} the values of the quality criterion at the previous step

I_m the values of the quality criterion at the current step

I_{m-1} the values of the quality criterion at the previous step

j the shock parameter of the pipeline

J_Σ the total inertia moment of the stopcock ED

K_h the weight coefficient by the head

K_n the weight coefficient by the power

K_g the weight coefficient by the power of recuperation

k_a the volume share of the unsolved air in the liquid

k_e, k_u proportionality coefficients

k_t the weight coefficient at the time component in the quality criterion

k_M the coefficient of the overloading ability of the electric motor by the moment

k_r reserve coefficient

l pipe length

L_1 the inductance of IM stator winding

L_2 the inductance of IM rotor winding

L_μ the mutual inductance of IM stator and rotor windings

$L_{1\sigma}$ the leakage inductance of IM stator winding

$L_{2\sigma}$ the leakage inductance of IM rotor winding

m liquid mass

m_w	the mass of the stopcock wedge
M	the electromagnetic moment of IM
M_1	the moment of friction occurring in the stopcock spindle thread
M_2	the moment of friction in the stopcock bush bearing
M_s	the moment of resistance created by the pump
M_{max}	the maximal allowable moment on the stopcock shaft
M_{tyr}	the torque developed by the turbine
M_v	the moment of the stopcock resistance
n	the coefficient of the intensity of the control of the pipeline valves
N	consumed power
N_L	the number of the sections
N_1	the reaction of the stopcock body sealing surface on the side of the medium input
N_2	the reaction of the stopcock body sealing surface on the side of the medium output
N_y	the number of breakdowns of the pipeline network during a year
N_{kav}	the hydraulic power of the steam-gas mixture
$N_{NK}(t)$	the current value of the power consumed by PC without cavitation processes in the pipeline
$N_{kav}(t)$	hydraulic power losses caused by the presence of cavitation processes in the pipeline
$N_{g\,max}(t)$	the maximal value of the recuperated power
$N_g(t)$	the current value of the recuperated power
$\Delta N_g(t)$	the discrepancy by the power of recuperation of the steam-gas mixture
P	the force of the liquid hydrostatic pressure
p	the absolute pressure in the pipeline
p_p	the number of IM pole pairs
p_{kav}	the pressure in the cavern
p_{pot}	the pressure in the liquid flow
p_{para}	the pressure of saturated steam

q_{im}	the specific load providing the leak-tightness of the stopcock sealing surface
Q	discharge
Q_c	the discharge in the consumer network
Q_{cn}	the rated discharge in the consumer network
Q_i	the discharge at the pipeline section output
Q_{i-1}	the discharge at the pipeline section input
Q_{tyr}	the discharge of the air turbine
Q_{kav}	the discharge of the steam-gas mixture
Q_{Σ}	gross axial thrust on the stopcock spindle
Q_{sb}	the axial effort component for overcoming the friction in the stuffing boxes
Q_w	the axial effort component for overcoming the forces acting on the wedge
Q_0	the axial thrust component for overcoming the internal pressure on the stopcock spindle butt
r_g	the gear reduction rate
r_{mc}	the average radius of the stopcock spindle thread
R	bubble radius
R_1	the resistive impedance of IM stator
R_2'	the reduced resistive impedance of IM stator
R_p	the internal resistance of the pump
R_{cn}	the rated hydraulic resistance of the consumer
R_{cb}	the radius to the center of the stopcock bearing balls
R_s	hydraulic network resistance
R_{VN}	the internal resistance of the pump
R_{pl}	the hydrodynamic resistance of the pipeline network
R_{kav}	the hydrodynamic resistance corresponding to the presence of cavitation processes
S	the cross section area of the pipeline
s	IM slid
S_0	pipeline resistivity

S_p	the wedge surface area subjected to the liquid pressure
t	the current time
t_{sh}	the time of the complete shutting of the pipeline valves
T	liquid temperature
T_{u}	the inertial time constant of the liquid column
T_e	the capacitive time constant of the water conduit
T_k	the inertial time constant of the cavitation cavern
T_{N}	the time constant of the pump
U_1	the current voltage supplied to IM stator winding
U_{1m}	the amplitude of the voltage supplied to IM stator winding
V_{kr}	the critical volume of the cavitation cavern
V_{res}	receiver volume
V_{kav}	the volume of the cavitation cavern
x	the coordinate along the pipeline axis
z	geodesic height
α	the relative current frequency of IM
α_w	the angle of the stopcock wedge
α_1	the angle of the lead of the stopcock spindle thread
β	the degree of opening of the regulating valve
γ	the specific weight of the liquid
γ_{IM}	the relative voltage of IM
δ	the thickness of the pipeline walls
η	efficiency
η_r	the efficiency of the gear
η_{ACE}	the efficiency of the hydraulic flow ADE
η_G	the efficiency of the electric generator
λ	friction hydraulic coefficient
λ_{pl}	the coefficient taking into account the friction along the pipeline length
ν_p	the relative rotation frequency of the pump
ν_v	the weight coefficient at the head component in the quality criterion

ν_{kr}	the relative critical rotation frequency of the pump ED, providing the non-cavitation operation of the hydrosystem
ξ_{pl}	the coefficient of the hydraulic resistance of the pipeline
ξ_{v}	the coefficient of the stopcock hydraulic resistance
ρ	liquid density
ρ'	the density of the liquid-air mixture
σ	surface tension coefficient
σ_{12}	the overall coefficient of IM dissipation
σ_1	the coefficient of dissipation of IM stator winding
σ_2	the coefficient of dissipation of IM rotor winding
υ	the velocity of liquid motion in the pipeline
υ_{kr}	the flow velocity corresponding to beginning of the cavitation process
φ	the angle of friction in the stopcock spindle thread
φ_{sp}	the current rotation angle of the stopcock spindle
χ	cavitation number
χ_{kr}	the critical number of cavitation
$\overrightarrow{\psi_1}$	the vector of the flux linkage of IM stator
$\overrightarrow{\psi_2}$	the vector of the flux linkage of IM rotor
Ψ_{pl}	the coefficient taking into account the deformability of the pipeline walls
ω	the angular rotation frequency of IM rotor
ω_{IM}	the current rotation frequency of IM
ω_0	the synchronous rotation frequency of IM
ω_{2e}	the angular slip frequency of IM
ω_p	the current rotation frequency of the pump
ω_{pn}	the rated rotation frequency of the pump
ω_{sp}	the current rotation frequency of the stopcock spindle

Chapter 1

THE CHARACTERISTIC OF THE NON-STEADY STATES OF THE PUMPING COMPLEX OPERATION

1.1. THE STRUCTURE AND CHARACTERISTIC OF PUMPING COMPLEX EQUIPMENT

The systems of public and industrial water and heat supply are among the biggest electricity users requiring up to 80% of the total energy consumption of the enterprise.

Pumping complexes (PC) represent a complicated electrohydraulic system including pumping plants (PP) with various patterns of pump units (PU) connection, a branched hydrodynamic network characterized by the presence of backpressure and hydraulic resistance [1–6]. Figure 1.1 shows a PC typical flowsheet.

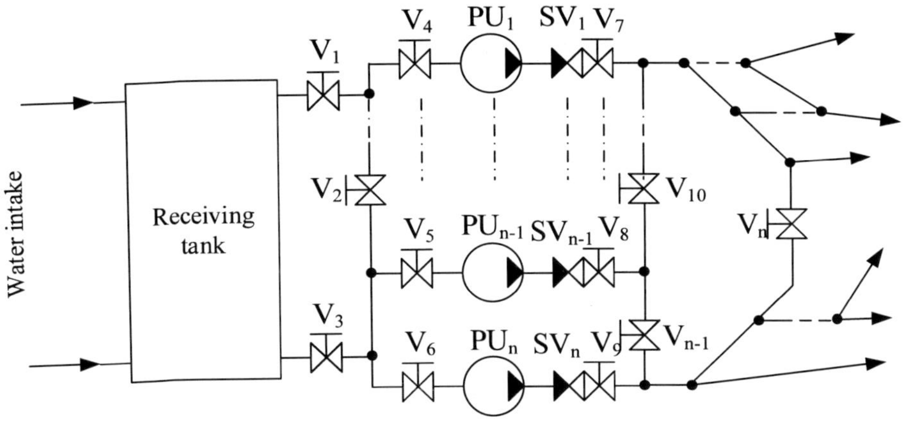

Figure 1.1. The structure of a pumping complex: $V_1 - V_n$ –shut-off and control valves; $PU_1 - PU_n$ – pump units; $SV_1 - SV_n$ – safety valves (check valves).

PC operation mode depends on many factors: the type and the published data of the used turbo-mechanism, PU connection pattern, the pipeline configuration and profile, the presence of shut-off and safety valves, the consumer's operation mode, the method of PP parameters regulation, etc. [1, 3–9]. The consumer's operation diagram providing an irregular time variation of the discharge in the system of municipal water supply confirms the above said (Figure 1.2).

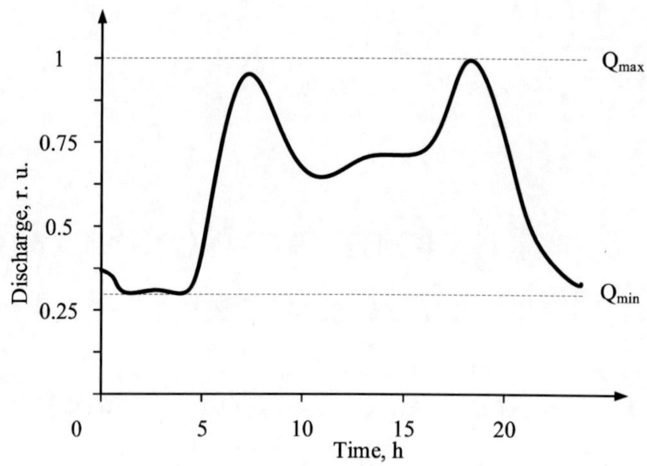

Figure 1.2. A typical 24-hour water consumption diagram.

In most water and heat supply pumping complexes designers use centrifugal pumps characterized by a number of parameters: discharge (Q, m^3/s), head (H, m), power consumption (N, kW), efficiency (η) and positive suction head (Δh_{dop}) [1, 3–9]. Dependences H–Q, N–Q, η–Q, $\Delta h_{dop–Q}$ refer to the pump published data and are cited for the rated rotation frequency and different diameters of the pump impeller.

The analysis of the pump published data [1–6] enabled revealing three basic types of the head-discharge characteristics: gently sloping characteristics (Figure 1.3, curve 1), steeply falling ones (Figure 1.3, curve 2) and characteristics with explicit maximum (Figure 1.3, curve 3).

Pumps with gently sloping characteristics and those having explicit maximum are used in PCs requiring a wide range of pump discharge control at insignificant variation of pressure. In pumps with explicit maximum, the characteristic part situated on the right of the maximum point is the operation zone. Pumps with steeply falling characteristics are used in PPs requiring pressure wide regulation at the insignificant change of discharge.

Centrifugal pumps have a growing power characteristic (Figure 1.3, curve 4), axial pumps have a decreasing characteristic (Figure 1.3, curve 5).

Curve η–Q characterizes the area of rational use of the turbo-mechanism at constant rotation frequency that is in the range of 15% of maximum value η_{max} (Figure 1.3, curves 6, 7).

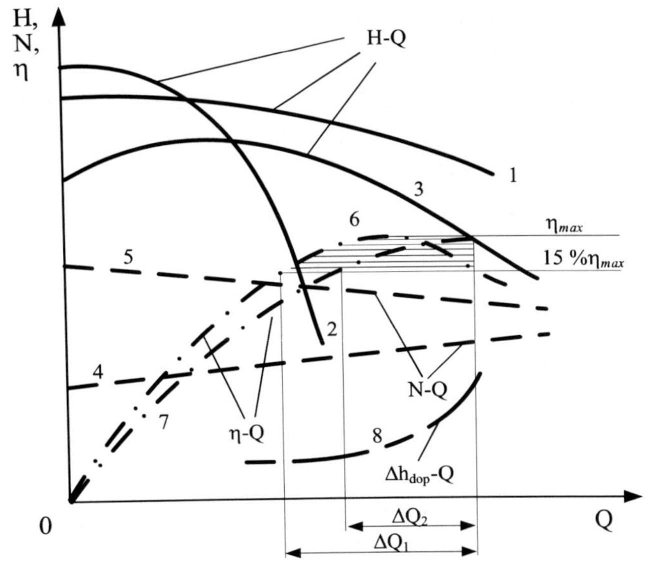

Figure 1.3. Pump head-discharge and power characteristics.

Curve Δh_{dop}–Q represents the pump intake pipe head loss dependence on discharge for the non-cavitation mode of the turbo-mechanism operation.

In practice, PP comprises several turbo-mechanisms operating simultaneously for a common pipeline. Sequential switching of pumps provides an increased head. In this case, the total head equals the sum of the heads created by operating units (Figure 1.4, a) at the same discharge value [1, 4–9].

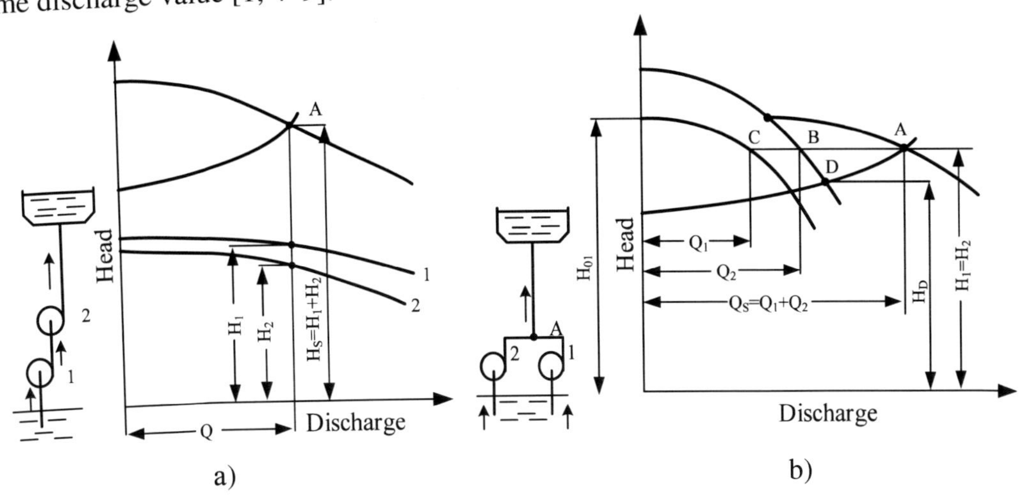

Figure 1.4. The diagrams of pump unit – pipeline network switching: a) sequential switching; b) parallel switching.

Pumps parallel switching (Figure 1.4, b) is usually used to increase discharge. To obtain total pump characteristic H–Q the values of the supply of the units operating at the same value of the head are added.

Alternating current induction and synchronous electric motors are commonly used as pump electric drives due to the compact design, simple connection to the turbo-mechanism, easy automation and relatively low operation cost [1, 2, 5, 6]. It should be noted that systems with pump variable-frequency electric drives (ED) have been widely used in practice of PP operation recently. They make it possible to change PP operation parameters gradually without unproductive loss of electricity with wide possibilities of the improvement of the flowsheet accuracy and efficiency. They allow increasing the durability of pipelines and equipment due to the reduction of the static and dynamic loads caused by high pulsations of pressure; reducing PC dimensions due to PUs enlargement and decreasing their number.

The hydrodynamic network of water and heat supply PC represents a complicated pipeline system, consisting of the pipeline parallel and sequential sections, each of which is characterized by its own diameter, length, geodesic position, presence of shut-off and control valves, etc. [1, 4–10].

The head-discharge characteristic of the hydraulic network is of a parabolic form (Figure 1.5) [1, 4-10] and is described by dependence:

$$H = H_{st} + R_s Q^2, \quad (1.1)$$

where H_{st} – static head caused by the difference between the geodesic marks of liquid discharge and intake, m; R_s – hydraulic resistance in the pipeline network, s^2/m^5.

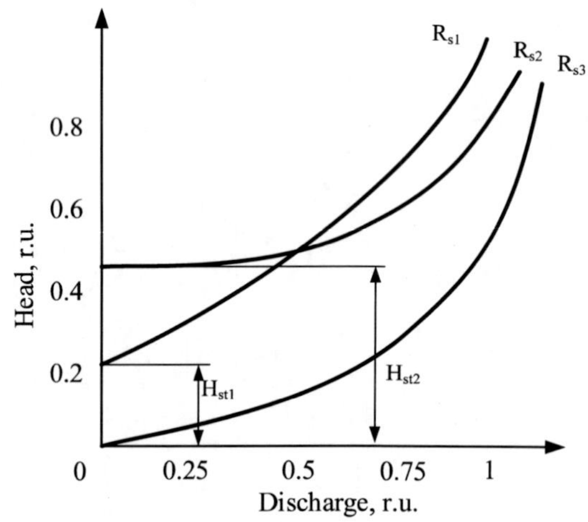

Figure 1.5. Head-discharge characteristics of the pipeline network.

The analysis revealed that in the systems of water and heat supply the static head makes (20-40)% of the PU rated head, which is caused by the high-rise of buildings and the geodesic position of the pipeline profile.

The second component in expression (1.1) characterizes the head losses in the pipeline length and at the local resistances [1–3, 10-13]. Knowing dynamic losses in the pipeline one can determine hydraulic resistance R_s, depending on the configuration, material and geometric dimensions of the pipeline, liquid motion mode, temperature, density and phases.

To determine the length head losses we use the Darcy-Weisbach equation, m [10–13]:

$$h_l = \lambda \frac{l}{d} \frac{\upsilon^2}{2g}, \tag{1.2}$$

where l – the pipe length, m; d – the pipe diameter, m; λ – hydraulic friction factor; $g = 9.81$ m/s^2 – acceleration of gravity; υ – liquid velocity in the pipe $\upsilon = Q/S = 4Q/\pi d^2$, m/s.

The analysis [1–3, 10–13] revealed that head losses along the length at turbulent motion, taking place in water and heat supply PC, (Figure A.7), depend on hydraulic friction factor λ:

$$\lambda = \frac{0.3164}{\mathrm{Re}^{0.25}}, \tag{1.3}$$

where $\mathrm{Re} = \dfrac{\upsilon d}{\nu}$ – the Reynolds number; $\nu = \dfrac{177.5 \cdot 10^{-8}}{1 + 0.0337t + 0.000221t^2}$ – kinematic viscosity (for water at $t = 20^0$C, $\nu = 101$ m^2/s).

Head losses at local resistances are caused by the sharp change of the flow area configuration, the change of the velocity mode, the flow transversal circulation, flows joint and division, the presence of shut-off and control valves. In the general case these head losses are determined by the Weisbach equation, m [10–13]:

$$h_m = \zeta \frac{\upsilon^2}{2g}, \tag{1.4}$$

where ζ – the local resistance coefficient dependent on the type of local resistance and the Reynolds number.

Table A.2 contains dependences for the determination of the head losses at the basic types of local resistances occurring in the practice of operation of water and heat supply PCs.

A real pipeline network represents a totality of the pipeline sections connected in series and in parallel; they can be united in a stub (branched) or a ring network (Figure 1.6).

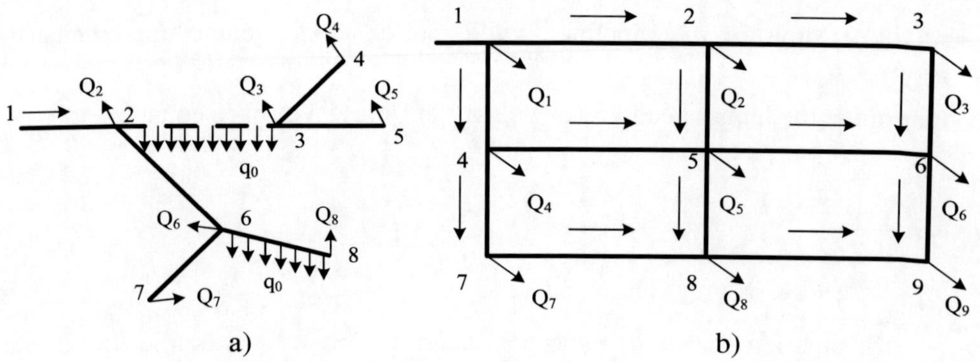

Figure 1.6. The pipeline networks configuration: stub (a); ring (b).

For the calculation of pipelines having a number of i-th sequential sections with different parameters the head-discharge characteristic of the pipeline is graphically created by adding, the ordinates (heads) at the same supply (Figure 1.7).

The head losses in such pipelines, m:

$$\Delta H = \sum_{i=1}^{n} h_i = \sum_{i=1}^{n} h_l + \sum_{i=1}^{n} h_m , \tag{1.5}$$

where $\sum_{i=1}^{n} h_l$ – total head losses along the length, m; $\sum_{i=1}^{n} h_m$ – total head losses at local resistances, m; n – the number of local resistance sections.

For the graphic creation of the head-discharge characteristics of the pipeline having parallel branches with common node points in their end and beginning one adds the abscissas (discharges) of H–Q characteristics at the same ordinates (heads) (Figure 1.7).

The head losses are analytically described by the expression of the following form, m:

$$\Delta H = \sum_{i=1}^{n} h_1 = \sum_{i=1}^{n} h_2 = \ldots = \sum_{i=1}^{n} h_n . \tag{1.6}$$

Taking into account the above said, the pipeline hydraulic resistance, s^2/m^5:

$$R_s = \frac{\Delta H}{Q^2} \text{ or } R_s = \sum S_0 l + \sum R_m , \tag{1.7}$$

where $S_0 = \dfrac{16\lambda}{g\pi^2 d^5}$ – the pipeline resistivity, s²/m⁶; $\sum R_m = \dfrac{0.81}{g}\sum \dfrac{\zeta}{d^4}$ – the sum of local resistances, s²/m⁵; l – the pipeline length, m.

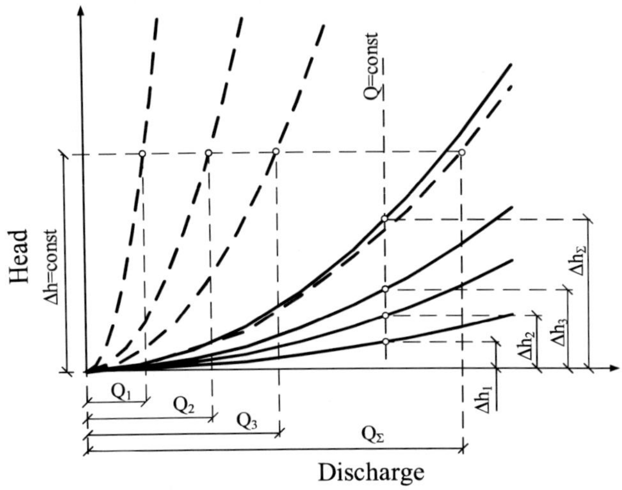

Figure 1.7. The pipeline head-discharge characteristics at sequential (—) and parallel (- - -) connection of the sections.

Stopcocks, back-pressure valves, check valves, etc. are used in PC as pipeline valves providing protective, preventive and control functions in pumping stations (PS). Valves control influences the character of transient processes in the hydraulic system both in usual modes (adjustment, start) and in emergency ones. The stopcocks, back-pressure valves, etc. are often shut (opened) without observing the required tempo and duration of control, which results in the occurrence of cavitation processes and surges in the communication network, increased vibrations of the walls and the flow path of the pump unit and the pipeline valves, oscillations of the moment on the pump shaft, etc.

1.2. The Dynamic Loads in the Working and Emergency Modes of the Pumping Complex Operation

When PC operation modes change, there appear stationary (operation) transient processes – PU start, stop, the start of adjacent pumps, supply or head regulation, etc. and non-stationary (emergency) transient processes. The latter are caused by sudden switch-off of all or a group of simultaneously operating pumps because of power supply outage; by the switch-off of one of the simultaneously operating pumps before shutting the stopcock on its head line; by the start of a pump with an open stopcock on a head line equipped with a check valve; by a mechanized shutting of the stopcock at the switch-off of the whole water conduit or its separate sections; by shutting or opening of quick-operating valves.

The analysis [2, 6, 9] revealed that the listed modes are characterized by high breakdown rate, low reliability indices, which is caused by the occurrence of essential dynamic loads in the form of pressure pulsations, surges, increased vibrations of the pipeline hydraulic system and PU. Figure 1.8 contains different modes and corresponding dynamic loads in PC.

The experience of PC operation reveals that the basic damages and emergencies occur at sudden disconnection of the electric motors from the power system, PU starts, stops taking place several times a day and during the regulation of the technological parameters [1-3, 6, 8].

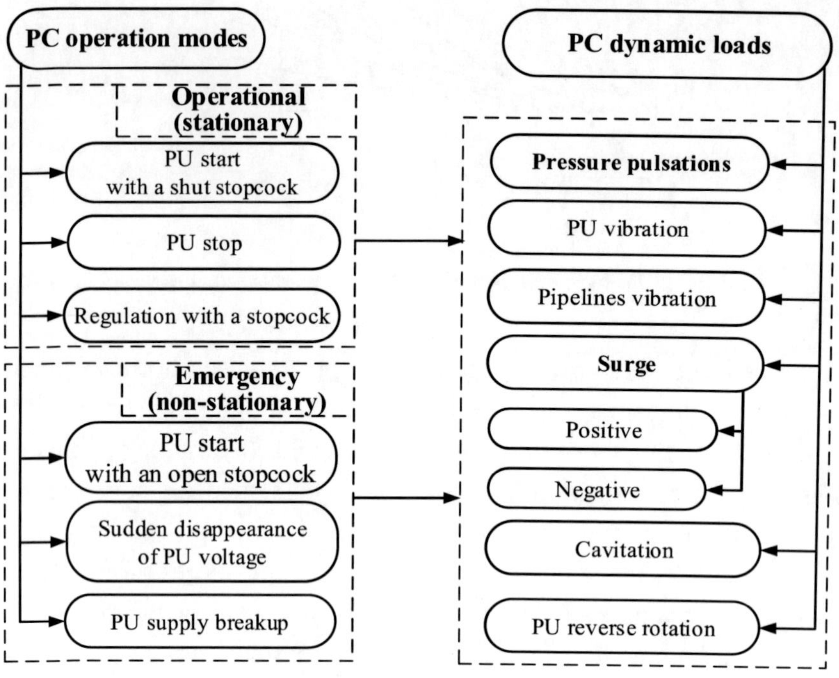

Figure 1.8. Dynamic loads in different modes of PC operation.

A typical feature of centrifugal pumps consists in the increase of the hydrodynamic moment on the shaft with the increase of discharge, so, to facilitate the electric motor operation conditions (to eliminate overloads); the pumps are started with a shut stopcock in the ascending pipe. When the pump reaches the set rotation frequency and head (time moment $t_{v=1}$) the output stopcock is slowly opened (time moment $t_{q=0}$) till the required discharge is achieved (Figure F1).

In some cases centrifugal PUs are started with an open stopcock with filled or/and empty head conduit after/before the check valve. The following cases are distinguished:

- the pipeline section from the pump to the check valve is filled (Figure F2) or not filled with liquid (Figure F3), and the value of the head on the side of the hydronetwork is in the domain of the operating heads of the pump (the pump start with a filled pipeline);

- the check valve is not waterlogged on the side of the pump and the pressure behind the check valve equals to zero (the pump start with an empty pipeline) (Figure F4).

In both cases in the beginning, the start takes place in the same way as with the shut back-pressure valve. When the pump reaches head h_p somewhat higher than the static one, the check valve opens (time moment $t_{q=0}$) and water motion begins ($q_p > 0$) with simultaneous increase of the rotation frequency until it becomes synchronous. Diagrams in Figures F2, F3 show that with such method of start the pump output parameters (q_p, ν) change without special fluctuations; the head transient processes are accompanied by insignificant increase of pressure at the moment of the check valve opening whose value is within (10-20)% of the pump rated head. In this case, one can observe insignificant short-term PU vibrations with the amplitude 1.5-2 times bigger than the values in a stationary mode [6].

At the pump start with an empty pipeline (Figure F4) the pressure in the ascending pipe starts growing and the section up to the check valve is filled with liquid ($t < t_{q=0}$). The air in the pipeline compresses to the value of the pressure behind the check valve, after that it opens (time moment $t_{q=0}$). Due to liquid inertia in the hydraulic network, there occurs a positive surge whereat the pressure at the pump increases by 5-7 times in comparison with the rated value. When the check valve opens, there occurs a short-term increase of the pump body vibration by 3-4 times in comparison with the values at the stationary modes of the unit operation. Then in the pipeline, there occurs a negative surge accompanied by a deep decrease of pressure at the pump output to the level of the admissible vacuum – h_{vak} and appearance of cavitation. In this case, the following phenomena take place: occurrence of a positive surge, a long-term (time interval $\Delta t_{q=0}$) decrease of the pump supply (practically to zero), significant increase of the pump vibration (by 6-8 times in comparison with the rated values) and pressure pulsations whose amplitude reaches (10-20)% of the pump rated head [9]. When cavitation disappears, the analyzed transient processes repeat several times with gradual attenuation.

PU is stopped by ED disconnection from the electricity network after preliminary shutting of the stopcock or the disk back-pressure valve in the pressure line. During the process of the pump stopping (Figure F5) one can observe a smooth decrease of supply to zero (time moment $t_{q=0}$), increase of the pressure in the head conduit without surge, increase of the dynamic loads on the unit in comparison with the values in the stationary mode. By the moment of the stopcock unit complete shutting, the pressure pulsations reach (15-20)% of the pump rated head and the vibrations amplitude increases by 6-8 times in comparison with the values in the stationary operation mode [9]. When the electric motor is disconnected from the electricity network (time moment t_{off}) the pressure pulsations and vibrations decrease proportionally to the squared rotation frequency. In this case, there occur vibration hits due to hydromechanical resonance – coincidence of the frequencies of hydrodynamic and mechanical oscillations caused by the generation of turbulent vortexes in the flow of the pumped medium and by PU impeller rotation.

Most dangerous dynamic loads in PC occur in the process of sudden disconnection of PU electric motor from the power network. The basic dynamic loads characterizing such process include the pressure in the head conduit during a surge, the discharge in the pipeline at the counterflow of the working medium, acceleration rotation frequency and the time of its achievement, pressure pulsations and vibration of the hydromechanical equipment. Depending on the type of the valves installed on the PC head conduit, the process develops in two directions:

1) a check valve is installed in the PU head line (Figure F6);
2) there is no check valve in the pump ascending pipe (Figure F7).

In the former case, when the electric motor is disconnected from the power network, the motor electromagnetic moment drops to zero and the unit rotation frequency gradually decreases by inertia. At the beginning of the counterflow mode (time moment $t_{q<0}$) the check valve shuts and the pump transfers into the pumping mode. The duration of the unit rotor direct rotation is determined by its inertia and the pressure in the head line. The water out of the conduit section from the check valve to the pump gradually flows down through it. PU sequentially goes through all the stopping modes (time moment $t_{\nu=0}$), and its output parameters (h_p, q_p, ν) smoothly decrease to zero. Due to the inessential reverse liquid discharge and low head the reverse rotation frequency, dynamic loads on the unit and the duration of the process are insignificant. In this case, as the diagrams show (Figure F6), at the check valve shut-in ($t_{q<0}$) in the head conduit there occurs a positive surge, whereat the pressure values are by 5-10 times higher in comparison with the rated value [4, 6, 9]. Hits overlapping may result in a repeated shut-in of the check valve and much higher increase of the pressure in the head line.

In the latter case, when there is no check valve in the head line, the process of sudden outage of electricity in PU is characterized by the following stages (Figure F7):

- the reduction of the rotation frequency in the pumping mode to the moment of the alteration of the water motion direction;

- the counterflow mode wherein, at the normal direction of the pump rotation, the flow moves from the head side to the suction side of the pump, which ends by the pump stop and the reverse of its rotation direction;
- the turbine mode wherein the pump rotates with acceleration in the opposite direction;
- the acceleration mode wherein the torque equals to the unit friction moment.

In comparison with the previous one, this mode is accompanied by the occurrence of a negative surge, water flowing down from the head conduit via the pump (counterflow), rotor rotation in the opposite direction and sharp increase of the dynamic loads on the unit, namely, by the increase of the pressure pulsations in the flow path and PU walls vibrations. So, during the transition from the stationary mode to the mode of liquid counterflow in the pump ascending pipe the amount of head pulsations is (20-40)% higher than in the stationary mode [6, 8, 9,14–23]. In this case, the PU wall vibrations increase and exceed the corresponding values in the rated mode by 5-8 times. In the turbine mode the vibromovements, as well as the pressure pulsations, at first reduce reaching the level of the stationary mode, then, after the achievement of the acceleration rotation frequency of the pump impeller, there occurs a vibration amplitude sharp maximum exceeding the admissible values by 15-25 times.

1.3. The Means of the Pumping Complex Hydrodynamic Protection

According to [14–23], to protect PC against surges the following is to be installed: valves for air input and jam on the water conduit; check valves dividing the water conduit into separate sections with a small static head at each of them and preventing the liquid reverse flow through PU; safety valves and damper valves; check valves with controlled opening and shutting on the pump head lines; blind diaphragms breaking when the pressure increases exceeding the admissible limit; water columns and air-water chambers (caps) cushioning the surge process; additional devices for reducing the dynamic loads in the hydrosystem (balance tanks, air pockets, interferential dampers, resilient members, etc.); regulating stopcocks providing both parameter regulation and main pipes hydraulic protection.

Such types of pipeline valves as shut-off and control stopcocks, taps, disk rotary shutters and hydraulic valves are often used in modern PCs of municipal and industrial water supply, sewage, heating, mining and quarry pumping plants, irrigation systems, etc. as the means of hydrodynamic protection. Table G1 in Appendix G contains their comparative characteristic.

Stopcocks, taps and back-pressure valves are installed in the suction, force and bypass lines at the output of single PU or a group of PUs operating in parallel, in PC discharge headers and are usually used for the regulation of the head and discharge.

Hydraulic valves providing both parameter regulation and PC protection are most widely spread and used devices. The valves are installed between PU and the regulating stopcock in the head or bypass conduit, in PU inclined or vertical mains. There are uncontrolled hydraulic valves opened/shut by the action of the current liquid flow due to the creation of the pressure difference on their plate, and controlled devices actuated by the action of an external effort of the drive on the locking body (diaphragm, plate, slide, piston). The analysis of the pipeline valve technical characteristics (Table G1) revealed that stopcocks and controlled hydraulic valves are mostly equipped with an electric or electromagnetic drive of the power up to 10–15 kW, more seldom with a hydraulic or pneumatic drive and are produced in a wide range of pipe sizes of 50-1200 mm. The specific features of hydraulic valves include: relatively low cost per unit (250 conventional units per 100 mm); the possibility of control by means of a drive; a short time of actuating of the uncontrolled valves (split seconds) and a long interval of shutting/opening (20-300 p) of controlled valves; a wide range of variation of the head and discharge; actuation under the pressure of the working medium; opening/shutting at the variation of the flow direction; a wide variety of designs.

The analysis of the conventional means of hydrodynamic protection of water supply and drainage system PCs during various emergencies (Figure 1.9) revealed that check valves and hydraulic locks preventing medium counterflow are used as the means of hydraulic protection at sudden power-off, abrupt stop of the pump resulting in the change of the flow direction. Safety, air and shutoff hydraulic valves are used for PC hydraulic protection at inadmissible excess of pressure in the pipeline network due to PU abrupt starts, stops, failure of the commutation system sections [15–50].

Safety valve devices have a number of typical drawbacks: a big difference of the valve opening and shutting pressures; a sharp shut of the back-pressure valve; a repeated actuation and generation of additional surges; difficulty of spring adjustment and frequent failures; incomplete attenuation of the surge, which caused their rather narrow use in hydrotransport systems. Damper valves representing a variety of hydraulic protection safety devices are characterized by dumping a big amount of water into the water header and by partial emptying of the force pipeline. A special feature of check valves operation consists in the plate sudden shut-in accompanied by a surge of a considerable force. The use of air valves or air-water chambers complicates the water supply equipment, and quick filling-in the chamber volume, at which the surge amplitude practically does not decrease, and the necessity of manufacture from anticorrosive materials limit the field of their application as pipeline protective valves.

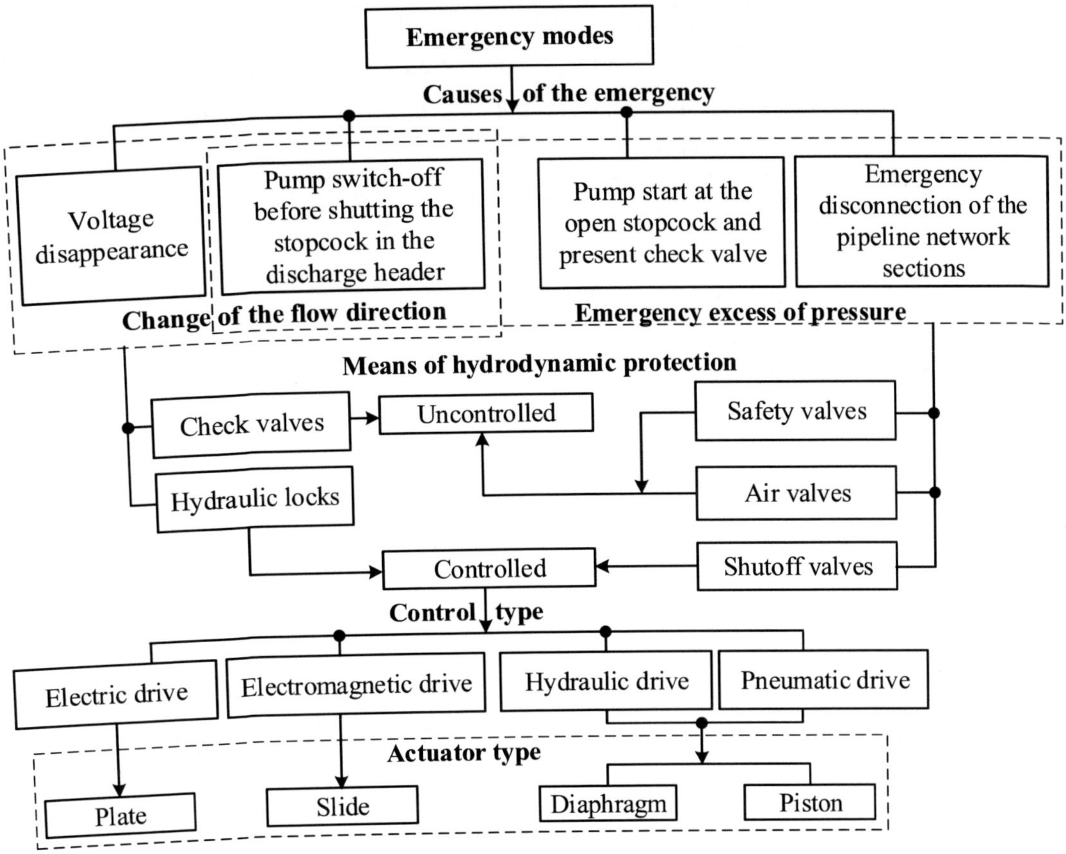

Figure 1.9. The classification of the means of PC hydrodynamic protection in emergency operation modes.

Thus, the existing PC hydraulic protection means are characterized by the following basic drawbacks (Figure 1.10): the lag of actuation performed when the emergency already occurred; an abrupt shut-in resulting in significant increase of the pressure in the pipeline network; occurrence of self-oscillations in locking devices, failures due to the presence of spring elements; uncontrollability resulting in repeated shut-in and increase of pressure in the hydronetwork.

Many scientists such as D. S. Begliarov, R. F. Ganiiev, Kh. N. Nizamov, D. Fox, K. P. Vishnevskii, V. D. Kilimnik, N. M. Koshkin, A. V. Shepelin. E. S. Shatalov et al. considered the problems related to the search of the methods and development of the systems providing minimization of the wave and vibration processes in PC pipeline networks. The analysis of their papers made it possible to single out the basic methods of PC hydrosystems protection against surges:

- the action on PU ED as the source of the occurrence of wave processes in pipeline networks;

- the action on the pumped flow of the working medium by installation of wave processes dampers, controlled and uncontrolled check valves, air caps, control by means of stopcock shutting, water or air inlet into the pipeline network.

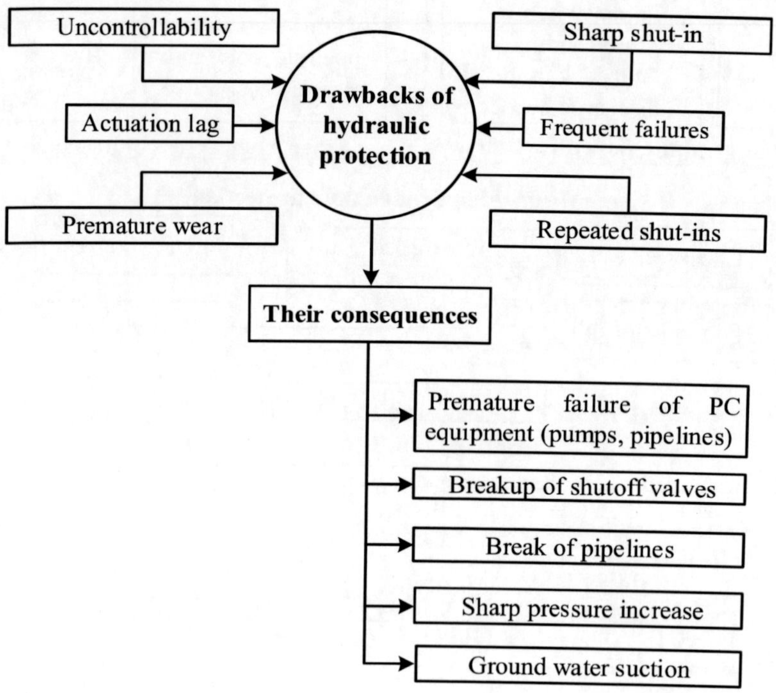

Figure 1.10. Basic drawbacks and consequences of the actuation of the means of PC hydrodynamic protection.

A. V. Shepelin, E. S. Shatalov and T. T. Tigist show in their papers and theoretically substantiate a possibility of the occurrence of head oscillations in the "controlled electric drive – turbomechanism – pipeline" system [37]. They determined that the optimization of the parameters of the regulators of the system of automatic control of the variable frequency electric drive of the turbomechanisms operating on the pipeline network enables the elimination of oscillations in the pipeline at a relatively high operation speed of the control system.

K. P. Vishnevskii, D. Fox, G. A. Mamedov in their papers consider a possibility of the pressure reduction in hydraulic transient processes, caused by shut-off valve shutting due to the increase of the time of its shutting and the use of unregulated electric drive with several rotation frequencies [15, 36]. However, the use of such systems does not provide considerable reduction of the amplitude of pressure oscillations in the pipeline due to the fact that the nonlinear hydraulic characteristics of the pipeline valves are not taken into account, and ED rotation frequency changes only at the last stage of its shutting.

R. F. Ganiiev, Kh. N. Nizamov, V. V. Sulimenko and other authors solve the problem of reducing dynamic loads in pipeline mains by means of various types of oscillation dampers and pressure stabilizers [34, 42]. The principle of such devices operation grounds on the distributed along the length of the pipeline dissipative and resiliently deforming action on the pulsating flow of the pumped medium. The best damping effect is achieved at the dissipation of the pulsation energy at the punched holes evenly distributed along the stabilizer length, and also due to damping caused by the pliability of the stabilizer resilient elements manufactured in the form of a gas cushion, chambers and bellows with walls made of springy and elastic materials. Additional effects of damping are provided when the flow widens in the antechambers and manifold of the stabilizer. The authors propose methods of assessment of the admissible values of wave and vibration loads for various types of media transported via the pipeline system with the use of pressure stabilizers. They prove that the installation of the pressure stabilizer in the pipeline allows reduction of the amplitude and pulsations of the pressure in the pipeline by more than two times and significant increase of the term of its trouble-free operation (by about eight times). However, the violation of the pipeline geometric form because of oscillator dampers and pressure stabilizers negatively influences the characteristics of the hydrosystem overall. Besides, due to the limited pliability of the stabilizer, it is impossible to reduce the level of the oscillation processes in the hydraulic main to the sufficient degree and a significant wear of damping elements results in the deterioration of the reliability of operation of the whole PC.

D. S. Begliarov in his papers experimentally confirms the necessity for taking into account the lag of the check valve plate actuation to determine the maximal increase of pressure in the head lines of pumping stations of closed irrigation systems [32]. The performed research revealed that, when a check valve with controlled shutting is installed on the pump head line, it is possible to use the calculation methods to determine its plate shutting mode providing the required reduction of pressure without the occurrence of an inadmissible reverse rotation of the pump unit rotor. This system did not develop as more efficient and reliable means of reduction of the dynamic loads in PC were found out.

N. M. Koshkin, V. D. Kilimnik researched the processes of opening/shutting the stopcocks equipped with a hydraulic drive under the conditions of closed irrigation systems. They theoretically substantiated the reduction of the value of the pressure in the pipeline network at an indirect surge by variation of the velocity of the shut-off valve shutting. The scientists substantiated the method for the calculation of a surge depending on the process of head conduit shutting taking into account different modes of shut-off valve movement at switching-on and –off the watering automation system, allowing one to set the head conduit shut-off mode reducing the increase of pressure in the hydrosystem by (20-40)%. They determined the regularities of the pressure value reduction during a surge depending on the time of hydraulic stopcocks shutting and the volume of the water discharged via the bypass after shutting the head conduit. It provides the decrease of the surge value by 1.5-2 times, which causes the improvement of the operation reliability and the efficiency of the operation of the closed irrigation network. However, the use of the hydraulic drive for the solution of the problems of reducing the PC dynamic loads is only reasonable under the conditions of closed irrigation systems, as PC water supply and sewage systems use an electric drive of pipeline valve.

Among the works of foreign researchers, the papers of Manlin Zhu [43] deserve attention, as the reduction of dynamic loads in PC during a surge is achieved due to the installation of a regulated air valve and a rotary disk back-pressure valve in the pipeline network. In this case, the variation of the cross-sectional area of the inlet and outlet holes of the air valve together with the step-by-step shutting of the disk back-pressure valve make it possible to decrease the high pressure in the water supply system PC hydraulic system by several times. The researcher has proved that the greatest reduction of the pressure in the hydrosystem is achieved at a certain relation of the cross-sectional area of the air valve inlet and outlet holes.

Mohammad Reza Bazargan-Lari [44] in his paper proposes a method for the determination of the non-linear trajectory of the hydraulic valve shutting with the use of a multi-object optimization model based on Bayesian network enabling the minimization of the pressure oscillations in the pipeline network. The application of Bayesian networks makes it possible to determine the optimal trajectory of valves shutting in the real time mode without starting the optimization models. The scientist demonstrated that, in comparison with the traditional step-by-step or linear control of the pipeline valve, the hydraulic valve shutting with the use of Bayesian networks allows considerable reduction of the range of pressure oscillations during surges in the pipeline network.

Jing-Yang Yu [45] considers the problem of the minimization of the dynamic loads in the hydrosystem during pipeline valves control with regard to long water supply systems PC. To solve the problem the author proposes the optimization of the traditional method of valves shutting in two stages with the use of the even search method. This optimization method provides the search of the optimal time of valves control by the examination of all options until the value corresponding to the minimal dynamic loads in PC is found.

1.4. The Cavitation Processes in the Pumping Complex

Cavitation is one of the most essential phenomena accompanying transient processes in PC. It is characterized by the occurrence of caverns filled with steam or gas causing breaks of the continuous flow of liquid, appearance of air locks in the hydraulic network, periodic oscillations of the pressure (Figure 1.11) and the discharge in the electro-hydraulic system, the increase of the head losses and growth of power consumed by liquid transportation [7–10, 22–49].

The cavitation processes caused by the liquid flow around the bodies are accompanied by the phenomenon of the carry-over of the steam-gas phase (the cavern content) down the flow with its following dissolution and condensation in the pressure space. In most cases, the amplitude of pressure oscillations (cavitation self-oscillations) is constant. Cavitation self-oscillations complicate the normal functioning of the hydrodynamic systems and sometimes make it impossible.

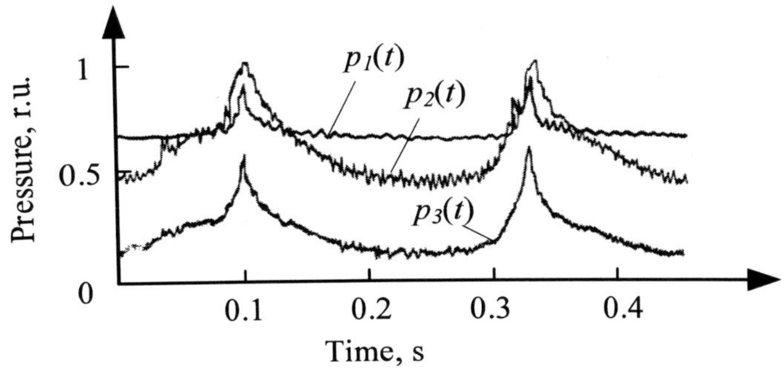

Figure 1.11. Liquid pressure variation curves at cavitation hollow pulsations: $p_1(t)$ – the pressure before the pump; $p_2(t)$ – the pressure at the pump output; $p_3(t)$ – the pressure in the pipeline.

1.4.1. The Essence of the Phenomenon and the Types of Cavitation

One can observe cavitation, as a hydrodynamic phenomenon, in pipelines, pump impellers, on hydraulic turbine blades; it is accompanied by increased vibration of equipment, material erosive destruction, the variation of the operation characteristics of the pumping and pipeline equipment [6, 10, 14, 15, 20, 22, 55–84].

To understand the cavitation phenomenon more clearly, we consider the process of the growth and destruction of a cavitation bubble filled with steam in detail. So, at the motion of liquid, free from impurities (Figure 1.12), with pressure p, equal to pressure p_{para} of saturated steam, it boils and the formed bubbles of radius R, taking into account surface tension forces $2\sigma/R$, are moved by the liquid flow into the pressure space where condensation takes place [55–57, 61–63]. When condition $p<(p_{para}-2\sigma/R)$ is met, the bubble grows, when $p>(p_{para}-2\sigma/R)$, the bubble contracts.

In most cases, the liquid flow contains a certain amount of not-dissolved gas wherein the smallest bubbles are of the radius of $R_0 \approx 10^{-9}$ m. When such liquid gets into the space of lower pressure, the bubbles grow and the gas diffuses through their surface in one or another direction. The equation of the static balance of such steam-gas bubble, taking into account the Boyle-Mariotte law, is of the form, Pa:

$$p = p_{para} + \left(p_0 - p_{para} + \frac{2\sigma}{R_0} \right) \frac{R_0^3}{R^3} - \frac{2\sigma}{R}, \tag{1.8}$$

where R – the bubble radius, m; σ – the coefficient of surface tension, N/m; p_0 – the initial pressure acting on the bubble, Pa; R_0 – the initial radius of the bubble, m.

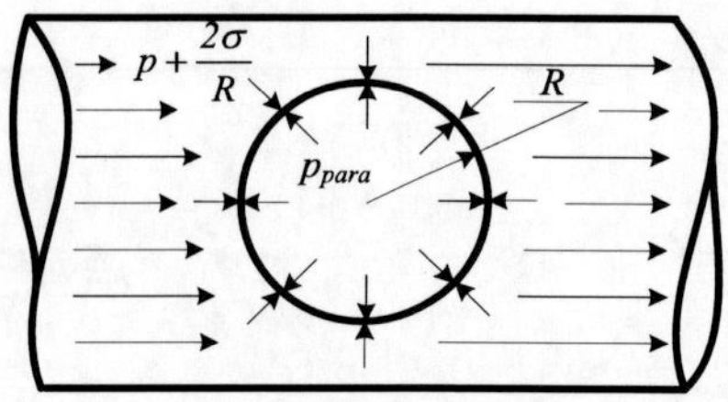

Figure 1.12. Forces acting on the surface of the steam bubble at static balance.

The formed steam-gas bubbles of the size of about tenth shares of a millimeter move together with the flow and, when they get into the pressure space, they collapse, giving rise to a shock wave developing huge pressure (of about 10^3 MPa). In this case, the gas that was in the bubbles dissolves and the formed emptiness is filled with liquid very quickly.

In most cases the processes of bubbles growth and collapse (destruction) include the following typical phases of the development of the cavitation cavern: T_1 – the phase of cavitation cavern growth; T_2 – the phase of cavern separation from the point of occurrence with creation of a new cavern at the same point, T_3 – the phase of destruction of the separated part of the cavitation cavern (Figure 1.13) [55–72]. In this case, the following cavern may appear both after and before the destruction of the previous one.

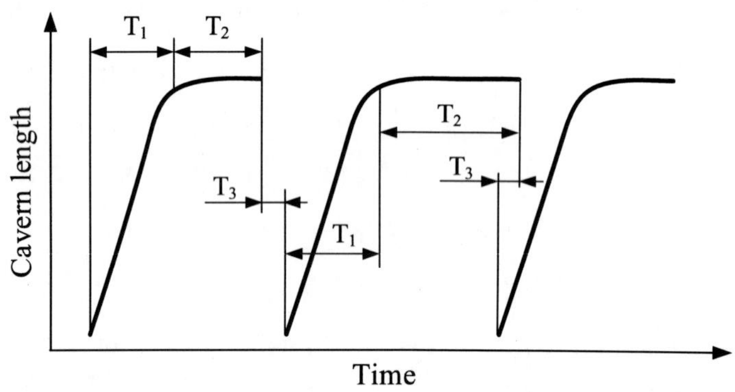

Figure 1.13. Variation of the length of the cavitation cavern.

Papers [55–75] demonstrate that a cavitation bubble may grow in 0.002 s up to 6 mm in diameter and completely collapse in 0.001 s, and under certain conditions of the cavitation process taking place in the area of 1 cm^2 more than 30 mln cavitation bubbles may appear and collapse in 1 s.

1.4.2. The Reasons for the Occurrence of Cavitation and Its Location in the Pumping Complex

The analysis of papers [6, 10–15, 20–23, 55–84] enabled singling out two types of cavitation processes: a hydrodynamic type and an acoustic (vibration) one.

In the former case the cavitation phenomenon is caused by narrowing of the flow or the presence of obstacles in its way at a high velocity of movement, which results in the local decrease of the pressure (proportionally to squared velocity), promoting the growth of the cavitation bubbles in the liquid flow. Acoustic cavitation develops if the liquid passes through acoustic oscillations subjecting the flow to tensile stress.

In the general case, cavitation represents a violation of the continuous character of the flow occurring at underpressure at the analyzed point of liquid, i.e., in the places where the liquid pressure becomes lower than a certain critical value – the pressure of saturated steam. In its turn, it results in the formation of cavitation bubbles filled with steam, gas or their mixture.

Figure 1.14 contains a diagram representing the classification of cavitation processes in PC.

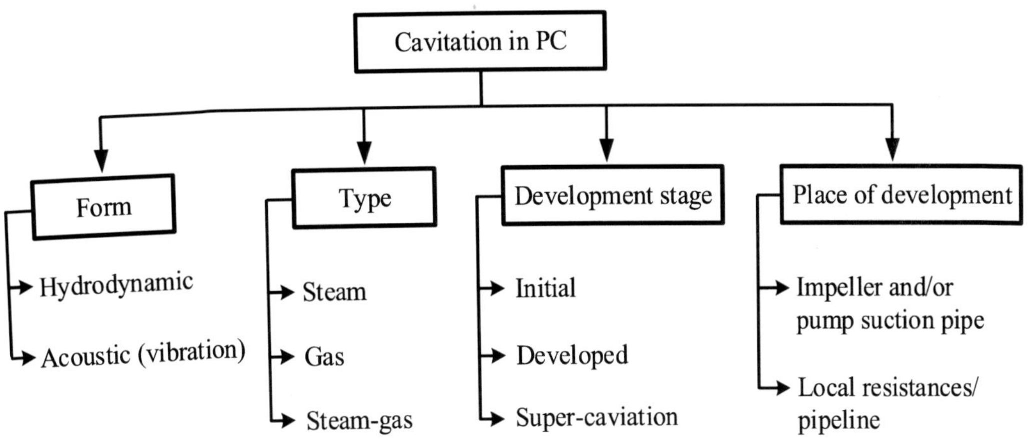

Figure 1.14. The classification of cavitation processes.

1.4.3. Cavitation Parameters

It should be noted that bubble collapse is accompanied by a sound impulse and the less gas the bubble contains the stronger the impulse is. If cavitation development reaches such a degree that a lot of bubbles appear and collapse, this process is accompanied by a loud noise with the frequency spectrum from several hundred Hz to several hundred kHz (Figure 1.15) [64, 65].

The basic parameters characterizing cavitation processes include [6, 14, 15, 55–84]:

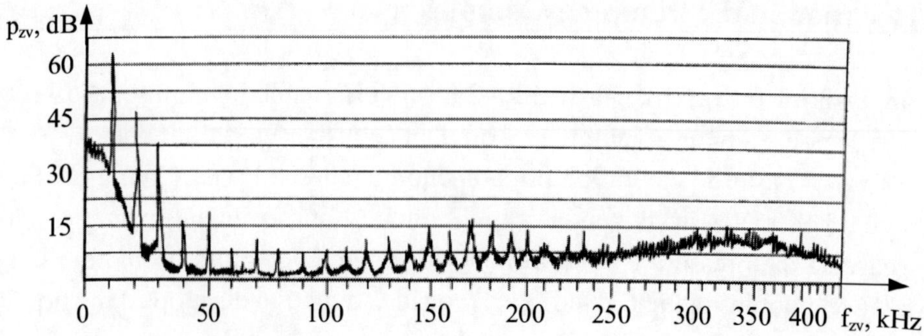

Figure 1.15. The spectrogram of cavitation noise.

- the cavitation number used for the quantitative assessment of the degree of the cavitation processes development

$$\chi = \frac{2(p_{pot} - p_{para})}{\rho \upsilon^2};\tag{1.9}$$

- positive suction head

$$\Delta h = \frac{p_{pot}}{\rho g} + \frac{\upsilon^2}{2g} - \frac{p_{para}}{\rho g};\tag{1.10}$$

- the volume of the break of the flow continuous character at the *i*-th point of the pipeline at moment *j*

$$V_{i,j} = V_{i,j-1} + \Delta Q_{i,j} \Delta t,\tag{1.11}$$

where υ – liquid flow velocity; p_{para} – the pressure of the liquid saturated steam, Pa; p_{pot} – the pressure in the liquid flow, Pa; $\rho = \frac{\rho_{20}}{1+\beta_1(T-T_0)}$ – liquid density, kg/m^3; T – liquid temperature, ^{0}C; ρ_{20} – liquid or gas density at temperature $T_0 = 20$ ^{0}C; β_1 – the coefficient of thermal expansion (for water $\beta_1 = 0.000014$); $\Delta Q_{i,j}$ – the average value of discharge at the adjacent sections; Δt – the time of cavitation existence.

The analysis [10–13] revealed that the pressure of the saturated steam of the transported liquid essentially depends on the temperature of the pumped liquid. Figure 1.16 contains a curve representing the saturated steam pressure dependence on temperature. It can be described with a sufficient degree of accuracy by approximation polynomial of the form:

$$p_{para} = 10\left(a + bT + cT^2 + dT^3 + eT^4\right), \tag{1.12}$$

where $a = 371.14$; $b = 55.15$; $c = 2.26$; $d = -0.017$; $e = 8.98 \cdot 10^{-4}$ – approximation coefficients.

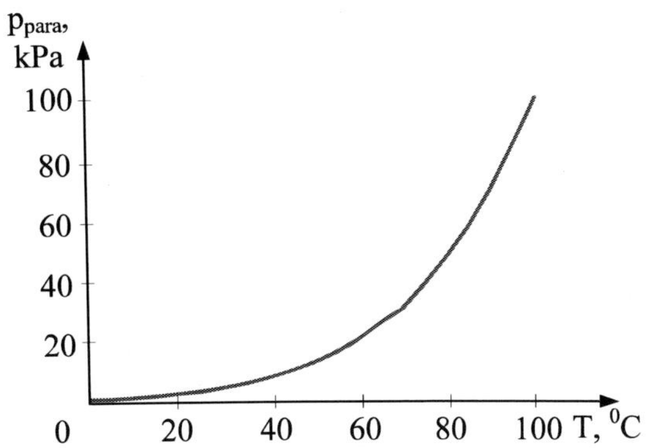

Figure 1.16. The saturated steam pressure dependence on temperature.

The value of the determination coefficient at the use of the polynomial (1.12) is 0.998.

It should be noted that at low values of the cavitation numbers the cavern size may significantly exceed the size of the bodies flowed round by the liquid, which results in super-cavitation [6, 14, 15, 55–84].

The analysis [6, 14, 15, 55–84] revealed that the pump cavitation properties are characterized by positive suction head representing the difference between the specific energy at the pump input and the energy corresponding to the pressure of steam generation. In this case, a sufficient positive suction head or the vacuum-metric height of the pump suction guarantees the non-cavitation operation of the turbomechanism.

1.4.4. Cavitation Process Influence on the Pumping Complex Characteristics

The analysis [6, 14, 15, 55–84] enabled singling out the main areas of cavitation location in PC (Figure 1.17):

- in the impeller and the pump suction pipe – as a result of the increase of the underpressure caused by the change of PC operation mode or the increase of liquid losses at the suction;

- in the siphon pipeline – due to the increase of the difference of the pressure and temperature at the pipeline suction and at the top point of the main, which causes the growth of the efforts stretching the liquid flow;
- at the local resistance – as a result of the increase of the head velocity component, which causes the local decrease of pressure in the liquid flow.

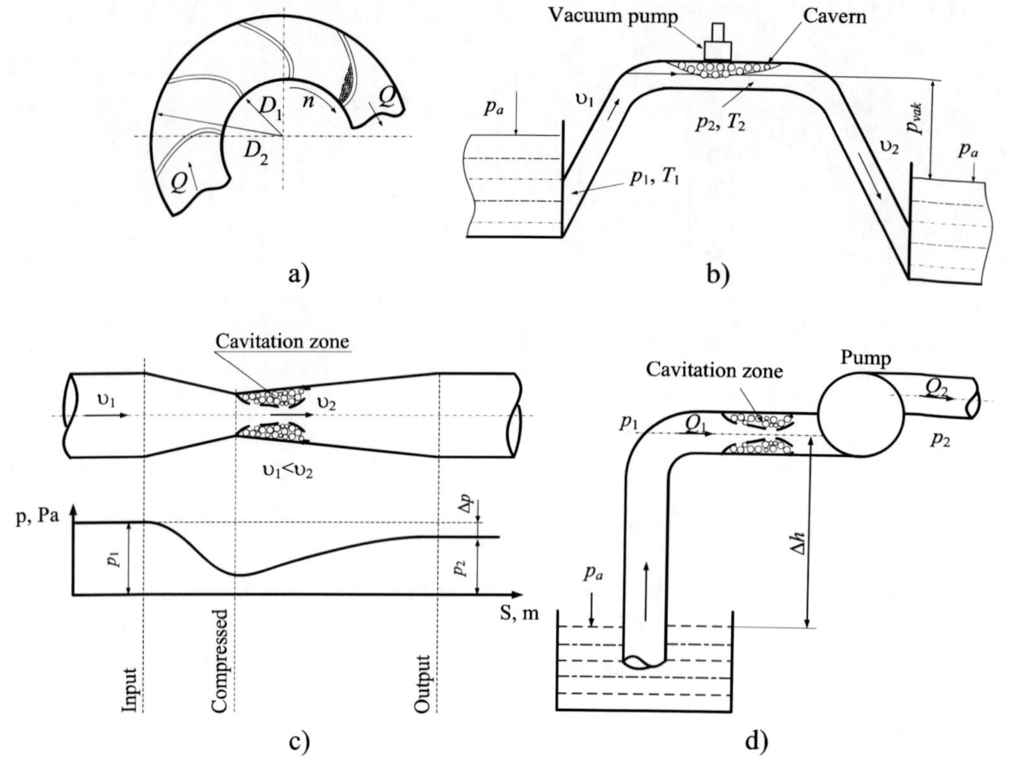

Figure 1.17. The cavitation location in PC: a) in the pump impeller; b) in the siphon pipeline; c) at the local resistance; d) in the pump suction pipe.

V. Ya. Karelin, L. P. Grianko, A. A. Lomakina in their papers [6, 14, 15, 55–84] consider the causes and features of cavitation processes in centrifugal and axial pumps, analyze the pump impeller cavitation processes influence on the energy and operation parameters of the turbomechanisms, give recommendations as to the provision of pump non-cavitation operation. Besides, the authors of papers [6, 14, 15, 55–84] state that the development of cavitation caverns on the impeller blades or in the pump, suction pipe results in a sharp reduction of the values of the head and efficiency of the turbomechanism. It should be noted that the growth of temperature of the pumped liquid results in the deterioration of the cavitation characteristics of the hydraulic machine and promotes the narrowing of the range of its non-cavitation operation (Figure 1.18).

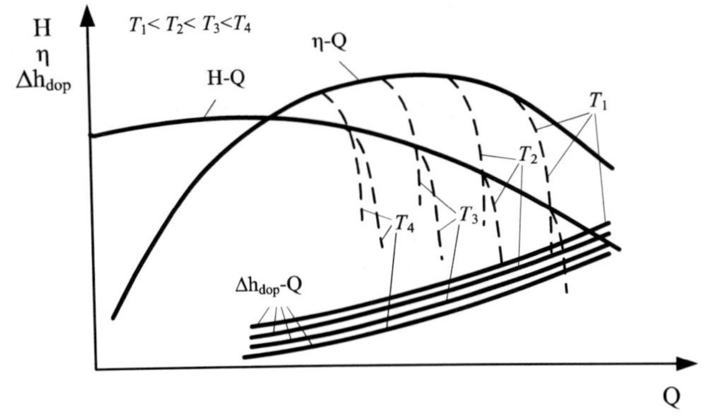

Figure 1.18. The head-discharge and energy characteristics of the pump at the occurrence of cavitation with different temperature of the transported liquid.

V.V. Rozhdestvenskii, A.N. Ivanov et al. [6, 14, 15, 55–84] in their papers studied different stages of cavitation development, the motion of the steam-gas bubble in the boundless liquid and near a hard wall of the body flowed or wetted by liquid. They found the empiric dependences of dimensionless coefficients of resistance, lifting force, gas moment and discharge depending on the cavitation number, which is confirmed by experimental research. However, the obtained results only touch upon the problem of cavitation occurrence when liquid flow passes around a certain object without taking into account PC operation mode changing in time.

A. D. Altshul, L. S. Zhivotovskii, A. S. Yuriiev et al. analyzed the cavitation processes influence on the value of head losses at local resistances [6, 10–15, 55–84]. The authors assess their cavitation characteristics by the critical value of the cavitation number obtained by experiment for different temperatures of the liquid. The latter makes it possible to determine the maximum velocity of liquid flow in the pipeline to provide the non-cavitation operation mode. Figure 1.19 contains curves representing the variation of the adjusting valve resistance coefficient depending on the cavitation number.

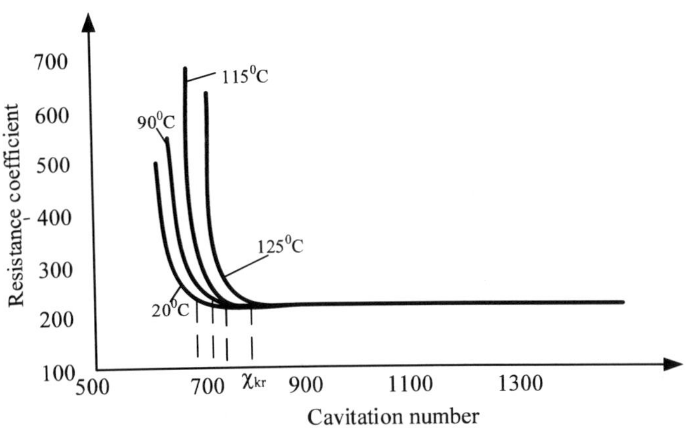

Figure 1.19. The adjusting of dependence of coefficient valve resistance on the cavitation number.

Their analysis revealed that cavitation development at the local resistances ($\chi \leq \chi_{kr}$), caused by the change of the temperature and velocity modes of the pipeline equipment operation, results in sharp increase of the coefficient of their resistance.

The phenomenon of cavitation in siphon pipelines is accompanied by the accumulation of air in the top point of the main, which results in the formation of an air damper decreasing the pipeline flow section [70].

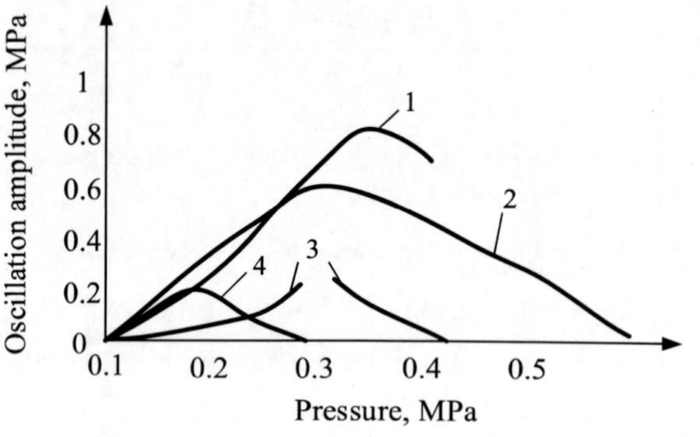

Figure 1.20. The dependence of cavitation oscillation amplitude on the pressure at the pump input at different values of the frequency of the pump shaft rotation: 1 – $n = 1.1n_{nom}$; 2 – $n = n_{nom}$; 3 – $n = 0.82n_{nom}$; 4 – $n = 0.7n_{nom}$.

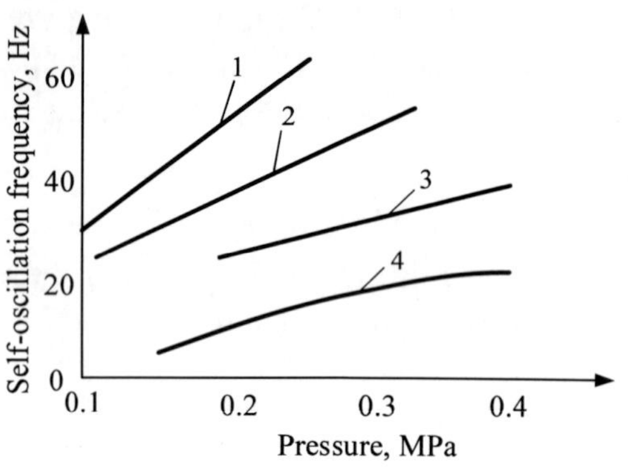

Figure 1.21. The self-oscillation frequency dependence on the pressure at the pump input at $Q = Q_{nom}$ at different lengths of the supply pipeline: 1 – $l = 0.08$ m; 2 – $l = 1.2$ m; 3 – $l = 3.15$ m; 4 – at $Q = 0.5Q_{nom}$ and $l = 1.2$ m.

Papers [6, 10–15, 55–84] deal with the issues of cavitation phenomena in the problems of the surge process in PC based on the research of N. E. Zhukovskii, A. I. Charnyi, L. Berzheronov, L. F. Moshnin and V. I. Blokhin. The authors studied the continuous flow break influence on the surge value and gave recommendations as to the research of non-stationary processes in PC.

V. V. Pilipenko, V. A. Zadontsev, M. S. Natanzon performed the analysis of the pressure and discharge oscillations caused by the presence of cavitation processes in PC [6, 10–15, 55–

84]. The amplitude of such self-oscillations may reach values that exceed the current head values by 2.5 times (Figure 1.20), and their frequency is within the range from 1 to 100 Hz (Figure 1.21). The curves experimentally obtained in [64–65] for different values of the input pressure and representing the variation of the pressure at the pump input and output confirm the above said (Figure 1.22). In one case the cavitation oscillation pressure frequency f_{kav} is 20 Hz (Figure 1.22, a), in the other case – $f_{kav} = 13.3$ Hz (Figure 1.22, b).

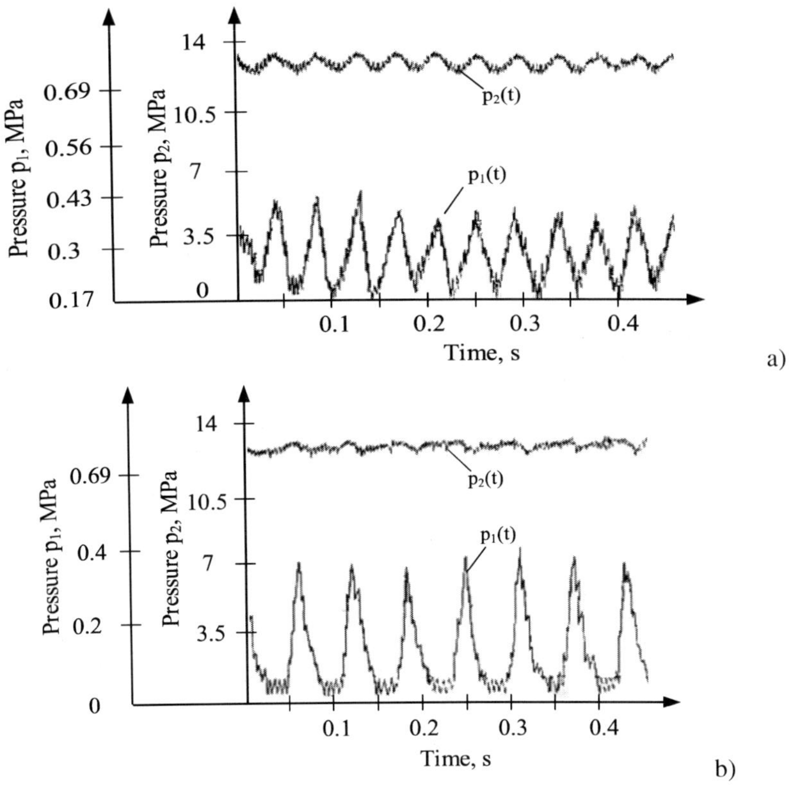

Figure 1.22. The pressure variation curves at the pump input $p_1(t)$ and output $p_2(t)$ at the shaft rotation frequency 300 s^{-1}: a) $p_2 = 13$ MPa, $p_1 = 0.3 \pm 0.122$ MPa, $f_{kav} = 20$ Hz; b) $p_2 = 13$ MPa, $p_1 = 0.2 \pm 0.18$ MPa, $f_{kav} = 13.3$ Hz.

Cavitation processes result in considerable deterioration of energy characteristics and reduction of the operation life of pumping equipment as the rate of its wear depends on the degree of cavitation development [6, 10–15, 55–84]. Figure 1.23 contains experimental curves representing cavitation wear rate I/I_{max} dependence on the cavitation development degree λ/h determined by the relative length L of the cavitation zone [64, 65].

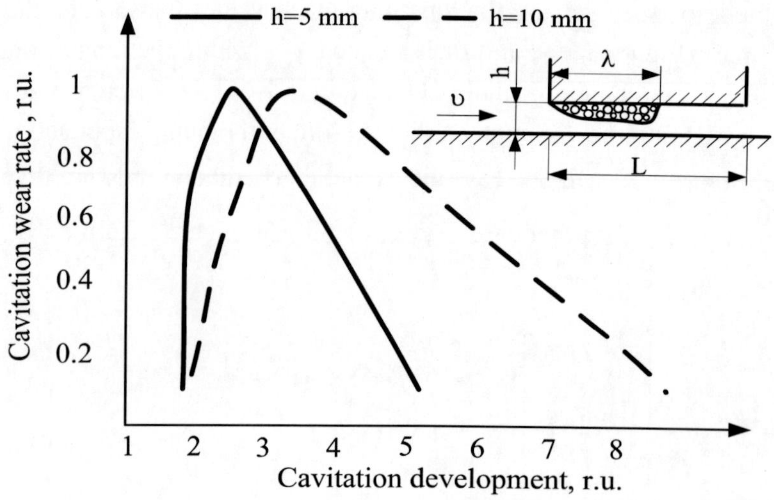

Figure 1.23. The cavitation wear rate dependence on the degree of cavitation development for $\upsilon = 18...20$ m/s.

The curves analysis (Figure 1.23) revealed that at first the cavitation wear rate grows reaching the maximum value, then decreases and may be completely absent at the super-cavitation modes. Therefore, when cavitation processes with $\lambda/h = (2.2...5)$ take place in PC, the equipment operation life reduces by 8-10 times and the efficiency value decreases by 10-12 % [41, 66–69].

Every cavitation bubble grows to the final size, and then collapses [6, 10–15, 55–84]. Surges occurring in this case, result in considerable pressure differences, which affects the technological equipment operation, causes continuous flow breaks, vibrations, noises, material erosive destruction (Figure A.7, Figure 1.24). It also calls forth the increase of the head losses and, correspondingly, the increase of the power consumed by liquid transportation. It causes the necessity for the search for solutions directed to the reduction of cavitation processes in the pumping complex.

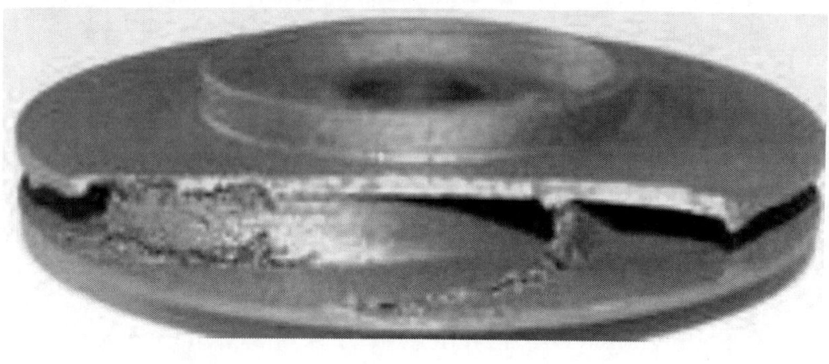

Figure 1.24. Erosion of the centrifugal pump impeller.

1.5. THE ANALYSIS OF THE EXISTING METHODS FOR THE REDUCTION OF THE CAVITATION PROCESSES IN THE PUMPING COMPLEX

The performed analysis [6, 10–15, 55–84] revealed that all the known measures of cavitation management are based on the following:

- the installation of pipeline protection appliances (water-air tanks, aeration valves, water inlet reservoirs, standpipes), intended not only for the prevention of the cavitation breaks of the continuous flows but also for the reduction of pressure pulsations in the hydrosystem. Water-air tanks (Figure 1.25) moderate the change of pressure in transient processes; reduce both increased and decreased pressures, preventing breaks of continuous flows in the pipeline. Aeration valves installed in the critical points of the pipeline profiles inject air into the cavitation zone, which results in the development of super-cavitation mode. Water inlet reservoirs and standpipes reduce underpressure in the hydrosystem preventing breaks of continuous flows;

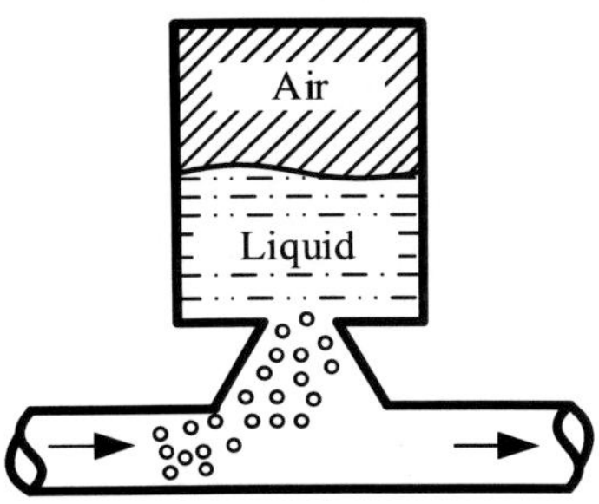

Figure 1.25. The diagram of installation of water-air tanks.

- the maintenance of the pump suction admissible height excluding occurrence of underpressure in the suction pipe and the pump impeller;
- the alteration of the angle of the flow around the hydromachine impeller blades, designing the blades in the shape of slightly bent profiles with rounded edges;
- the cover of the hydraulic surfaces with material more resistant to cavitation destruction;
- degassing the liquid flow in the process of water preparation by a chemical method (using additives, Figure A.8) or a physical method (installation of de-aerators, Figure A.9), separators, air vents).

The above analysis of the methods and devices for cavitation reduction revealed that most of them are expensive and material consuming; they are directed to the decrease of the destructive action of cavitation, the increase of the range of non-cavitation mode of PC operation by the alteration of the configuration of the pipeline network or structural parameters of hydraulic equipment.

The papers by V. Ya. Karelin, L. P. Grianko, A. D. Altshul, V. V. Pilipenko, V. A. Zadontsev, M. S. Natanzon deserve special attention [6, 10–15, 55–84] as they contain the analysis of the issues of the pump electric drive rotation frequency influence on the losses, the amplitude and the frequency of pressure oscillation in PC caused by the presence of cavitation processes. Figure 1.17 contains dependence of cavitation oscillation amplitude on the pressure at the pump input for different values of the frequency of the pump shaft rotation. They are of explicitly non-linear character with maximum at the pressure exceeding the pressure of cavitation break. As the pressure at the pump input and output depends on the rotation frequency of its shaft, when the value of the latter decreases, the amplitude of pressure self-oscillations in the elctro-hydraulic system reduces (Figure 1.17) or they stop completely. So, the decrease of the pump impeller rotation frequency by 30% of the rated value (Figure 1.17) results in the decrease of the maximum value of pressure cavitation oscillations in the electro-hydraulic system by three times. The frequency of the cavitation oscillations of the pump input pressure decreases with the increase of the length of the supply pipeline, of the rotation frequency of the hydraulic machine shaft and with the reduction of the discharge via the pump. However, later this method of cavitation reduction was not developed.

Thus, the use of a variable-frequency ED in water- and heat-supply PC for the regulation of the technological parameters and creation of the control system on its basis will enable the control of occurrence of the zones with low pressure. It will also allow the alteration of the pump operating velocity to correct the parameters (pressure or discharge) aiming at the provision of the cavitation protection of the hydraulic equipment. This approach provides the improvement of the energy efficiency and reliability of the operation of the elctrohydraulic equipment by means of the decrease of the pump electric drive power nonproductive losses caused by the presence of cavitation processes in the pipeline.

Conclusion

Generalizing the results presented in this chapter, one can come to the following conclusions.

1. The performed analysis of the modes of operation of modern pumping complexes revealed, that they are characterized by low indices of reliability and high breakdown rate. During the process of pumping complexes operation, the alteration of the operation modes in the hydrosystem there occur wave phenomena of various nature (surges, pressure pulsations, etc.). They are caused by a number of factors: emergency outage of the electricity supply of operating pump units; check valve

reaction; quick shutting or opening the safety or shut-off valves; erroneous actions of the maintenance staff; switching-on or -off of a big consumer, etc.

2. It is determined that the existing approaches to the solution of the problem of PC hydrodynamic protection do not consider the problems of control of the pipeline valves with a variable frequency ED aiming at minimization of the dynamic loads in the hydrosystem. Step-by-step or linear control of pipeline valves is mostly used but it does not provide significant decrease of the amplitude of pressure oscillations in the hydrosystem. In this relation, it is expedient to develop and research the systems of dynamic loads in PC based on a variable-frequency ED of the shut-off regulating pipeline valves.
3. It is shown that at the decrease of the pressure in the discharge header, the alteration of the temperature of the pumped liquid, the presence of local resistances along the length of the pipeline in the hydrosystem there occur cavitation processes characterized by the appearance of periodic oscillations of pressure and discharge. They result in the decrease of the efficiency of the pump unit and the growth of non-productive losses of power for the transportation of the operation product. In this case, the frequency of cavitation self-oscillations is within (1-100) Hz; the pulsations amplitude may exceed the current values of pressure and discharge by (2-2.5) times; power non-productive losses caused by cavitation reach 30% of the determined power of the pump electric drive.
4. Practical methods and technical means of cavitation decrease based on the installation of pipeline protective appliances, using additives in the liquid, injections of liquid or air into the cavitation area are directed to the reduction of the destructive action of the cavitation processes and are characterized by considerable material costs.
5. The performed analysis of the influence of the frequency of the pump shaft rotation on the cavitation processes in the electrohydraulic complex revealed that with the decrease of the rotation frequency of the pump shaft there occurs reduction of losses, the amplitude and frequency of pressure oscillations caused by the presence of cavitation processes or their complete termination. Thus, the use of a variable-frequency electric drive and a control system on its basis will make it possible to improve the energy efficiency and reliability of the operation of the electromechanical system for liquid transportation.

Chapter 2

Working Out the Models of a Pumping Complex with a Variable-Frequency Electric Drive at the Development of Non-Steady Processes in the Pipeline Network

2.1. The Approximation of the Hydraulic Characteristics of the Pumping Complex Pipeline Valves

The analysis [26] revealed that the pipeline valves are characterized by the throughput coefficient K_v and cavitation coefficient K_c dependence on hydraulic resistance coefficient ξ (Figure 2.1), head losses Δh_{rv} dependence on the variation of the flow velocity υ, hydraulic resistance coefficient ξ dependence on the degree β of opening and flow velocity υ (Figures 2.2, 2.3) etc.

The valves cavitation characteristics enable the determination of the range of variation of the control valves hydraulic resistance coefficient ξ whereat the continuous liquid flow does not break because of the occurrence of cavitation caverns (cavities filled with gas or air). The occurrence of cavitation negatively affects the valves operation: the throughput decreases, the hydraulic resistance coefficient grows, a moment when the discharge will be equal to zero may come.

The valves basic hydraulic characteristic influencing the parameters of hydraulic transportation consists in hydraulic resistance coefficient ξ dependence on height h of the sliding disk rise (for stopcocks and valves), the rotating disk rotation angle α_{bv} (for disk back-pressure valves), the seat cross section free area S_h for the control valves) or liquid motion velocity υ (for the check valves), (Figures 2.1, 2.2).

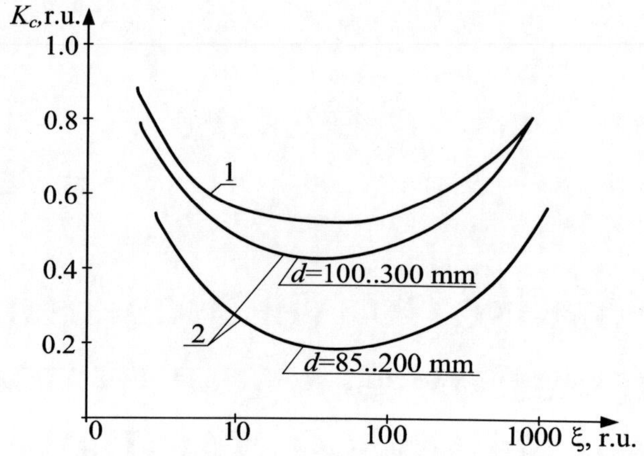

Figure 2.1. Cavitation coefficient K_c dependence on hydraulic resistance coefficient ξ for the control single-seat valves (1) and double-seat valves (2) of different diameters d .

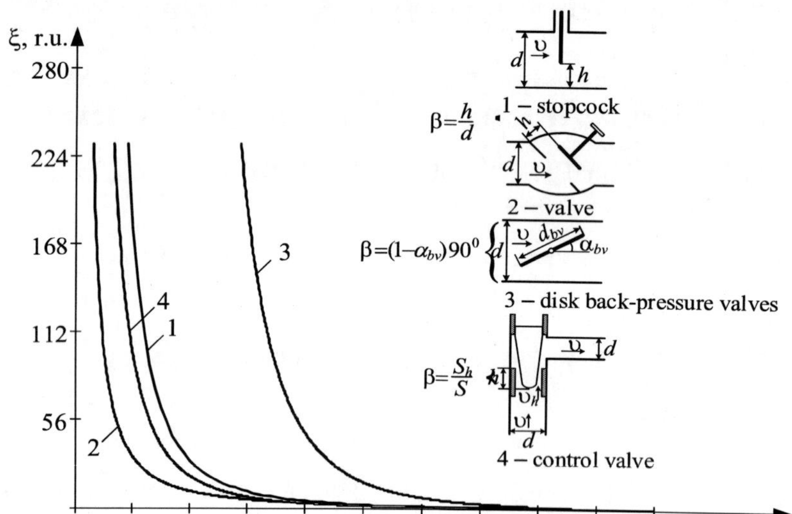

Figure 2.2. The dependence of the hydraulic resistance coefficient ξ on the relative degree β of opening the pipeline valves.

The variation of these characteristics is nonlinear: hydraulic resistance coefficient ξ grows significantly at a certain critical value of β_{cr}, which equals 0.25 for the stopcocks, for valves – 0.3, for disk back-pressure valves – 0.6, for control valves – 0.2, and at υ_{cr}, equal to 1 m/s for the check valves.

The analysis [10–13] revealed that there is no universal analytical dependence relating hydraulic resistance coefficient ξ and relative degree β of opening, required to the determination of the optimal laws of control of the pipeline valves. In this regard, an approximation function of the following form is proposed:

$$\xi = A\left(\left(1/\beta\right)-1\right)^{C} + B\left(\left(1/\beta\right)-1\right)^{D} + \xi_0, \quad (2.1)$$

where A, B, C, D – approximation coefficients depending on the type of the pipeline; ξ_0 – hydraulic resistance coefficients at full opening of the valves (β=1).

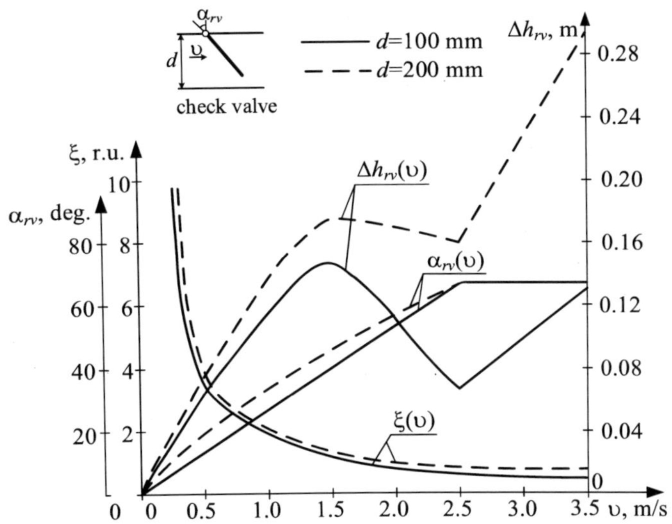

Figure 2.3. The hydraulic resistance coefficient ξ, plate rotation angle α_{rv} and the check valve head losses Δh_{rv} dependence on the liquid flow velocity υ.

Table 2.1 contains the values of approximation coefficients for various pipeline valves whose hydraulic characteristics are given in Figure 2.2.

Table 2.1. The approximation coefficients for various pipeline valves

Coefficient	Stopcock		Valve	Disk back-pressure valve	Single-seat control valve
	wedge	parallel			
A	0.505	1.019	2	1.7	1.8
B	1.868	1.115	1.2	3.8	2
C	2.35	2.5	0.129	10.628	0.664
D	1.4	1.1	1.555	5.652	0.794
ξ_0	0.05	0.1	0.5	0.24	0.18

The choice of the approximation dependence type (2.1) is caused by the following feature of the pipeline valves operation – the reduction of the relative degree β of opening results in the increase of hydraulic resistance coefficient ξ Dependence (2.1) is characterized by a high value of determination coefficient R^2 that is higher than 0.92 for various types of pipeline valves.

In expression (2.1) parameter β depends on the law of valves control, which significantly influences the character of transient processes in PC: the occurrence of pressure inrushes, head losses, cavitation phenomena resulting in the break of the continuous liquid flow, etc.

That is why it is reasonable to research the pipeline valves control laws in order to decrease the dynamic loads in PC and improve the reliability of the technological equipment operation.

When the law of valves control is not observed, there occurs a surge in the hydraulic system. The surge phase is its basic parameter [10-15, 26]:

$$T_{ph} = 2L/c, \tag{2.2}$$

where L – the pipeline length, m; $c = c_a / \sqrt{1 + (E_a/E_{pl})\Psi}$ – the velocity of surge wave propagation; E_a – the module of the liquid volume elasticity; E_{pl} – the module of elasticity of the pipe walls material; $c_a = \sqrt{E_a/\rho}$ – the velocity of sound propagation in the liquid, m/s; ρ – the liquid density, kg/m^3; Ψ – the coefficient taking into account the pipeline walls deformability; ΔH_k – the increase of the head in every phase of the surge, m.

When the valves actuate, there occur a direct and an indirect surges, and at the direct one time t_{sh} of the full shutting of the valves is shorter than the surge phase T_{ph}.

Various laws of the pipeline valves control can be described by the expression of the form:

$$t = (1-\beta)^n t_{sh} \tag{2.3}$$

where β – the relative degree of opening of the pipeline valves ($\beta = h/d$ for the stopcocks and valves; $\beta = (1-\alpha)90^0$ – for the disk back-pressure valves; $\beta = S_h/S$ – for the control valves).

Having transformed (2.3), we get:

$$\beta = 1 - (t/t_{sh})^{1/n},\ n \geq 1. \tag{2.4}$$

Figure 2.4 shows the curves representing the variation of the relative degree h/d of the stopcock wedge rise for different coefficients n of intensity of control of the shut-off and regulating valves at $t_{sh} = 5T_{ph}$.

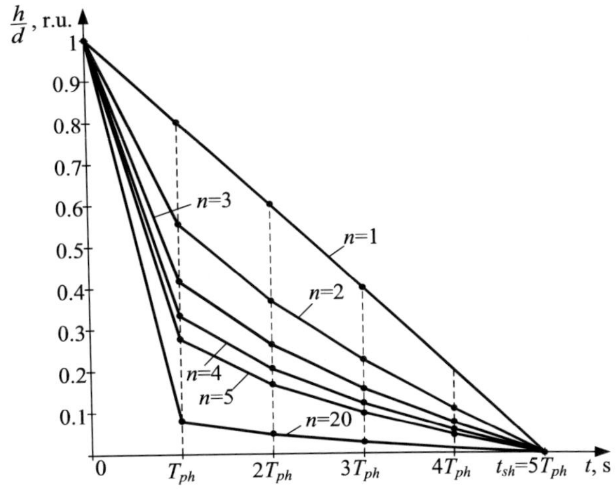

Figure 2.4. Curves representing the variation of the relative degree h/d of the stopcock wedge rise depending on the time for different coefficients n of the intensity of valves control.

Figures 2.5, 2.6 contain the graphs of the variation of the head absolute and relative increase $\Delta h = \Delta H/\Delta H_{dir}$ respectively, where ΔH_{dir} – the increase of the head at direct surge for the hydrosystem with the parameters: $\upsilon_0 = 1$ m/s, $\Delta h_{fr} = 7.13$ m, $L = 3000$ m, $\xi_{pl} = 140$, $c = 1000$ m/s, $T_{ph} = 6$ s. A stopcock installed in the consumer communication network and closed with different intensity coefficients n and control time t_{sh} equal to $5T_{ph}$ is used as the pipeline valve in the hydrosystem.

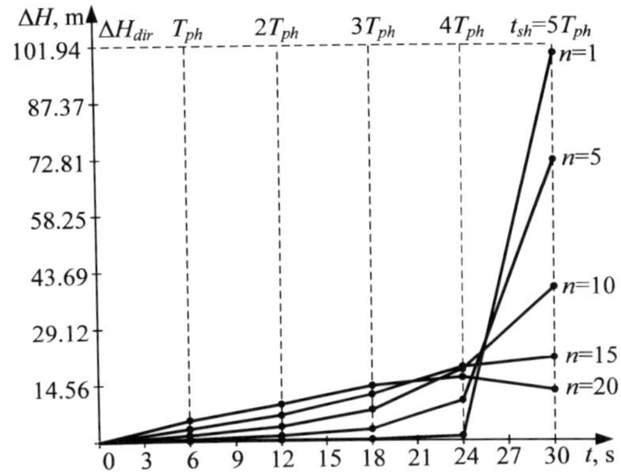

Figure 2.5. The graphs of the variation of the head absolute increase ΔH at the surge for different intensity coefficients n and valve control time t_{sh}.

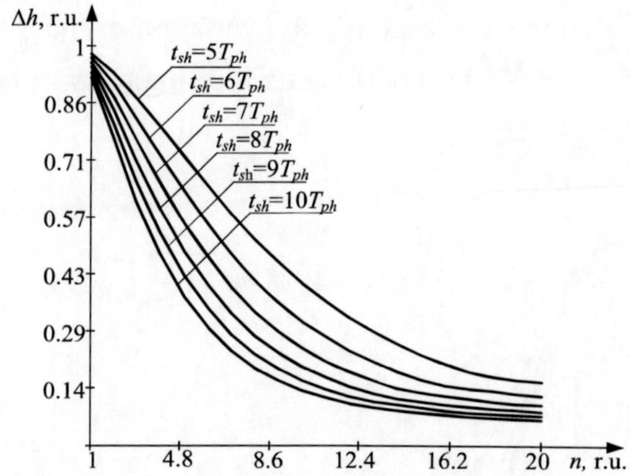

Figure 2.6. The graphs of the variation of the head relative increase ΔH at the surge for different intensity coefficients n and valve control time t_{sh}.

The analysis of graphs in Figure 2.6 reveals that an essential decrease of the head growth at surge can be achieved by increasing control intensity coefficient n and stopcock shutting time t_{sh}.

2.2. The Mathematical Description of a Pumping Complex with Adjustable Pipeline Valves

Linear/nonlinear differential and ordinary equations or their systems describing particular physical processes in the system are the basis for the existing approaches to PC mathematical description.

Three components are inherent elements in any PC: a pump unit, a pipeline network and a consumer forming a single production string for pumping and providing the consumer with the working product. Figure 2.7 shows a functional diagram of a PC with adjustable pipeline valves.

The figure demonstrates that the pipeline network represents a system of interconnected sections of the pipeline where the output parameters (head H, discharge Q or medium motion velocity υ) of the previous section are the input parameters for the following section and vice versa. In this case, the first section of the pipeline system is the pump ascending pipe. The last section of the pipeline network and the consumer form an interconnected subsystem. Shut-off and control valves can be installed at any section of the pipeline network. EDs are used for the control of the pump and the shut-off and control valves.

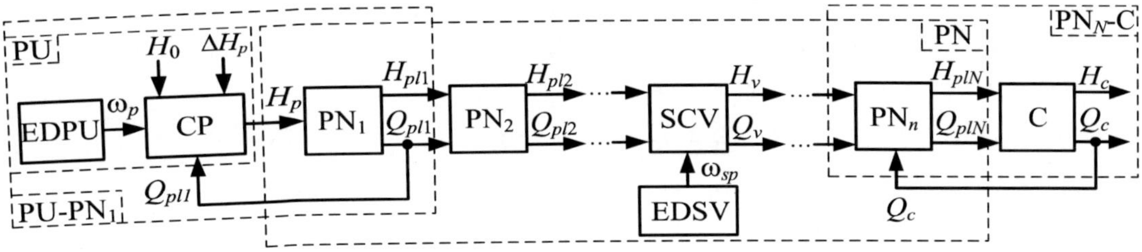

Figure 2.7. A functional diagram of a PC with adjustable pipeline valves: CP – centrifugal pump; SCV – shut-off and control valves; EDPU, EDSV – electric drive of the pump unit and of the shut-off and control valves, respectively; PN1-PNn – the sections of the pipeline network PN; C – consumer.

Telegraph equations known from [14, 15, 83–98] are used for the description of the pipeline network and the research of the dynamic processes occurring in it:

$$\begin{cases} \dfrac{\partial H}{\partial x} - \dfrac{1}{gS}\dfrac{\partial Q}{\partial t} - \dfrac{\lambda_{pl}}{d}\dfrac{1}{S^2}\dfrac{Q|Q|}{2g} = 0; \\ \dfrac{\partial H}{\partial t} + \dfrac{c^2}{gS}\dfrac{\partial Q}{\partial x} = 0, \end{cases} \tag{2.5}$$

where $H = p/(\rho g)$, Q – piezometric head and discharge of liquid in the current pipeline cross section, respectively, m, m^3/s; p – the pressure in the current pipeline cross section, Pa; $S = \dfrac{\pi d^2}{4}$ – the pipeline cross section area, m^2; d – the pipeline diameter, m; λ_{pl} – the dimensionless coefficient of resistance of the pipeline; c – the velocity of propagation of the surge wave or sound in the working medium, m/s; ρ – the density of the working medium, kg/m^3; g=9.81 m/s^2 – the acceleration of gravity. Equations of the following form represent a centrifugal pump:

$$H_p = H_0 \nu_p^{\ 2} - R_p Q^2, \tag{2.6}$$

where H_0, H_p – the head at zero feed and at the output of the pump impeller, respectively, m; R_p – the pump internal resistance, s^2/m^5; $\nu_p = \omega_p / \omega_{pn}$ – the relative rotation frequency of the pump impeller; ω_p, ω_{pn} – the current and rated pump rotation frequencies, respectively, s^{-1}.The regulating stopcock is described by expression (2.1), in which the relative degree of opening of the stopcock is equal to $\beta = \varphi_{sp}/(2\pi n_{sh})$, where

$\varphi_{sp} = \int_0^{t_{sh}} \omega_{sp}(t)dt$ – the current rotation angle of the valve spindle, rad; $\omega_{sp} = \omega_{IM} / r_g$ – the current rotation frequency of the valve spindle, s^{-1}; ω_{IM} – the current rotation frequency of the shaft of the drive electric motor, s^{-1}; r_g – the gear ratio of the stopcock ED reducer; n_{sh} – the number of turns required for the full shutting of the stopcock.

The stopcock drive IM is represented by differential equations in a synchronous coordinate system «uυ0» [83-100, 143]:

$$\begin{cases} d\psi_{1u}/dt = -R_1\psi_{1u}/(L_2\sigma_{12}) + R_1L_\mu\psi_{2u}/(L_1L_2\sigma_{12}) + p_p\omega_0\psi_{1\upsilon} + U_{1m}; \\ d\psi_{1\upsilon}/dt = -R_1\psi_{1\upsilon}/(L_2\sigma_{12}) + R_1L_\mu\psi_{2\upsilon}/(L_1L_2\sigma_{12}) - p_p\omega_0\psi_{1\upsilon}; \\ d\psi_{2u}/dt = -R_2'\psi_{2u}/(L_1\sigma_{12}) + R_2'L_\mu\psi_{1u}/(L_1L_2\sigma_{12}) + \omega_{2e}\psi_{2\upsilon}; \\ d\psi_{2\upsilon}/dt = -R_2'\psi_{2\upsilon}/(L_1\sigma_{12}) + R_2'L_\mu\psi_{1\upsilon}/(L_1L_2\sigma_{12}) - \omega_{2e}\psi_{2u}; \\ M = (3/2)p_p L_\mu/(L_1L_2\sigma_{12})(\psi_{1\upsilon}\psi_{2u} - \psi_{1u}\psi_{2\upsilon}); \\ \omega_0 = 2\pi f_0; \quad \omega_{2e} = \omega_0 - \omega_{IM}; \\ J_\Sigma\, d\omega_{IM}/dt = M - M_s; \\ \vec{i_1} = \left(L_2\overrightarrow{\psi_1} - L_\mu\overrightarrow{\psi_2}\right)/\left(L_1L_2 - L_\mu^{\,2}\right); \\ \vec{i_2} = \left(L_1\overrightarrow{\psi_2} - L_\mu\overrightarrow{\psi_1}\right)/\left(L_1L_2 - L_\mu^{\,2}\right), \end{cases} \tag{2.7}$$

where ψ_{1u}, $\psi_{1\upsilon}$, ψ_{2u}, $\psi_{2\upsilon}$ – flux linkages of the electric machine stator and rotor by coordinates u and υ respectively, Wb; U_{1m} – the amplitude value of the supply voltage of the stator winding, V; R_1, R_2' – the stator resistive impedance and the rotor reduced resistive impedance, respectively, Ohm; L_1, L_2 – the inductance of the stator and rotor phase winding, respectively, H; L_μ – mutual inductance of the stator and rotor windings, H; M – IM electromagnetic moment, N·m; $M_s = M_v/r_g$ – the static moment on IM shaft, N·m; $M_v = f(p, \beta)$ – resistance moment created by the stopcock and depending on pressure p at the stopcock and its opening relative degree β, N·m; ω_0 – the synchronous rotation frequency of IM stator magnetic field, s^{-1}; ω_{2e} – the IM angular slip frequency, s^{-1}; J_Σ – the total inertia moment of the stopcock ED, kgm^2; f_0 – the frequency of supply voltage of IM stator, Hz; p_p – the number of IM poles pairs; σ_{12} – dissipation coefficient; $\overrightarrow{\psi_1}, \overrightarrow{\psi_2}$ – the vector of the flux linkage of the stator and rotor, respectively, Wb; $\vec{i}_1, \vec{i}_2$ – the vector of the current of the stator and rotor phase, respectively, A.

The consumer is described by a hydraulic characteristic of the form [1, 2, 6, 10]:

$$H_c - h_s = R_{cn} Q_c^2 , \tag{2.8}$$

where H_c, Q_c – the head and discharge in the consumer network, respectively, m, m³/s; h_s – the static head in the pipeline system, m; $R_{cn} = (H_{cn} - h_s)/Q_{cn}^2$ – the consumer rated hydraulic resistance, s²/m⁵; H_{cn}, Q_{cn} – the rated head and discharge in the consumer network, respectively, m, m³/s.

Solution of (2.5) by means of the final elements method enables the representation of the pipeline network by a final number of sections with equal parameters and passing from partial derivatives to difference analogs:

$$\frac{\partial H}{\partial x} = \frac{H_i - H_{i-1}}{l} ; \frac{\partial Q}{\partial x} = \frac{Q_i - Q_{i-1}}{l} , \tag{2.9}$$

where $l = \frac{L}{N_L}$ – the length of the pipeline section, m; L – the length of the pipeline network, m; N_L – the number of the sections into which the pipeline is divided; H_i, Q_i, H_{i-1}, Q_{i-1} – the head and discharge at the output and the input of the *i*-th section of the pipeline, respectively, m, m³/s.

Then for the *i*-th section, the head and discharge equations will be of the form:

$$\begin{cases} H_i - H_{i-1} + l_0 l \dfrac{dQ_i}{dt} + r_0 l_{уч.} Q_i |Q_i| = 0; \\ \dfrac{dH_i}{dt} + c_0 \dfrac{1}{l} (Q_i - Q_{i-1}) = 0, \end{cases} \tag{2.10}$$

where $r_0 = \lambda_{pl} / (2gS^2 d)$, $l_0 = 1/(gS)$, $c_0 = c^2/(gS)$ – the specific parameters of the pipeline section, s²/m⁶, s²/m³, m⁻¹; $H_i = h_i + h_{s_i}$, $H_{i-1} = h_{i-1} + h_{s_{i-1}}$ – the head at the output and input of the *i*-th section of the pipeline, respectively, m; $\lambda_{pl} = 0.11 (k_{pl}/d)^{0.25}$ – the coefficient of resistance of the pipeline; k_{pl} – the pipe roughness, m; $h_{s_i}, h_{s_{i-1}}$ – the height of the location of the beginning and the end of the *i*-th section in relation to the pump axle, respectively, m; $h_i = p_i/(\rho g)$, $h_{i-1} = p_{i-1}/(\rho g)$ – the absolute head at the output and input of the i-th section of the pipeline, respectively, m; p_i, p_{i-1} – the absolute pressure at the output and input of the i-th section of the pipeline, respectively, Pa.

This representation of the pipeline network makes it possible to install the regulating stopcock and to research the dynamic processes at its any section.

Based on the above given mathematical expressions, we made a block diagram of the PC model (Figure 2.8) for the research of the transient processes during the stopcock control.

In this case, the head at the stopcock is determined according to the formula:

$$H_v = (R_v + R_{cn})Q_{cn}^2, \tag{2.11}$$

where $R_v = \xi_v(\beta)/(2gS^2)$ – the stopcock hydraulic resistance, s²/m⁵; $\xi_v(\beta)$ – the regulating stopcock hydraulic resistance coefficient ξ_v dependence on the relative degree β of its opening.

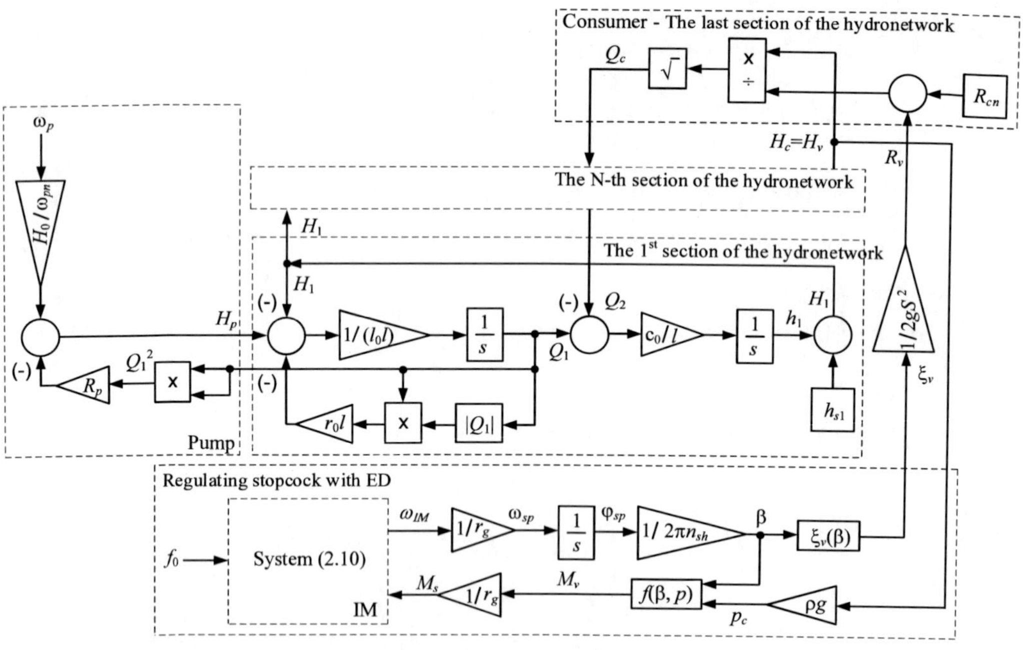

Figure 2.8. A block diagram of the model of PC with adjustable pipeline valves.

The developed model of PC with adjustable pipeline valves enables the formation of the required trajectories of its actuator motion by assigning the frequency f_0 of the voltage supplied to the electric motor stator. The obtained PC model, unlike the conventional ones, represents a pump, a consumer and a pipeline network in the form of inseparable interconnected systems, wherein the output parameters of one of them are the input parameters for another one and vice versa.

The hydronetwork multilink structure allows the research of the dynamic processes at every point of the hydrotransport system, and the variation of the network configuration (length, diameter, the section height (static head) etc.), the parameters of the stopcock, the consumer and the analysis of these factors influence on the transient processes. In addition, there is a possibility to install the stopcock at any section of the pipeline system.

2.3. The Dynamic Processes in a Pumping Complex with Adjustable Pipeline Valves

To research the dynamic loads we analyze PC with the pump rated head $H_{pn} = 60$ m, the working medium flow velocity $\upsilon_n = 1$ m/s and the consumer hydraulic resistance $R_{cn} = 40.65$ s²/m⁵ respectively; the pipeline length $L = 3000$ m and diameter $d = 1.2$ m respectively; the number of the pipeline sections $N = 20$; the velocity of the propagation of the surge wave $c = 1000$ m/s. The values of the specific parameters of every section: $r_0 = 0.002$ s²/m⁶, $l_0 = 0.0901$ s²/m³, $c_0 = 90132$ m⁻¹. The full shutting of the used stopcock requires $m = 50$ turns of its spindle, which provides the induction ED with the power of $P = 2$ kW and the reducer with the gear ration $r_g = 28$.

Figures 2.9–2.20 contain the graphs of time-variable relative degree β of valve opening, head H_p at the pump output, head H_v in the pipeline network at the stopcock, discharge Q_c in the consumer network at linear (a, c) and nonlinear (b) laws of its shutting with the constant and variable frequency of ED supply voltage, respectively.

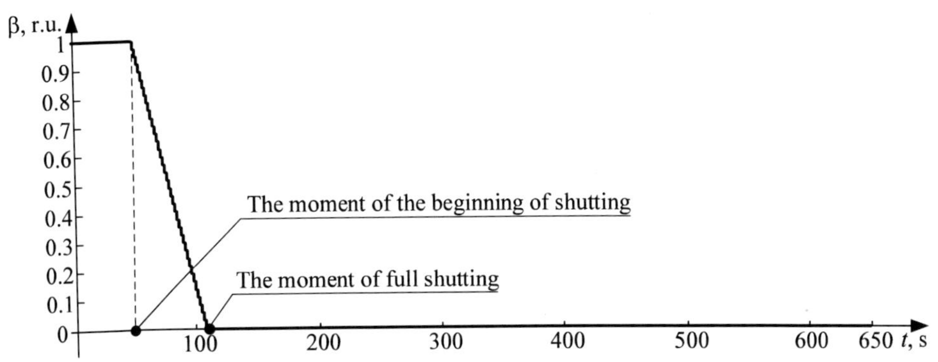

Figure 2.9. A graph of the time-variable relative degree β of the stopcock opening at the linear law of its shutting with the ED supply voltage frequency equal to 50 Hz.

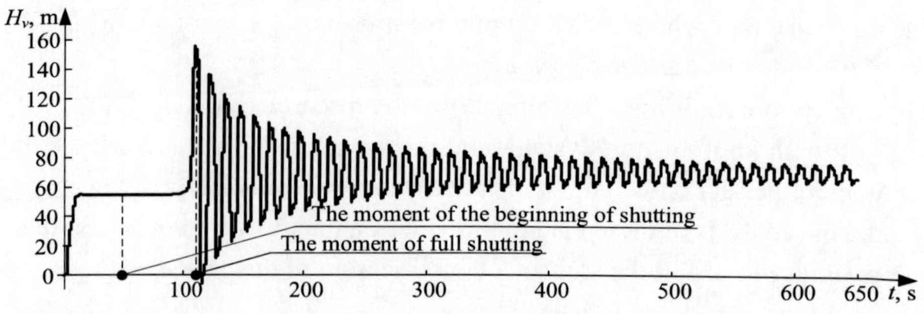

Figure 2.10. A graph of the time-variable head H_v in the pipeline network at the stopcock at the linear law of its shutting with the ED supply voltage frequency equal to 50 Hz.

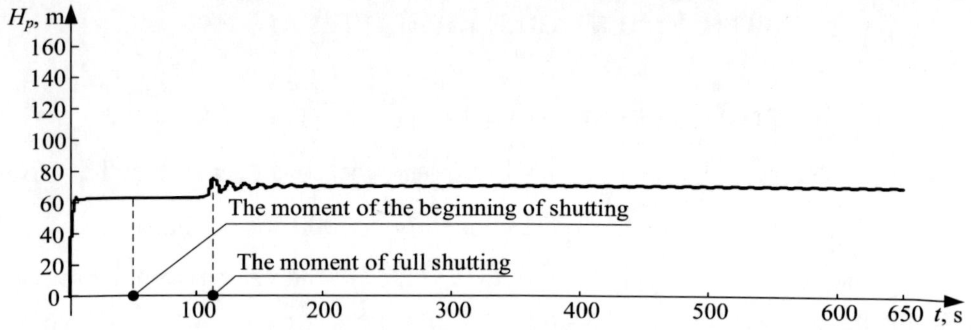

Figure 2.11. A graph of the time-variable head H_p at the pump output at the linear law of the stopcock shutting with the ED supply voltage frequency equal to 50 Hz.

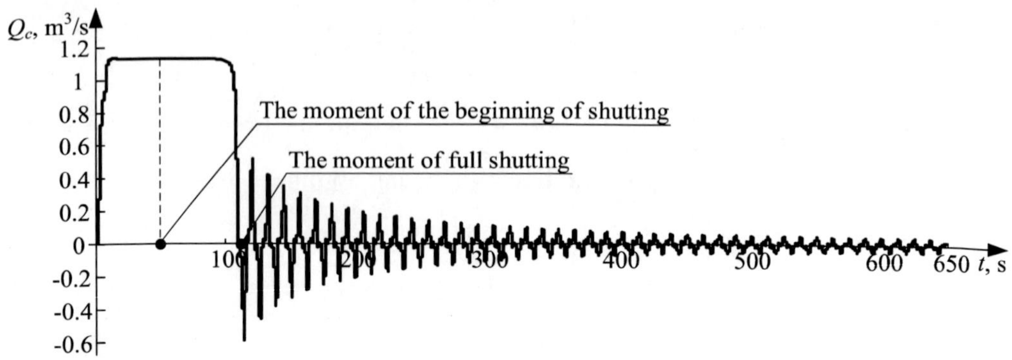

Figure 2.12. A graph of the time-variable discharge Q_c in the consumer network at the linear law of the stopcock shutting with the ED supply voltage frequency equal to 50 Hz.

The stopcock linear shutting is performed at constant frequency of the supply voltage of 50 Hz and 5 Hz, respectively. The stopcock nonlinear shutting is performed with the frequency of 50 Hz on the section at $\beta > 0.2$ and 5 Hz – at $\beta \leq 0.2$ respectively.

The graphs demonstrate that linear shutting of the stopcock with the supply voltage frequency equal to 50 Hz (Figure 2.9–2.12) is accompanied by a quick increase of pressure at the stopcock; its value exceeds the pump rated head by 2.5 times. The reduction of the supply voltage frequency to 5 Hz results in the decrease of the head to the value of 0.7 of the rated value and significant increase of the valve control time (Figure 2.13–2.16).

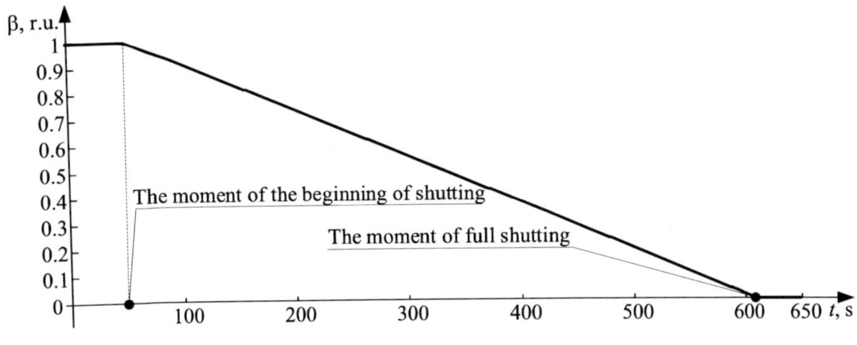

Figure 2.13. A graph of the time-variable relative degree β of the stopcock shutting at the linear law of its shutting with the ED supply voltage frequency equal to 5 Hz.

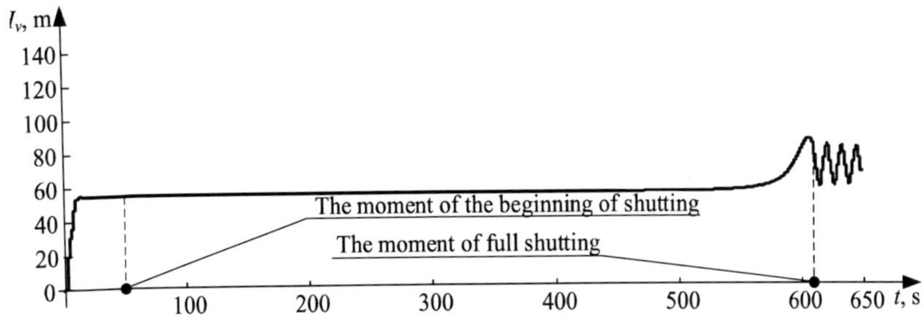

Figure 2.14. A graph of the time-variable head H_v in the pipeline network at the stopcock at the linear law of its shutting with the ED supply voltage frequency equal to 5 Hz.

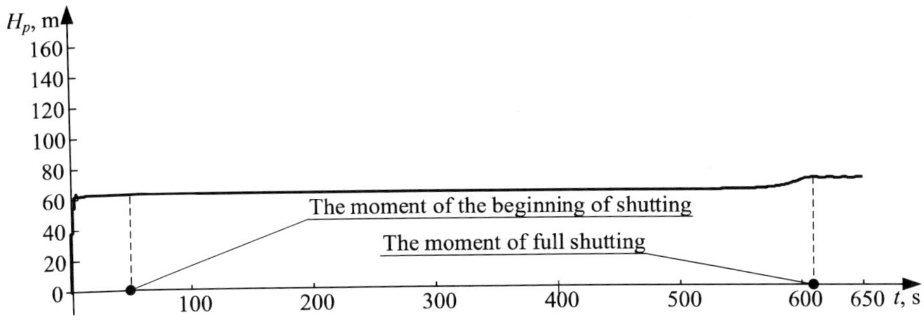

Figure 2.15. A graph of the time-variable head H_p at the pump output at the linear law of the stopcock shutting with the ED supply voltage frequency equal to 5 Hz.

A possibility of the pipeline valve shutting by the nonlinear law is worth considering (Figure 2.17–2.20). The analysis of the curves representing the variation of the head at the stopcock revealed that significant decrease of dynamic loads in the pipeline is observed both at the reduction of the value of the supply voltage frequency and at the formation of the nonlinear law of shutting. Besides, the increase of pressure in both cases is of the same value, and the time of valve shutting decreases by dozens of times, which is important for valve control in emergency modes caused by sudden outage of electricity supply in PC and occurrence of liquid counterflow.

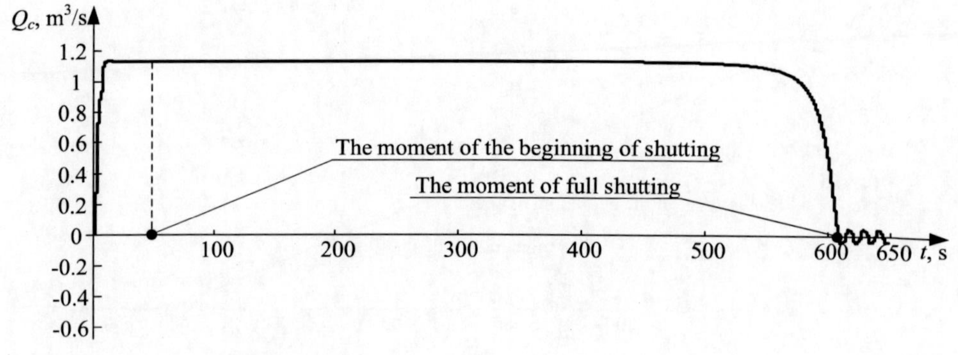

Figure 2.16. A graph of the time-variable discharge Q_c in the consumer network at the linear law of the stopcock shutting with the ED supply voltage frequency equal to 5 Hz.

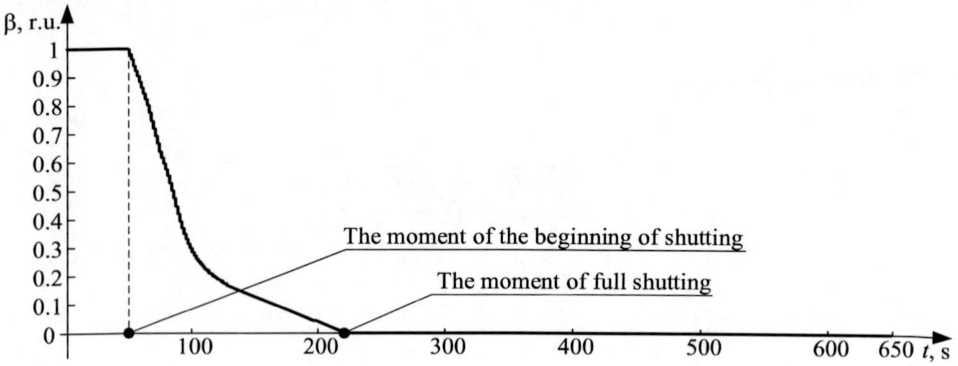

Figure 2.17. A graph of the time-variable relative degree β of the stopcock opening at the nonlinear law of its shutting with the ED supply voltage frequency equal to 50 Hz on a section at $\beta > 0.2$ and 5 Hz – at $\beta \leq 0.2$.

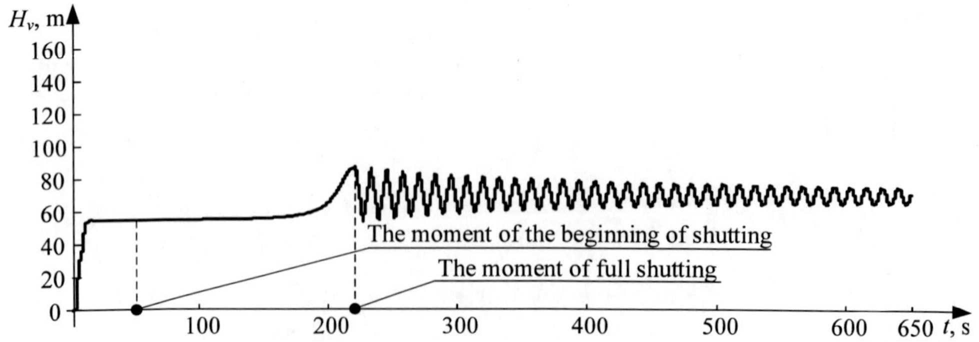

Figure 2.18. A graph of the time-variable head H_v in the pipeline network at the stopcock at the nonlinear law of its shutting with the ED supply voltage frequency equal to 50 Hz on the section at $\beta > 0.2$ and 5 Hz – at $\beta \leq 0.2$.

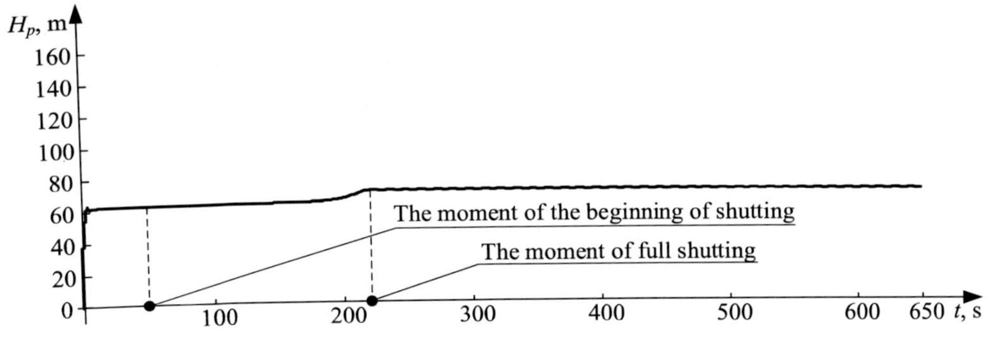

Figure 2.19. A graph of the time-variable head H_p at the pump output at the nonlinear law of its shutting with the ED supply voltage frequency equal to 50 Hz on the section at $\beta > 0.2$ and 5 Hz – at $\beta \leq 0.2$.

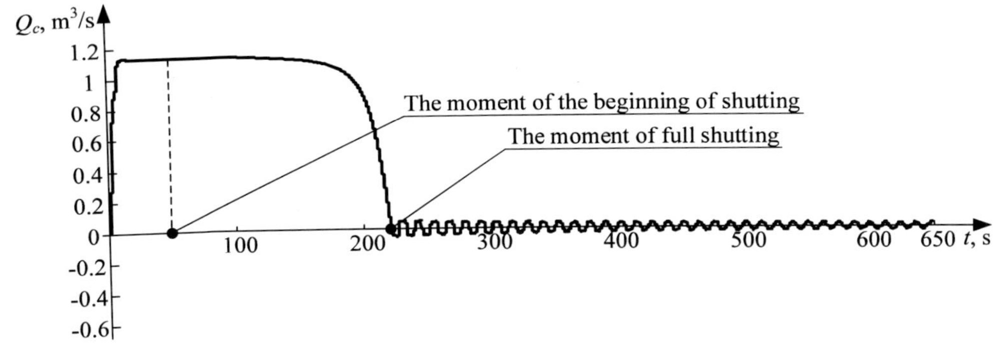

Figure 2.20. A graph of the time-variable discharge Q_c in the consumer network at the nonlinear law of its shutting with the ED supply voltage frequency equal to 50 Hz on the section at $\beta > 0.2$ and 5 Hz – at $\beta \leq 0.2$.

2.4. The Block Diagrams of the Models of a Pumping Complex with a Variable-Frequency Electric Drive and a Cavitation Channel in the Pipeline Network

2.4.1. The Power Components of the Liquid Head Motion in the Pumping Complex

The analysis of the power balance of the real liquid head motion is performed in relation to a functional diagram of PC (Figure 2.21), including a pump and a pipeline network consisting of three sections.

Let liquid with mass m and velocity υ move under pressure along a pipeline elementary section of length l and cross section area S (Figure 2.22).

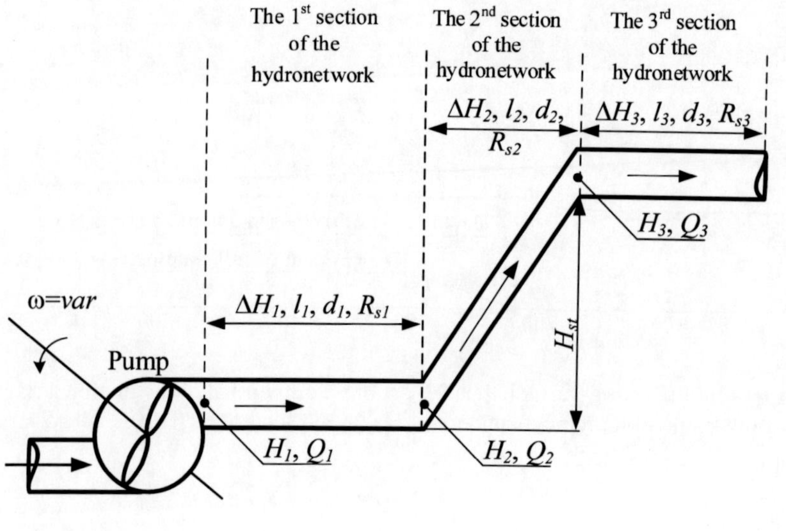

Figure 2.21. PC functional diagram.

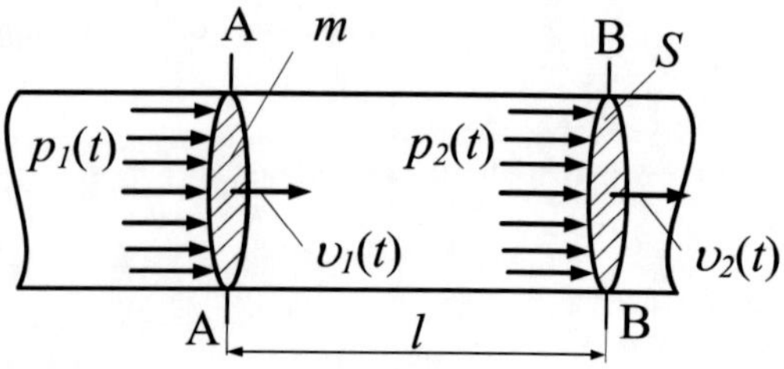

Figure 2.22. An elementary section of the pipeline.

According to Newton's law, to move incompressible non-viscous liquid of mass m from one point of the pipeline (cross section A-A) to another one (cross section B-B) with velocity υ it is necessary to apply force:

$$F = F_{kin} + F_{tr} = m\frac{d\upsilon}{dt} + \Delta pS\,, \tag{2.12}$$

where $F_{kin} = ma = m\frac{d\upsilon}{dt}$ – the force required to move the liquid, N; $F_{tr} = \Delta pS$ – resistance force, caused by friction at liquid motion, N; m – liquid mass, kg; a – acceleration, m/s^2; υ – the velocity of the liquid flow motion, m/s; Δp – pressure losses, Pa; S – the cross section area of the liquid flow, m^2.

On the other hand, according to the laws of hydrostatics, the force action at the pipeline section [10–15, 101–105]:

$$F = (p_1 - p_2)S\,, \tag{2.13}$$

where p_1, p_2 – the pressure at the pipeline section input and output, Pa.

Then, jointly solving equations (2.12) and (2.13) we get:

$$(p_1 - p_2)S = m\frac{d\upsilon}{dt} + \Delta pS . \tag{2.14}$$

Taking into account mass $m = \rho V$ and velocity $\upsilon = \frac{Q}{S}$ of the liquid flow at volume $V = lS = const$, we obtain:

$$\frac{\upsilon S}{p_2}(p_1 - \Delta p) = \frac{\rho l \upsilon}{p_2}\frac{dQ}{dt} + Q , \tag{2.15}$$

where ρ – the density of the liquid flow, kg/m^3; l – the length of the pipeline section, m; Q – liquid discharge, m^3/s.

Then the transfer function describing the liquid flow motion along the pipeline is:

$$W(p) = \frac{Q(p)}{p_1(p) - \Delta p(p)} = \frac{k_u}{T_u p + 1} , \tag{2.16}$$

where $k_u = \frac{\upsilon S}{p_2}$ – the coefficient of proportionality, m^3/s·Pa; $T_u = \frac{\rho l \upsilon}{p_2}$ – the inertia constant of the liquid column, s.

Taking into account the mass conservation law in the elementary volume, the liquid flow motion is to be supplemented with the continuity equation:

$$\rho(Q_1 - Q_2) = \frac{dm}{dt} , \tag{2.17}$$

where $Q_1 = Q$, $Q_2 = \sqrt{\frac{\rho g \Delta p}{R_s}}$ – the liquid discharge at the pipeline section input and output, respectively, m^3/s; R_s – hydrodymanic resistance, s^2/m^5.

Assuming that $m = \rho V$ at volume $V = lS = const$, and the squared velocity of the sound in the still liquid $c^2 = \frac{dp}{d\rho}$, we obtain a differential equation of the form:

$$\frac{c\rho}{S}(Q_1 - Q_2) = \frac{l}{c}\frac{dp}{dt}, \tag{2.18}$$

where p_0 – the steady-state value of the pressure in the stationary mode, Pa; Q_0 – the steady-state value of the discharge in the stationary mode, m³/s.

Then the transfer function describing the operation of the pipeline as a reservoir is:

$$W(p) = \frac{\Delta p(p)}{Q_1(p) - Q_2(p)} = \frac{k_e}{T_e p}, \tag{2.19}$$

where $T_e = \frac{l}{c}$ – the capacity time constant of the water conduit, s; $k_e = \frac{\rho c}{S}$ – the coefficient of proportionality, kg/s·m⁴; Δp – the pressure action at the pipeline network section (pressure losses), Pa.

Thus, the analysis of the headed motion of the liquid made it possible to single out three energy components of the liquid flow:

- an active component – related to head losses in the pipeline caused by the presence of the hydronetwork resistance, W:

$$N_R(t) = R_s Q^3(t), \tag{2.20}$$

where $Q(t)$ – the liquid discharge at the hydronetwork section, m³/s;

- a kinetic component – caused by the motion of liquid of mass m with velocity υ, W:

$$N_L(t) = \frac{dW_L(t)}{dt} = \frac{d\left(\frac{m\upsilon^2(t)}{2}\right)}{dt} = \frac{d\left(C_1 Q^2(t)\right)}{dt} = C_2 Q(t), \tag{2.21}$$

where $W_L(t)$ – the kinetic energy at the hydronetwork section, W·s; $C_1 = \frac{m}{2S^2}$, $C_2 = 2C_1$ – constants, respectively, kg/m⁴·s, kg/m·s².

- a potential component corresponding to the pipeline filling or rise of the liquid to a certain level, W:

$$N_C(t) = p_{pot}(t)Q(t) = \frac{k_{tr}}{T_{tr}} Q(t) \int_0^t Q(t)dt\,, \tag{2.22}$$

where $p_{pot}(t)$ – the pressure acting at the hydronetwork section, Pa; k_{tr} – the pipeline structural coefficient, kg/m^4·s; T_{tr} – the pipeline filling time constant, s.

Such representation makes it possible to describe the energy processes at each of the n-th sections of the communication network by an RLC-contour and to take into account the density, temperature, liquid heterogeneity changing in time.

2.4.2. The Model of a Pumping Complex with One Section of the Hydraulic Network

The simplest PC consists of a pump with gently sloping *H-Q* characteristics, operating for a pipeline network with constant parameters without counter-pressure ($H_{st} = 0$) and with time-constant hydrodynamic resistance of the pipeline ($R_s = const$).

The differential equations of the pump induction ED at variable frequency, represented in the form of flux linkages for the synchronous coordinate system rotating with the velocity of the stator field, and provided that $\omega_k = \omega_{0e}$, $u_2 = 0$, are of the form (2.7).

In this case, the power supplied to the machine from the stator is:

$$N_d = u_{1u}i_{1u} + u_{1v}i_{1v}\,, \tag{2.23}$$

where u_{1u}, u_{1v}, i_{1u}, i_{1v} – the vectors of the voltages and currents of the corresponding coordinates.

Equations (2.7) are obtained provided that the real axis of the coordinate system (u–v), rotating synchronously, is aligned with the stator voltage vector u_1.

The resistance moment created by the pump [1-6], N·m:

$$M_s = \frac{N_N}{\omega} = \frac{\rho g H_N Q}{\eta_N \omega}\,, \tag{2.24}$$

where N_N – the useful power of the pump, W; ρ – the density of the pumped liquid, kg/m^3; $g = 9.81$ m/s^2 – the acceleration of gravity; H_N – the head developed by PU, m; Q – PC discharge, m^3/s; η_N – the pump efficiency.

Taking into account (2.7), the block diagram of the induction motor model is shown in Figure 2.23.

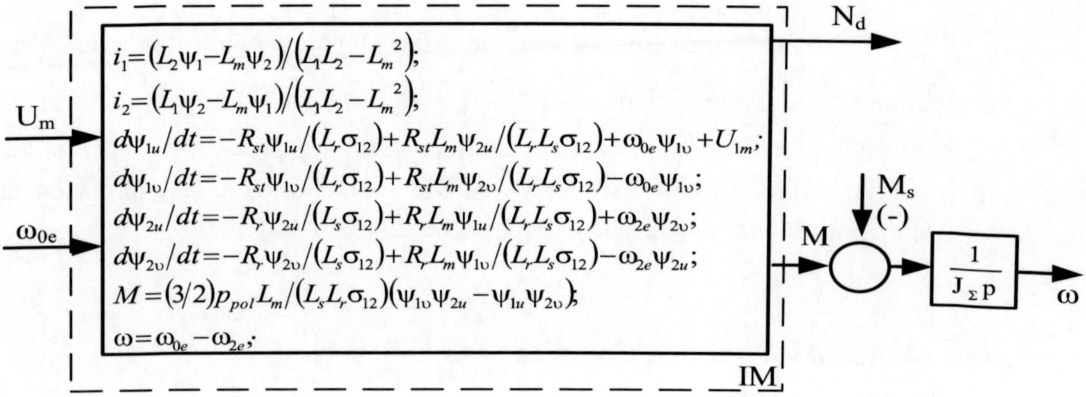

Figure 2.23. The block diagram of the induction motor model.

The head developed by PU at the constant rotation frequency of the pump ED is determined according to expression, m:

$$H_N = H_0 - \Delta H_N = H_0 - R_{VN} Q^2 , \tag{2.25}$$

where H_0 – the head of the pump at zero discharge, m; ΔH_N – the head losses in the pump, m; R_{VN} – the internal resistance of the pump, s^2/m^5.

The pump is an inertial mechanism whose time constant T_H is determined by the time of liquid motion from the intake pipe into the head pipe [101–108], and the latter is an inseparable part of the pump and represents an elementary section of the pipeline. Then, taking into account the above analyzed and the relations for the head $H_{pot} = p_{pot}/\rho g$, discharge $Q = \upsilon S$, the physical processes in the pump and the adjacent pipeline are described by transfer functions of the form:

$$W_{N1}(p) = \frac{Q(p)}{H_N(p) - \Delta H_\Sigma(p)} = \frac{k_u}{T_\Sigma p + 1} ; \tag{2.26}$$

$$W_{N2}(p) = \frac{\Delta H_N(p)}{Q^2(p)} = R_{VN} ; \tag{2.27}$$

$$W_{N3}(p)=\frac{\Delta H(p)}{Q(p)-Q_{OS}(p)}=\frac{k_e}{T_e p};\tag{2.28}$$

$$W_{N4}(p)=\frac{Q_{OS}^2(p)}{\Delta H(p)}=R_{s\Sigma},\tag{2.29}$$

where $k_u=\frac{Q_{nom}}{H_{nom}}$, $k_e=\frac{c}{gS}$ – the coefficient of proportionality, respectively, m²/s, s/m²; H_{nom} – the rated value of the pump at the output of the pump/section of the pipeline, m; Q_{nom} – the discharge of the liquid at the pipeline section at the head H_{nom}, m³/s; $T_\Sigma=T_N+T_u$ – the time of the liquid motion through the pump and along the adjacent pipeline, s; $T_u=\frac{lQ_{nom}}{gSH_{nom}}$, $T_e=\frac{l}{c}$ – the inertia and capacity time constants of the pipeline, respectively, s; $T_N=4(d_2-d_1)/\bar{\upsilon}z_p\,ln(d_2/d_1)$ – the pump time constant, s; d_1, d_2 – the input and output diameters of the centrifugal pump circular grid, m; $\bar{\upsilon}$ – the velocity of liquid flow in the inter-blade space, m/s; z_p – the number of blades in the circular grid; ΔH_Σ – the total head losses in the hydrodynamic network (in this case, $\Delta H_\Sigma=\Delta H$), m; ΔH – the head losses in the adjacent section of the pipeline, m; $R_{s\Sigma}$ – the hydrodynamic resistance of the pipeline (in the considered case – of the adjacent section of the hydronetwork R_s), s²/m⁵.

Figure 2.24 shows a block diagram of the pump with a gently sloping *H–Q* characteristic and an adjacent section of the hydronetwork.

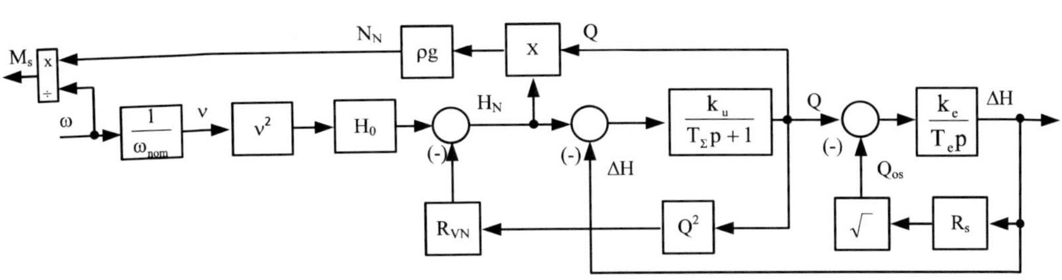

Figure 2.24. A block diagram of the model of the pump with an adjacent section of the hydronetwork.

In the operation practice, one can often deal with pumps whose head-discharge characteristic has a maximum. Figure 2.25 shows a block diagram of such turbomechanism and an adjacent section of the hydronetwork.

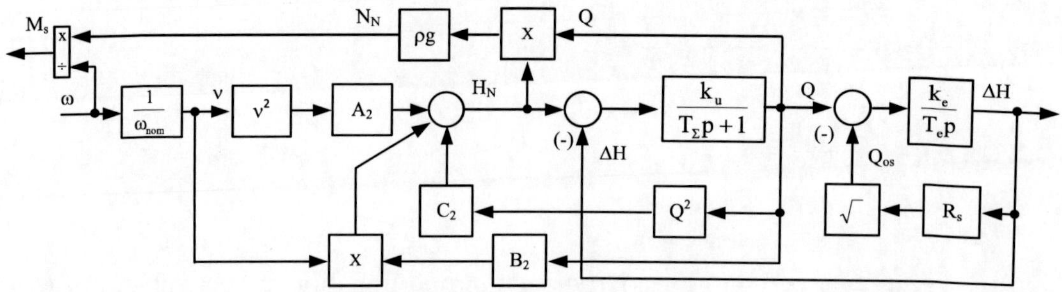

Figure 2.25. A block diagram of the model of the pump unit whose head-discharge characteristic has a maximum.

A generalized structural block diagram of the model of the IM-pump-hydronetwork section system has the form shown in Figure 2.26.

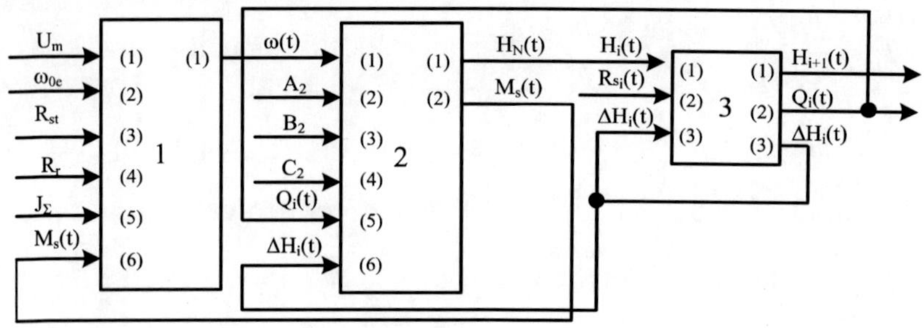

Figure 2.26. A block diagram of the model of the IM-pimp-hydronetwork section system: 1 – IM; 2 – pump; 3 – hydronetwork section.

2.4.3. The Model of a Cavitation Channel

Papers [55-70, 78-82] reveal that provided the current cavitation number χ equals or is less than its critical value χ_{kr}, determining the boundary of the development of the cavitation processes, PC operation mode is accompanied by the occurrence of cavitation processes resulting in the growth of unproductive power losses caused by the variation of the pipeline hydraulic resistance. The latter results from periodic occurrence of cavitation caverns whose volume essentially depends on the pressure and velocity of motion of the liquid flow. Taking into account the research described in papers [55-66, 78-80, 101-108], the value of volume V_{kav} of the cavitation cavern is determined according to expression:

$$V_{kav} = \frac{1}{T_k} \int \left(Q_1(t) - Q_2(t)\right) dt, \tag{2.30}$$

where $T_k = \frac{V_0}{G}\gamma$ – the inertia time constant of the cavitation cavern, s; G – the weight discharge of liquid at V_0, N/s; V_0 – the initial volume of the cavitation cavern, m^3; $\gamma = \rho g$ – the liquid specific weight, N/m^3.

Taking into consideration the relation between the weight and mass discharges $G = \rho g Q$, the specific weight and density $\gamma = \rho g$ the inertia time constant of the cavitation cavern is:

$$T_k = \frac{V_0}{Q}. \tag{2.31}$$

We assume that the cavitation cavern has the form of a sphere. Then, taking into account expression (1.9), the value of the critical volume of the cavitation cavern corresponding to its destruction is determined by the expression of the form:

$$V_{kr} = \frac{4\pi}{3}\left(\frac{2\sigma}{p_{kavi} - p_{poti}}\right), \tag{2.32}$$

where p_{kavi} – the pressure in the cavern, Pa; p_{poti} – the pressure at the *i*-th section of the pipeline, Pa.

With the assumptions that the cavitation cavern has the form of a sphere with constant volume, does not move along the length of the pipeline and represents a local narrowing at the network section, the head losses, caused by the presence of cavitation processes in the hydronetwork, m:

$$\Delta h_{kav} = \zeta_{kav}\frac{\upsilon^2}{2g}, \tag{2.33}$$

where $\zeta_{kav} = \left(\frac{d^2}{\varepsilon d_p^2} - 1\right)^2$ – the coefficient of local resistance of the flow narrowing caused by the cavitation processes in the pipeline; $\varepsilon = 0.57 + \frac{0.43}{\left(1 - \left(\frac{d_p^2}{d^2}\right)\right)}$ – the coefficient of flow contraction; d – the pipeline diameter before the narrowing, m; $d_p = d - \sqrt[3]{\frac{6V_{kav}}{\pi}}$ – the diameter of the flow in the narrow part, m.

Then, the hydrodynamic resistance corresponding to the presence of cavitation processes, s²/m⁵:

$$R_{kav} = \frac{\zeta_{kav}}{2gS^2}. \tag{2.34}$$

Taking into account expression (2.23), the hydraulic resistance at the i-th section of the pipeline network is:

$$R_{s\Sigma\, i} = R_{s\, i} + R_{kav\, i}. \tag{2.35}$$

Then the block diagram of the model of the cavitation channel is of the form shown in Figure 2.27.

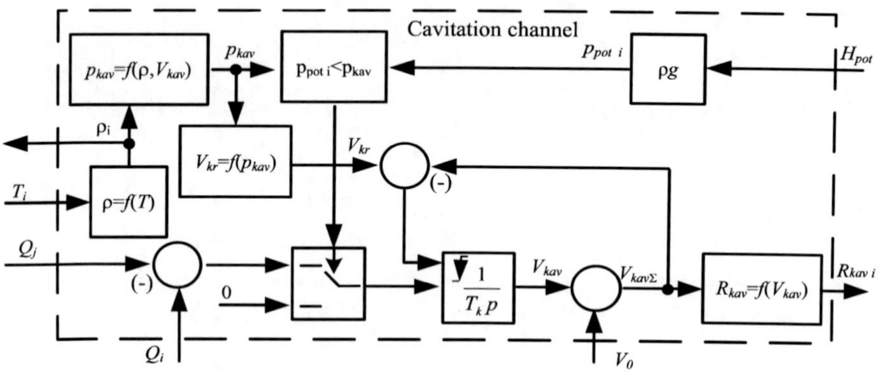

Figure 2.27. A block diagram of the model of the cavitation channel.

Such representation of the cavitation channel enables its connection to any section of the pipeline where cavitation processes develop.

A generalized structural block diagram of the model of the IM-pump-hydronetwork section system taking into consideration the cavitation processes in the pipeline is of the form shown in Figure 2.28.

When researching a mathematical model of PC with the parameters: $H_{nom} = 75$ m, $Q_{nom} = 630$ m³/h, $H_0 = 92$ m, $\omega_{nom} = 151.8$ s⁻¹, $N_{nom} = 176$ kW, $\eta_{nom} = 0.75$; $T_V = 25$ °C, $l = 1000$ m, $d = 0.4$ m, $V_{kr} = 0.015$ m³ we obtained the characteristics of the pumping complex without taking into account (Figure 2.29) and taking into account (Figure 2.30) the cavitation processes in the pipeline. Their analysis revealed that the development of the cavitation processes (switching-on the cavitation channel at the time moment t_1) at the hydronetwork section is accompanied by PC technological parameters pulsations caused by periodical collapse of the cavitation caverns in the liquid flow with the frequency of 10 Hz.

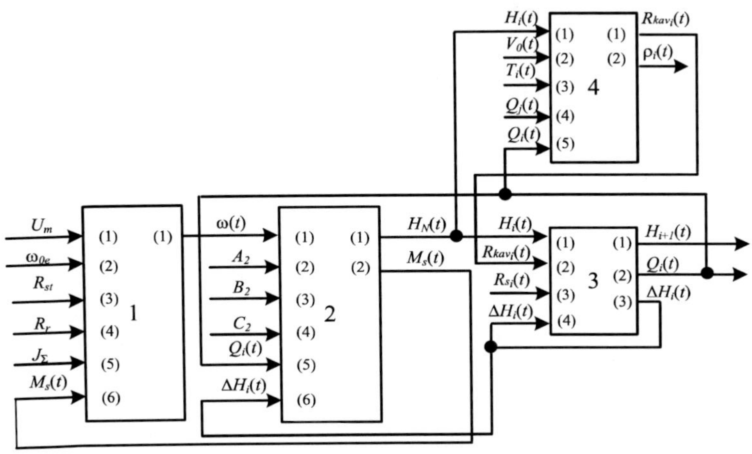

Figure 2.28. A structural block diagram of the model of IM-pump-hydronetwork system taking into account the cavitation processes in the pipeline: 1 – IM; 2 – pump; 3 – a hydraulic network section; 4 – a cavitation channel.

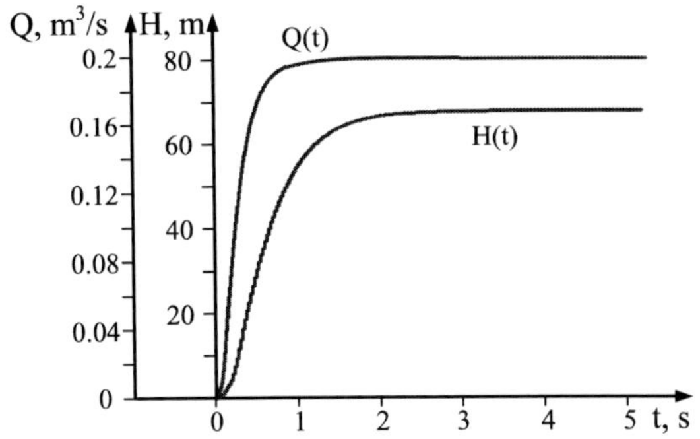

Figure 2.29. The dynamic characteristics of PC without cavitation processes in the hydronetwork.

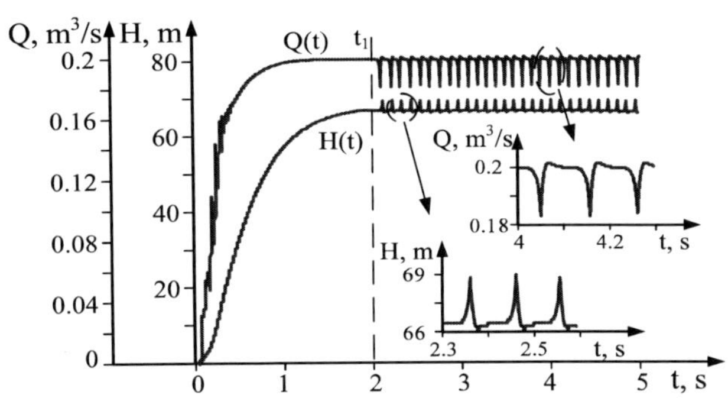

Figure 2.30. The dynamic characteristics of PC at the development of cavitation processes in the pipeline.

The proposed method of describing the pumping complex with cavitation processes in the hydrodynamic network makes it possible to research the dynamic modes of operation in the system with time-variable hydraulic resistance, water consumption mode, temperature, liquid density, etc.

2.4.4. The Model of a Pumping Complex Operating for the Communication Network of Complicated Configuration

It is known that the communication network of water- and heat-supply PC has a complicated configuration consisting of pipeline sections connected sequentially or in parallel. Each of them is characterized by its own diameter, length, the value of counter-pressure, the presence of local resistances (stopcocks, check valves, etc.), different properties of the transported medium (the degree of aeration, the temperature of the liquid flow), and by the time-variable mode of water consumption.

The free head at the end of the *i*-th section of the pipeline is:

$$H_{i+1} = H_i - \Delta H_i = \Delta H_{i+1} + \ldots + \Delta H_n . \tag{2.36}$$

Then the *i*-th elementary section of the hydrodynamic network is described by transfer functions of the form:

$$W_{i1}(p) = \frac{Q_i(p)}{H_i(p) - (\Delta H_i(p) + \Delta H_{i+1}(p) + \ldots + \Delta H_n(p))} = \frac{k_{u\,i}}{T_{u\,i}p + 1}; \tag{2.37}$$

$$W_{i2}(p) = \frac{\Delta H_i(p)}{Q_i(p) - Q_{OS\,i}(p)} = \frac{k_{e\,i}}{T_{e\,i}p}; \tag{2.38}$$

$$W_{i3}(p) = \frac{Q^2_{OS\,i}(p)}{\Delta H_i(p)} = R_{s\Sigma\,i}, \tag{2.39}$$

where in (2.37 – 2.39): $\Delta H_\Sigma = \Delta H_i + H_{st\,i} + \Delta H_{i+1} + \ldots + \Delta H_n$ – the total head losses in the hydrodynamic network; ΔH_i, ΔH_{i+1}, ΔH_n – the head losses at the *i*-th, (*i*+1)-th and *n*-th sections of the pipeline, respectively; $R_{s\Sigma\,i}$, $R_{s\Sigma\,i+1}$, $R_{s\Sigma\,n}$ – the hydrodynamic resistance of the *i*-th, (*i*+1)-th and the *n*-th sections of the pipeline, respectively.

A generalized structural block diagram of the model of PC with a complicated configuration of the pipeline is of the form shown in Figure 2.31.

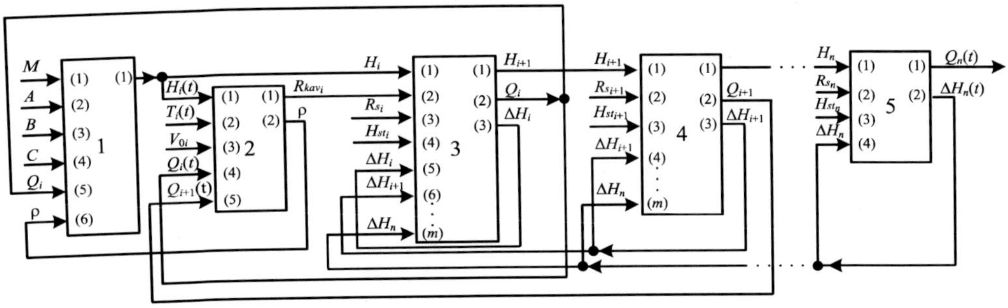

Figure 2.31. A block diagram of the model of PC with complicated configuration of the pipeline: 1 – PU; 2 – cavitation channel; 3 – 5 – *i*-th (adjacent to the pump), (*i*+1)-th and the *n*-th section of the hydronetwork, respectively.

The number *n*, of the sections into which the hydrodynamic network is conditionally divided depends on the posed tasks of the research. It should be noted that if there is an excessive number of the hydronetwork sections, the time spent on mathematical modeling will increase. Besides, it will be one of the sources of accumulation of the error.

Figure B.1 contains a block diagram of the model of the pumping complex with a variable-frequency ED and a cavitation channel consisting of a frequency converter represented by a first-order aperiodic link with the law of frequency control $U/f^2 = K_{pr}$; an induction motor with a squirrel-cage rotor whose mathematical description is performed based on a model in «*u*, *v*, 0» – coordinates [99, 100]; a pump unit with a gently sloping head = discharge characteristic; a pipeline network represented by three sequentially connected pipeline sections of different configuration; a cavitation channel taking into account cavitation processes at the second section of the hydronetwork.

When researching a PC mathematical model with the parameters: $H_{nom} = 75$ m, $Q_{nom} = 630$ m³/h, $H_0 = 92$ m, $\omega_{nom} = 151.8$ s⁻¹, $N_{nom} = 176$ kW, $\eta_{nom} = 0.75$, $l = 1000$ m, $d = 0.4$ m, we obtained curves representing the variation of the discharge and head at each of the pipeline sections (Figure 2.32, a), the temperature, the volume of the cavitation cavern and the pipeline head losses (Figure 2.32, b).

The analysis revealed that the pump operation in the interval $0 - t_1$ is characterized by technological parameters invariable in time. At the time moment $t_1 - t_2$ the liquid temperature *T* grows by 10°C, which is accompanied by the variation of the head and discharge (curves $H_1(t) - H_3(t)$ and $Q_1(t) - Q_3(t)$ Figure 2.32, a).

When the pressure at the 2-nd section becomes less than the pressure of the saturated steam (time moment t_2), cavitation cavities of volume V_{kav} generate (Figure 2.32, b), and collapse when they reach a certain size. It results in periodic oscillations of hydraulic resistance and, accordingly, in the alteration of the head losses (curves $\Delta H_1(t) - \Delta H_3(t)$), of free head (curves $H_1(t) - H_3(t)$) and discharge (curves $Q_1(t) - Q_3(t)$) (Figure 2.32).

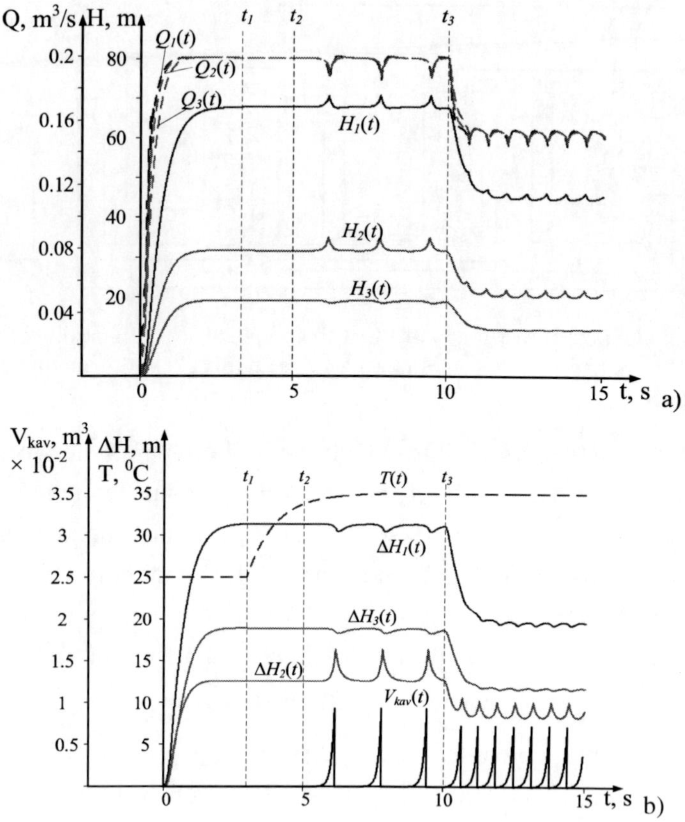

Figure 2.32. The curves representing the variation of the technological parameters of the pumping complex with a variable–frequency ED at the development of cavitation processes in the pipeline: $T(t)$ – temperature; $H_1(t)$ – $H_3(t)$ – head at the input of every section of the pipeline, respectively; $\Delta H_1(t)$ – $\Delta H_3(t)$ – head losses in the pipeline at every section; $Q_1(t)$ – $Q_3(t)$ – liquid discharge at every section of the pipeline; $V_{kav}(t)$ – the volume of the cavitation cavern.

Therefore, at the time interval $t_2 - t_3$ the change of temperature results in the growth of the head losses ΔH_2 in the pipeline by 30%. It should be noted that the presence of static head at the second section of the pipeline reduces the cavitation influence on the following section, which is accompanied by insignificant oscillations of the head at the third section of the hydronetwork.

Time moment $t_3 = 10$ s is characterized by the decrease of the pump ED rotation frequency by 20% down from the rated value, which results in the alteration of the parameters of the whole PC. Obviously, in this case, the cavern volume V_{kav} decreases by 40% and, accordingly, hydraulic resistance decreases too. For the analyzed PC the reduction of the rotation frequency results in the decrease of the amplitude of the wave processes (curves $H_2(t)$, $Q_2(t)$, $\Delta H_2(t)$) by two times.

2.4.5. The Transient Processes in a Pumping Complex with a Cavitation Channel

When researching a mathematical model of PC with the parameters given in Table 2.2, we obtained curves representing time-variable head (Figure 2.33), discharge (Figure 2.34), head losses (Figure 2.35), hydraulic power (Figure 2.36) and cavitation caverns volume (Figure 2.37) for different values of cavitation cavern time constants $T_{k1} = 2$ s; $T_{k2} = 0.2$ s.

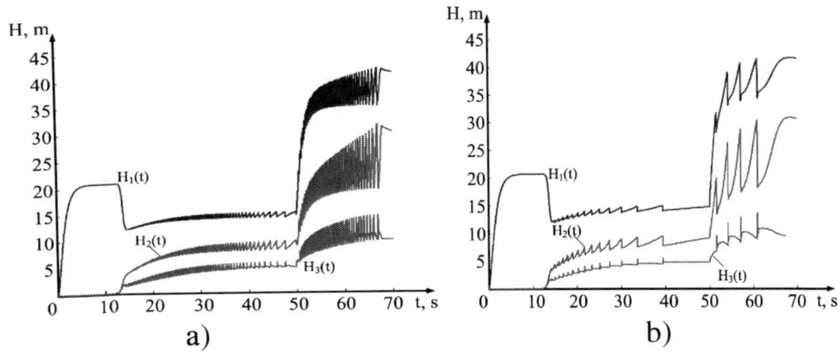

Figure 2.33. PC head curves at $T_{k1} = 0.5$ s (a) and $T_{k2} = 5$ s (b).

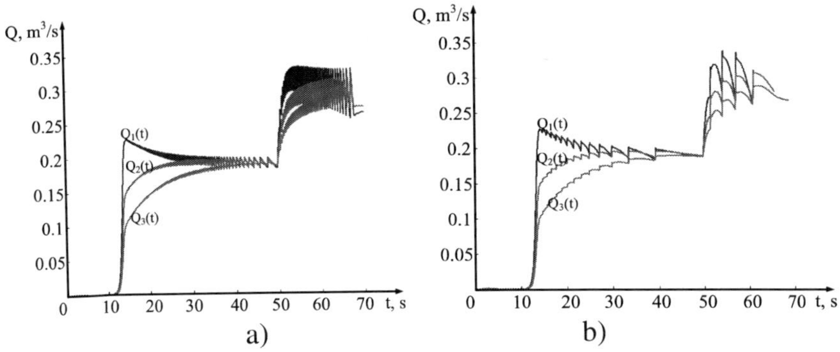

Figure 2.34. PC discharge curves at $T_{k1} = 0.5$ s (a) and $T_{k2} = 5$ s (b).

The analysis of the curves (Figures 2.33–2.36) revealed that in the interval of $0-10$ s the pump starts on the shut stopcock. In this case, the head at the output of the turbomechanism corresponds to the relative rotation frequency of the pump ED $\nu = 0.5$. The discharge, PC hydraulic power and the volume of cavitation caverns equal to zero.

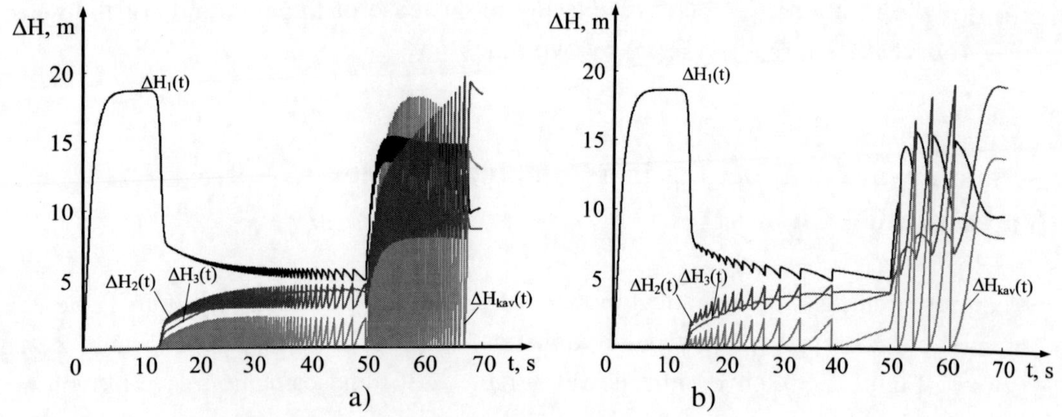

Figure 2.35. PC head losses curves at $T_{k1} = 0.5$ s (a) and $T_{k2} = 5$ s (b).

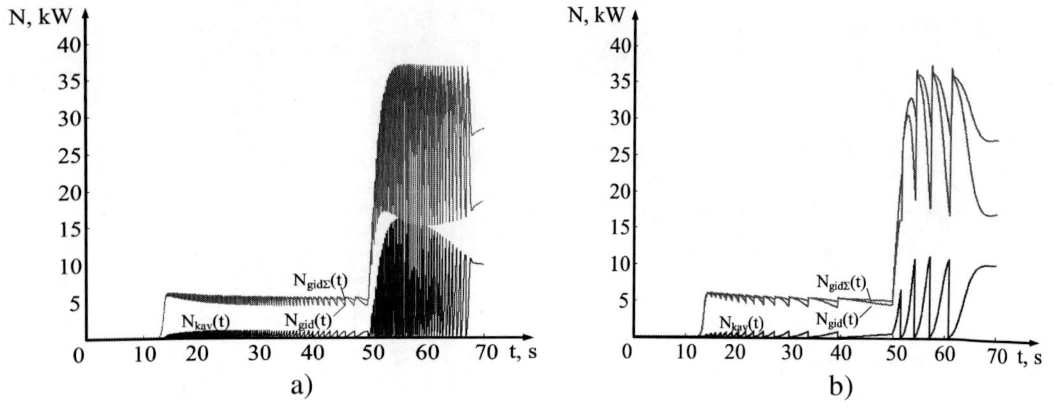

Figure 2.36. PC hydraulic power curves at $T_{k1} = 0.5$ s (a) and $T_{k2} = 5$ s (b).

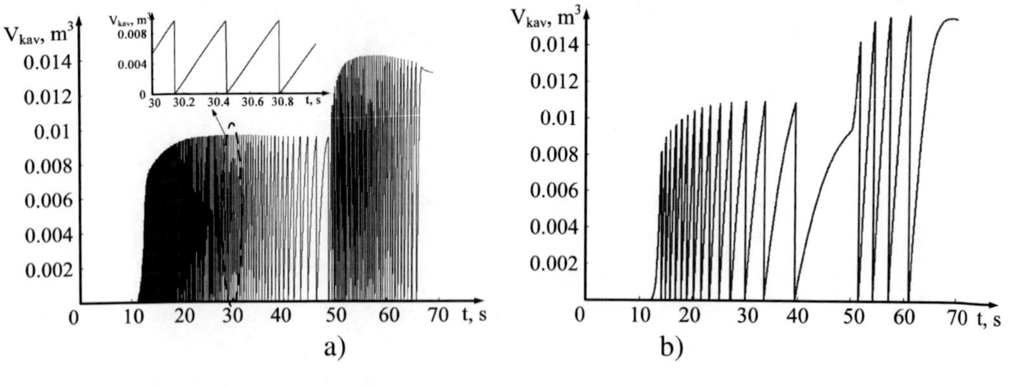

Figure 2.37. The curves representing the alteration of the volume of the cavitation cavern in the pumping plant $T_{k1} = 0.5$ s (a) and $T_{k2} = 5$ s (b).

Time moment $t = 10$ s is characterized by opening of the stopcock at the pipeline section adjacent to the pump. In this case, the value of head H_1 at the pump output decreases (Figure 2.33), and at the second H_2 and the third H_3 pipeline sections the head value grows with simultaneous increase of the liquid discharge at each section (curves $Q_1(t)$ – $Q_3(t)$ Figure 2.34) and PC hydraulic power $N_{gid\Sigma}$ (Figure 2.36). As the head value H_2 at the second section of the hydronetwork is less than the critical value H_k, determining the boundary of the cavitation operation mode, the volume V_{kav} of the cavitation cavern grows at the second section (Figure 2.37). In this case, when every cavitation cavern reaches the maximum value V_{kr} determined by expression (2.33), it collapses. This process repeats, which is accompanied by a sharp alteration of the pipeline hydraulic resistance and occurrence of pulsations of PC technological and energy parameters (Figures 2.33-2.36).

Table 2.2. PC parameters

Name	Value	Name	Value
Head developed by the pump	$H_{nom} = 62.5$ m	Cavern maximal volume	$V_{kr} = 0.0513Q$ m³
Discharge developed by the pump	$Q_{nom} = 1250$ m³/h	Length of the first pipeline section	$l_1 = 1000$ m
Head developed by the pump at zero supply	$H_0 = 83$ m	Length of the second pipeline section	$l_2 = 500$ m
Rated rotation frequency of the pump ED	$\omega_{nom} = 151.8$ s⁻¹	Length of the third pipeline section	$l_3 = 800$ m
Power of the pump electric drive	$N_{nom} = 250$ kW	Pipeline diameter	$d = 0.4$ m
Supply voltage	$U = 380$ V	Cavitation critical number	$\chi_{kr} = 350$
Pump efficiency	$\eta_{nom} = 0.86$	Velocity of sound in the liquid	$a = 1450$ m/s
Temperature of the pumped liquid	$T_V = 25$ °C	Radius of the air bubbles unsolved in the liquid	$R_0 = 10^{-4}$ m
Number of the hydronetwork sections	$n = 3$	Initial volume of cavitation caverns	$V_0 = 10^{-7}$ m³
Pressure in the cavern	$p_{kav} = 10$ kPa	Pump time constant	$T_N = 0.005$ s
Inertia time constants of the hydronetwork sections	$T_{u1} = 4.05$ s	Capacity time constants of the hydronetwork sections	$T_{e1} = 5.86 \cdot 10^{-4}$ s
	$T_{u2} = 6.75$ s		$T_{e2} = 8.79 \cdot 10^{-4}$ s
	$T_{u3} = 10.35$ s		$T_{e3} = 0.00135$ s

The analysis of the curves (Figures 2.35, 2.36) revealed that the losses of the head (curve $\Delta H_{kav}(t)$ Figure 2.35) and the hydraulic power (curve $N_{kav}(t)$ Figure 2.36), caused by cavitation processes, at $\nu = 0.5$ make about 40 and 20%, respectively. It should be noted that the amplitude of pulsations of the head and discharge at the pump output is higher than behind the cavitation section of the pipeline.

Time moment $t = 50$ s is characterized by the growth of water consumption, which is achieved by the increase of the relative rotation frequency $\nu = 0.8$ of ED turbomechanism. In this case, the PC technological and energy parameters grow (Figures 2.33-2.36), and the value of the maximal volume V_{kr} of cavitation caverns (Figure 2.37) increases by 1.4 times. The latter results in the growth by 40, 18, 60 and 30% of the amplitude of head H_2 pulsations, discharge Q_2, head losses ΔH_{kav} and hydraulic power N_{kav} at the second section of the pipeline, which are caused by cavitation processes, respectively.

Thus, the tenfold increase of the cavitation process time constant results in considerable decrease of the frequency of pulsations of PP technological and energy parameters and the resistance of the PC mathematical model to self-oscillations. The reduction of the pump ED rotation frequency is accompanied by cavitation oscillations with lower amplitude.

2.5. The Substantiation of the Possibility of the Use of the Gas-Vapor Mixture Energy in the Liquid Flow

The analysis [55-66, 78-80, 101-108] revealed and the results of mathematical modeling performed in 2.4 confirmed, that cavitation processes have their own energy characteristics and result in the growth of total losses of power in the pipeline when the value of cavitation number $\chi = \dfrac{2(p - p_{para})}{\rho \upsilon^2}$ decreases below the critical value χ_{kr}. We performed a quantitative assessment of the gas-vapor mixture at the headed motion of the liquid. For the analysis of the cavitation processes in PC the following parameters are adopted: head $H = 38$ m, supply $Q = 0.188$ m³/s, consumed power $N = 148$ kW, liquid temperature $T_V = 26$ ⁰C, saturated steam pressure $p_{para} = 3361.06$ Pa, pipeline diameter $d = 0.4$ m, liquid density $\rho = 1000$ kg/m³, cavitation critical number $\chi_{kr} = 700$ for the pump ascending pipe.

Figure 2.38 shows the head-discharge and energy characteristics of PC consisting of a centrifugal pump with a regulating electric drive operating on the pipeline network with the maximal value of discharge corresponding to point A of the operation mode.

For the considered case, the curve representing the variation of the cavitation number is of the form $\chi - Q$ (Figure 2.38). Comparing the cavitation number corresponding to the current operation mode (by curve $\chi - Q$), with the critical number χ_{kr}, we come to the conclusion about the presence of cavitation processes in the flow of liquid.

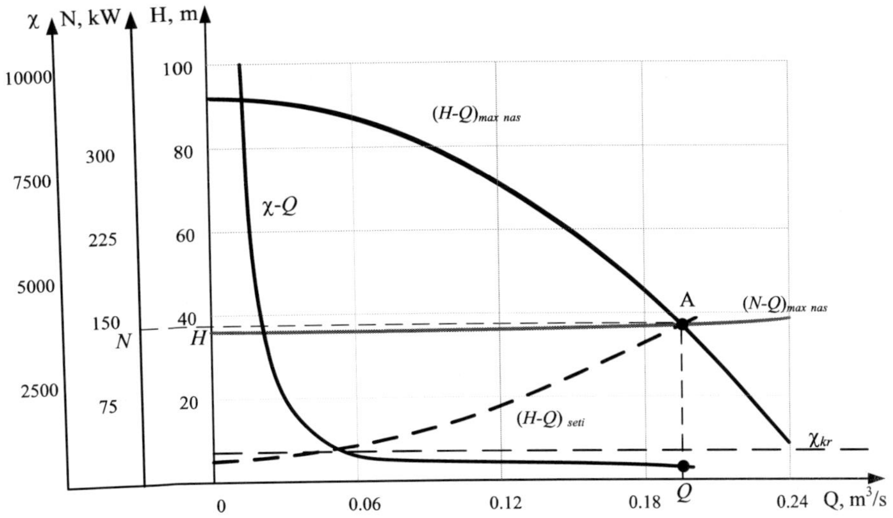

Figure 2.38. PS technological and energy characteristics.

The operation mode of PC changes in accordance with the consumption schedule characterized by maximal Q_{max} and minimal Q_{min} values of discharge.

Let us assume that at PC discharge equal to Q the gas-vapor mixture discharge in the liquid is $Q_{kav} = 0.25Q$ [55-66, 78-80]. Then the gas-vapor mixture hydraulic power, W:

$$N_{kav} = p_{pot} Q_{kav} . \tag{2.40}$$

For the considered system at the pipeline pressure $p_{max} = 372780$ Pa, discharge $Q_{max} = 0.188$ m³/s, the discharge and power of the gas-vapor mixture is $Q_{kav\ max} = 0.047$ m³/s and $N_{kav\ max} = 16.8$ kW (23% of the liquid flow power), respectively.

At PP operation mode with pressure $p_{min} = 176666$ Pa and discharge $Q_{min} = 0.12$ m³/s, the gas-vapor mixture discharge is $Q_{kav\ min} = 0.03$ m³/s. In this case, the value of the gas-vapor mixture power corresponds to $N_{kav\ min} = 5.3$ kW (25% of the liquid flow power).

The analysis of the obtained results revealed that cavitation processes in PC are characterized by considerable power of the gas-vapor mixture making (23-25)% of the liquid flow power. Thereby, it is reasonable to develop devices and systems of cavitation processes control by means of the efficient use of cavitation caverns energy.

CONCLUSION

Generalizing the results presented in this chapter, one can come to the following conclusions.

1. We have determined that the dynamic loads in PC pipeline network depend on velocity υ of liquid flow, the pipeline parameters (material, diameter d, length L, head losses Δh_{fr} along the length, the pumped medium characteristics); hydraulic characteristics $\xi(\beta)$ of the pipeline valves determined by its structural features; the type of control characterized by shutting time t_{sh} and coefficient n of the intensity of valves control, which are assigned by the alteration of the supply voltage frequency of the shut-off an control valves ED. Thus, the development of the system of the pipeline valves adjustable ED will make it possible to eliminate inadmissible increase of pressure in the PC pipeline network and, consequently, to improve its reliability in emergency and operation modes, to increase the service life of the technical equipment.
2. The proposed mathematical model for the description of PC with adjustable pipeline valves, operating for a complicated hydronetwork with static head, makes it possible to research the equipment operation emergency modes caused by the non-observance of the law of the control of shut-off and control pipeline valves at starts/stops and the regulation of the pump unit discharge or head; sharp alteration of the pump operation mode at regulation of the discharge or head by a stopcock in the ascending pipe; abrupt outage of the pump electricity supply at sharp collapse of the check valves, etc.
3. We have demonstrated that during the description of the hydrodynamic processes in the pipeline network it is necessary to take into account the energy aspects of liquid transportation: the active component caused by the presence of the hydronetwork resistance; the kinetic component caused by the transfer of the liquid mass along the pipeline length; the potential component corresponding to the pipeline filling or the rise of the liquid to a certain level. The above said enabled the presentation of the pipeline network as a multilink structure consisting of RLC-strings whose parameters reflect the conditions of the flow and the properties of the pumped liquid.
4. We have developed a cavitation channel mathematical model taking into account the presence of cavitation processes in the hydrosystem at the alteration of the temperature mode of liquid motion and the pumping plant technological parameters. We have demonstrated that the cavitation processes time constant depends on the degree of cavitation development, the motion velocity mode and the liquid flow aeration. The reduction of the cavitation channel time constant

results in the growth of the frequency of pulsations of the pressure and discharge in the hydronetwork, caused by the presence of cavitation processes and, correspondingly, in the increase of power losses in the pipeline during liquid transportation.

5. We have proposed the block diagrams of the models of pumping complexes with different operation characteristics of the pumping and pipeline equipment, with the possibility of operation of pump units equipped with a variable-frequency electric drive for a pipeline network of complicated configuration with installed shut-off and control valves.
6. The performed calculation revealed that the cavitation processes in the pumping complex are characterized by considerable power of gas-vapor mixture making (23-25)% of the power of the liquid flow, which confirms the reasonability of the efficient use of the energy of the cavitation caverns.

Chapter 3

The Methods and Principles of Making the Control Systems of the Electric Drive of the Pumping Complex in Non-Steady Operation States

3.1. Working out the Electromechanical System of the Dynamic Loads Decrease in a Pumping Complex

When the stopcocks or hydraulic valves shut/open sharply, significant dynamic loads occur in the pipeline network. They negatively influence the operation of PU and the pipeline equipment [6, 14, 15, 21–23]. To decrease them it is necessary to provide such control of the shutoff valves at which their opening will be minimal and the pressure in the hydrosystem will not exceed the maximum permissible value. It is possible to solve the problems related to the control of PU dynamic loads by means of a variable-frequency ED of the pipeline valves, wherefore one must know its power characteristic – the efforts and moments acting during its shutting or opening.

Paper [86] contains a theoretically substantiated possibility for the decrease of the dynamic loads in PU both in the stationary and emergency modes by means of the control of the stopcock shutting taking into account the complex nonlinear character of the alteration of hydraulic resistance coefficient ξ_v depending on relative degree β of its opening.

The analysis of curve $\xi_v(\beta)$ typical of shutoff-control stopcocks, shown in Figure 2.2, (curve 1) revealed that there is an essential growth of the hydraulic resistance coefficient at section $\beta \le 0.2$.

The stopcock resistance moment M_v dependence on relative degree β of its opening and its spindle rotation frequency ω_{sp} is not less important and takes into account the nonlinear character of $\xi_v(\beta)$ alteration.

3.1.1. The Mechanical Characteristics of the Pipeline Valves with a Variable-Frequency Electric Drive

The analysis of forces (Figure 3.1), acting on the stopcock wedge, revealed that, to shut it, it is necessary to apply the following torque to its spindle:

$$M_v = M_1 + M_2, \tag{3.1}$$

where $M_1 = Q_\Sigma r_{mc} tg(\alpha_1 + \varphi)$ – the friction torque occurring in the thread, N·m; $M_2 = Q_\Sigma f_b R_{cb}$ – the friction torque in the bearing bushing, N·m; r_{mc} – the average radius of the thread, m; R_{cb} – the radius to the center of the bearing balls, m; α_1 – the thread helix angle, grad; φ – the friction angle in the spindle thread, grad; f_b – the coefficient of friction in the bearings (for the sliding-contact bearing – $f_b = 0.1 \div 0.15$; for the rolling-contact bearing – $f_b = 0.01$).

Table 3.1 contains calculation expressions for the determination of the acting efforts, wherein the following designations are assumed: p – the working pressure of the medium, Pa; $S_p = \pi(d_{in} + 2/3b)^2/4$ – the area of the wedge surface subjected to the action of liquid pressure p, m^2; d_{in} – the internal diameter of the sealing ring, m; b – the width of the sealing ring, m; $q_{im} = (0.25 \div 0.5)p$ – the specific load providing the leak tightness at the sealing surface, Pa; d_{out} – the external diameter of the sealing ring, m; f_c – the coefficient of friction on the sealing surface ($f_c \approx 0.15$); d_{sp} – the spindle diameter, m; h_{sb} – the height of the stuffing box, m; f_{sb} – the coefficient of friction in the stuffing boxes ($f_{sb} \approx 0.1$); α_w – the wedge angle, grad; m_w – the wedge mass, kg; g – the acceleration of gravity, m/s^2.

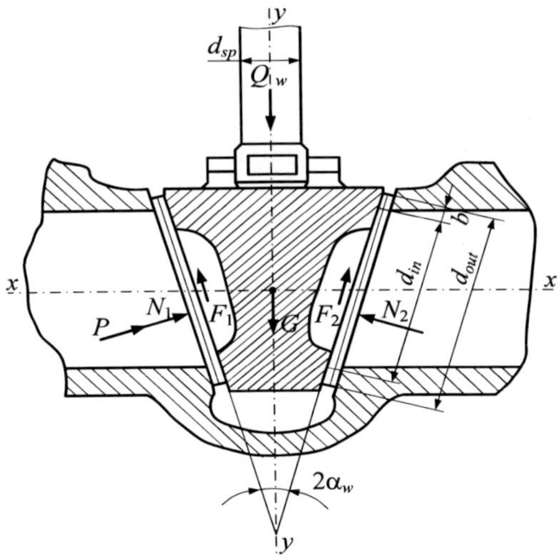

Figure 3.1. The diagram of the forces acting on the stopcock wedge.

Table 3.1. The expressions for the determination of the forces acting on the stopcock wedge

Effort type	Analytical expression
Total axial effort, N	$Q_\Sigma = Q_w + Q_{sb} + Q_0$
The axial effort component for overcoming the forces acting on the wedge, N	$Q_w = P(2\sin\alpha_w + f_c\cos\alpha_w) + 2N_1(\sin\alpha_w + f_c\cos\alpha_w) - G$
The axial effort component for overcoming the friction in the stuffing boxes, N	$Q_{sb} = 4\pi d_{sp} f_{sb} h_{sb} p$
The axial effort component for overcoming the internal pressure on the spindle butt, N	$Q_0 = p\left(\pi d_{sp}^2\right)/4$
The force of the medium hydrostatic pressure, N	$P = pS_p$
The response of the sealing surface of the body on the side of the medium input, N	$N_1 = q_{im}\pi\left(d_{out}^2 - d_{in}^2\right)/4$
The response of the sealing surface of the body on the side of the medium output, N	$N_2 = \dfrac{(P+N_1)\cos\alpha_w - F_1\sin\alpha_w}{\cos\alpha_w - f_c\sin\alpha_w}$
Friction forces, N	$F_2 = N_2 f_c$, $F_1 = N_1 f_c$
Gravity, N	$G = m_w g$

For uncontrolled valves, torque M_v is determined from the condition of the constancy of values S_p and p, in this case, p is assumed equal to the medium working pressure in the steady condition. When using a variable-frequency ED of the stopcock it is necessary to take into account the variation of area S_p and pressure p of the liquid. Accordingly, there is a necessity for the determination of the pipeline valves resistance moment M_v analytical dependence on relative degree β of its opening and its spindle rotation frequency ω_{sp}.

Figure 3.2 shows for the case of a cylindrical pipeline how the surface area S_p of the stopcock sluice changes when liquid pressure p acts on it and when relative degree $\beta = h/d_{in}$ of the pipeline valves opening changes, where h – the height of the sluice rise, m.

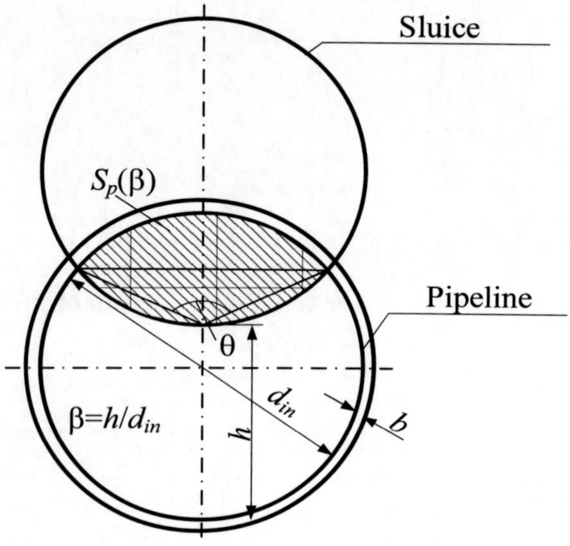

Figure 3.2. The dependence of area S_p of the stopcock sluice surface, subjected to liquid pressure, on relative degree β of its opening.

Area S_s of the circular segment formed by an arc of θ degrees is equal to:

$$S_s = \frac{d_{in}^{\ 2}}{8}\left(\frac{\pi\theta}{180} - sin\theta\right). \tag{3.2}$$

The change of the relative degree β of the stopcock opening results in the change of angle θ:

$$\theta = 2arccos\,(\beta). \tag{3.3}$$

Then:

$$S_p(\beta) = \left(\pi(d_{in} + 2b)^2/8\right)\left(2arccos(\beta) - sin(2arccos(\beta))\right). \tag{3.4}$$

Based on PC mathematical model worked out in [84] we obtained curves H_v of head variation at the pipeline section with a stopcock, depending on the relative degree β of opening, for different frequencies ω_{sp} of spindle rotation at a linear law of its shutting with constant frequency f_0 of the supply voltage of the drive electric motor (Figure 3.3).

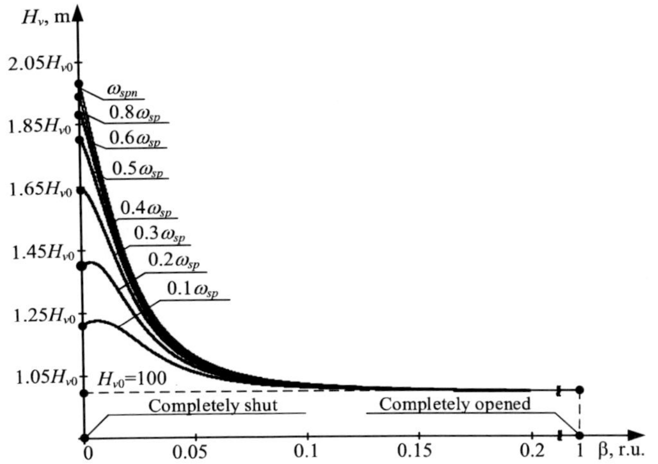

Figure 3.3. The dependence of head H_v at the pipeline section with a stopcock at its shutting on relative degree β of its opening for different frequencies ω_{sp} of rotation of the spindle.

PC parameters: the pump rated head $H_{pn} = 100$ m; the velocity of the working medium flow $\upsilon_n = 1.4$ m/s; the hydraulic resistance at the consumer $R_{cn} = 52$ s^2/m^5; length $L = 5000$ m and diameter $d = 1.2$ m of the pipeline; the number of the pipeline sections $N = 20$; the velocity of shock wave propagation $c = 1000$ m/s; the specific parameters of each section: $r_0 = 0.000162$ s^2/m^6, $l_0 = 0.0901$ s^2/m^3, $c_0 = 90132$ m^{-1}. As PC shutoff-regulating pipeline valves we used a wedge stopcock 30ch930br with a non-retractable spindle and the pipe size $d = 1.2$ m, designed for the pipelines transporting water or steam at the temperature of 120°C. According to the designations adopted in Table 3.1, the stopcock parameter values are: $d_{out} = 1.25$ m, $d_{in} = 1.2$ m, $b = 0.01$ m, $d_{sp} = 0.032$ m, $h_{sb} = 0.0084$ m, $f_s = 0.1$, $r_{mc} = 0.013$ m, α_1=60^0, $R_{cb} = 0.038$ m, $f_b = 0.12$.

The analysis of curves $H_v(\beta, \omega_{sp})$ revealed that the decrease of ED supply voltage results in the lower values of the pressure at the pipeline section with a stopcock at its shutting.

By the approximation of curves $H_v(\beta, \omega_{sp})$ and taking into account dependence $p = \rho g H$ we obtain:

$$p(\beta, \omega_{sp}) = \rho g\left(H_{v0} + \left(a_1 + a_2\omega_{sp}^{\ 3} + a_3\sqrt{\omega_{sp}}\right)e^{-\beta\left(b_1 + b_2\omega_{sp} + b_3\sqrt{\omega_{sp}}\right)}\right), \tag{3.5}$$

where H_{v0} – the head at the stopcock before the beginning of shutting (at $\beta = 1$), m; $a_1, a_2, a_3, b_1, b_2, b_3$ – the coefficients of approximation; ρ – the density of the medium, kg/m^3.

Then, using expressions (3.1), (3.4), (3.5) and formulae in Table 3.1, the dependence of the stopcock resistance moment on the relative degree β of opening and frequency ω_{sp} of its spindle rotation will be of the form:

$$M_v(\beta, \omega_{sp}) = p(\beta, \omega_{sp})\left(S_p(\beta)f_c + \frac{k\pi\left(d_{out}^2 - d_{in}^2\right)}{4} - m_w g + \right.$$
$$\left. + \pi d_{sp} 0{,}4 h_{sb} + \frac{\pi d_{sp}^2}{4}\right)\left(r_{mc} tg(\alpha_1 + \varphi) + f_b R_{cb}\right). \tag{3.6}$$

Using (3.6), we obtained the plots of alteration of moment M_v of the stopcock resistance depending on the relative degree β of its opening for a number of frequencies ω_{sp} of its spindle rotation, shown in Figure 3.4.

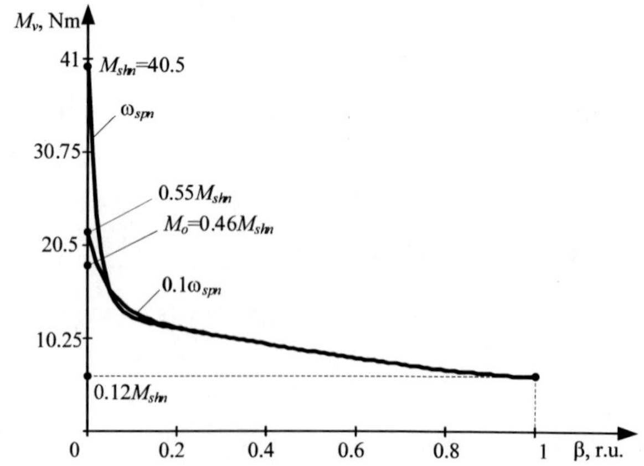

Figure 3.4. The dependence of the stopcock resistance moment M_v on the relative degree β of its opening at different frequencies ω_{sp} of rotation of the spindle.

The analysis of the curves revealed that at the section $0.2 \leq \beta \leq 1$ moment M_v changes linearly; at the section where $\beta \leq 0.2$, there is a considerable growth of moment M_v at the decrease of the relative degree β of the stopcock opening.

The plots demonstrate that at the stopcock shutting with rated frequency ω_{sp} of rotation of the spindle at ED supply voltage rated frequency $f_{0n} = 50$ Hz moment M_v assumes the maximal value $M_{shn} = 40.5$ N·m at the complete shutting of the stopcock $\beta = 0$, caused by a sharp increase of pressure at the stopcock in the pipeline network (Figure 3.3).

When the spindle rotation frequency ω_{sp} decreases to $0.1\omega_{spn}$ due to the reduction of the supply voltage frequency to $f_{01} = 5$ Hz, moment M_v decreases to value M_{sh1}, which is by 1.8 times less than M_{shn}.

In the shut position, head H_0, developed by the pump at zero discharge, acts on the stopcock working element. In this case, area S_p of the stopcock wedge surface, subjected to the working medium pressure, is maximal and is determined from (3.4) at $\beta = 0$. Substituting the quantities of the mentioned values into expression (3.6) one can determine the stopcock resistance moment M_o, which is to be overcome at its opening. The value of this moment is by 0.46 times lower than resistance moment M_{shn}, occurring at complete shutting of the stopcock ($\beta = 0$) with the rated frequency ω_{spn} of its spindle rotation (Figure 3.4).

3.1.2. The Determination of the Power and the Choice of a Variable-Frequency Electric Drive of the Pipeline Valves

The analysis of papers [42, 44, 46, 49, 98] revealed that for the solution of the problems of the decrease of dynamic loads in the PC pipeline network in both emergency and adjustment modes it is reasonable to use a frequency-variable ED of the stopcock. In this case, more attention is paid to the problems of the technical realization of the power converter and the creation of the microprocessor systems of the control of pipeline valves ED, the algorithms of the efficient control of power convertors. However, some problems remained uncovered. These are problems concerning the determination of the electric motor power and the gear reduction rate, the valves variable-frequency electric drive taking into account the nonlinear character of its hydraulic and mechanical characteristics.

The pipeline valves serial unregulated induction ED is designed for the control of the valves of various types and designs meant for the operation in the wide ranges of variation of the pipeline diameter and the pressure of the working medium. The valves ED power is calculated according to expression:

$$P_{IM} = \frac{k_r M_{max} n_{IM}}{9550 \eta_r r_g k_M}, \tag{3.7}$$

where M_{max} – the maximal moment at the ED shaft, N·m; n_{IM} – the rotation frequency of the electric motor shaft, rev/min; r_g – the gear reduction rate; η_r – the efficiency of the gear; k_M – the coefficient of the overload capacity of the motor by the moment; k_r – the reserve coefficient assumed to be equal to 1.1–1.5.

When moment M_{max} is determined, the alteration of the working medium pressure p and area S_p of the working element surface subjected to the action of the flow are neglected and assumed to be constant.

The basic catalog parameters of the stopcock ED include rated power P_{IMn} of the electric motor and time t_{sh} of valves complete opening/shutting (Table 3.2).

Table 3.2. The basic parameters of the stopcock ED

Pressure p = 2.5 kG/cm^2			
Diameter d, mm	Electric motor	Power P_{IMn}, kW	Time t_{sh} of opening or shutting, min
600	AOC2-11-4	0.6	2
800	AOC2-22-4	2.2	2
1000	AOC2-31-4	3	2.1
1200	AOC2-41-4	5.2	2.6
1400	AOC2-42-4	7.5	4.5
1600	AOC2-42-4	7,5	5.2
2000/1800	AOC2-41-4	5.2	5.8
Pressure p = 10 kG/cm^2			
600	AOC2-31-4	3	2.4
1000	AOC2-41-4	5.2	3.2
1200	AOC2-41-4	5.2	4.6
1400	AOC2-41-4	5.2	4.6
1600	AOC2-41-4	5.2	4.6

Table 3.2 demonstrates that electric motors AOC2 with increased sliding and the poles pairs number $p_p = 2$ are used as stopcock EDs. They are meant for the drive of the mechanisms characterized by the presence of relatively big flyweights, an irregular blast nature and pulsations of the load, a high frequency of starts and reverses.

ED is chosen based on the pipeline network diameter and the working pressure in the system. For the analyzed PC with the pump rated head $H_{pn} = 100$ m and the pipeline diameter $d = 1.2$ m, using the data in Table 3.2, we chose a stopcock ED with drive IM (AOC2-41-4) of the power 5.2 kW, and the opening/shutting time equal to 4.6 min.

The decrease of the supply voltage frequency of the ED with such parameters, in order to reduce the dynamic loads in PC, results in considerable increase of the stopcock shutting time. So, for the analyzed ED, when the supply voltage frequency reduces from 50 to 5 Hz, the shutting time increases by tenfold and is 46 min, which is caused by a great gear reduction rate r_g equal to 80. The mentioned value of the stopcock shutting time is inadmissible in both stationary and emergency PC operating conditions. That is why there is a necessity for the specification of the ED gear reduction rate whereat the minimal time of the pipeline valves control is provided.

The alteration of the ED gear reduction rate causes the change of moment M_v of resistance on the shaft of the drive electric motor. In this case, ED with a new reduction rate is to provide:

- the motor starting torque overcoming moment M_o of the stopcock resistance occurring at its opening ($\beta = 0$);
- the maximal speed of response of the valve electric drive at the section of the shutting trajectory, where $0.2 < \beta \leq 1$;
- the electric motor moment with the value exceeding moment M_{vmax} at full shutting ($\beta = 0$) and low frequencies of supply voltage at $\beta \leq 0.2$.

The above stated requirements can be met due to the use of the law of frequency control with the maintenance of constant power, which is implemented by means of contemporary serial frequency convertors (FC):

$$\gamma_{IM}(\alpha) = \sqrt{\alpha}, \qquad (3.8)$$

where γ_{IM} – the stator relative voltage; α – the stator current relative frequency.

The plots in Figure 3.3 demonstrate that at the stopcock spindle rotation frequency equal to $0.1\omega_{spn}$, provided at the supply voltage frequency f_0 equal to 5 Hz, the increase of head ΔH in the pipeline network equals to 24 m. In this case, the maximal head in the hydrosystem is 124 m, which is by 3% higher than head H_0 at the zero discharge, equal to 120 m for the analyzed PC and being its maximally admissible value at the fully shut stopcock. In this connection, there is no necessity for the reduction of the stopcock ED supply voltage frequency lower than 5 Hz.

The stopcock IM stator electromagnetic moment M and current I_1 dependences on sliding s at the law of frequency control (3.8) are respectively described by expressions of the form:

$$M(s,\alpha)=\frac{3\left(U_1\sqrt{\alpha}\right)^2}{\omega_0}\frac{R_2's}{A(s,\alpha)}; \tag{3.9}$$

$$I_1(s,\alpha)=U_1\sqrt{\alpha}\sqrt{\frac{C(s)}{A(s,\alpha)}}, \tag{3.10}$$

where s – the motor sliding; U_1 – IM stator phase voltage, V; ω_0 – the synchronous rotation frequency of IM stator magnetic field, s^{-1}; R_1, R_2' – the stator resistance and the rotor reduced resistance, respectively, Ohm; $L_{1\sigma}$, $L_{2\sigma}$, L_μ – the leakage inductance of the stator, rotor and magnetization circuit, respectively, H; X_1, X_2, X_μ – the inductive reactances of the stator, rotor and magnetization circuit, respectively, Ohm; $\sigma_{12}=\sigma_1+\sigma_2+\sigma_1\sigma_2$ – the total leakage coefficient; $\sigma_1=L_{1\sigma}/L_\mu$, $\sigma_2=L_{2\sigma}/L_\mu$ – the leakage coefficient for the stator and rotor, respectively; $v=R_1/X_\mu$; $w=(1+\sigma_1)$; $r=R_1(1+\sigma_2)$; $q=X_\mu\tau'$; $a(\alpha)=\sqrt{r^2+q^2\alpha^2}$; $b(\alpha)=\sqrt{v^2+w^2\alpha^2}$; $C(s)=R_2'/X_\mu^2+(1+\sigma_2)^2s^2$; $A(s,\alpha)=a(\alpha)^2s^2+2R_1R_2'\alpha s+b(\alpha)^2(R_2')^2$.

Figure 3.5 shows the electromagnetic moment M dependence on the stopcock AOC2-41-4 IM rotor frequency ω at the supply voltage frequency equal to 50 Hz and 5 Hz, respectively.

The plot in Figure 3.5 shows that at the linear law of the stopcock shutting with the supply voltage frequency equal to 50 Hz, in accordance with (3.6), the moment at IM shaft changes in the range from M_{min50} (at $\beta=1$) to M_{max50} (at $\beta=0$). At the nonlinear law of the stopcock shutting with the supply voltage frequency equal to 50 Hz, at the section where

$\beta > 0.2$ and 5 Hz – at $\beta \leq 0.2$ the moment on the IM shaft changes in the range from M_{min50} (at $\beta = 1$) to M'_{max5} (at $\beta = 0$).

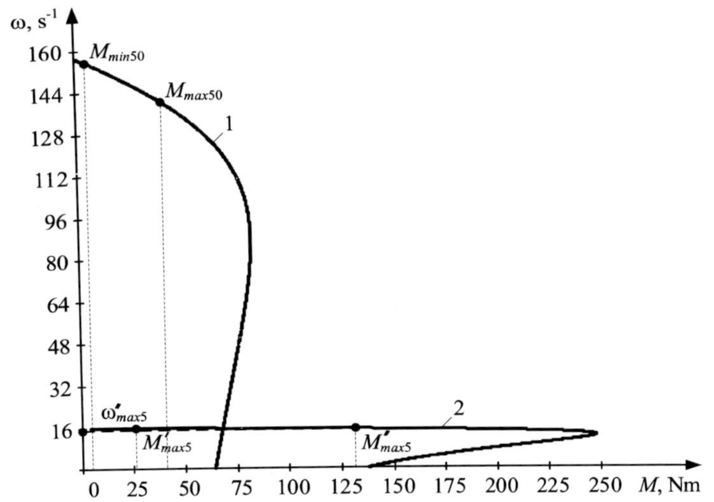

Figure 3.5. The stopcock AOC2-41-4 IM electromagnetic moment M dependence on the rotation frequency of its rotor at the supply voltage frequency equal to 50 Hz (1) and 5 Hz (2).

The frequency convertor changes the frequency and maintains the constant value of the voltage/frequency relation. In this case, the maximal current of the invertor overload is assumed to be equal to $(1.2-1.5)I_{1n}$, where I_{1n} – the rated current of IM stator.

Figure 3.6 shows the stopcock AOC2-41-4 IM stator current I_1 dependence on rotation frequency ω of its rotor at the supply voltage frequency equal to 50 Hz and 5 Hz, respectively.

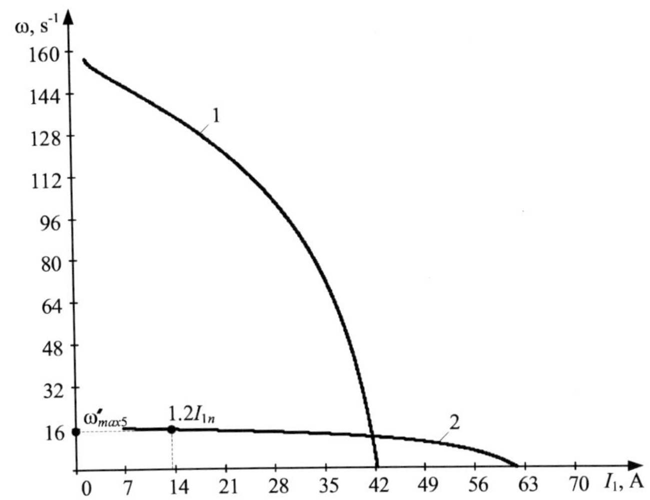

Figure 3.6. The stopcock AOC2-41-4 IM stator current I_1 dependence on rotation frequency ω of its rotor at the supply voltage frequency equal to 50 Hz (1) and 5 Hz (2).

The figure shows that at the limitation of the stator current I_1 by value $1.2I_{1n}$ and the supply voltage frequency equal to 5 Hz IM rotor rotation frequency must not be lower than the value of ω''_{max5}. IM electromagnetic moment corresponding to the rotor rotation frequency equal to ω''_{max5}, represents the maximal admissible electromagnetic moment of the considered IM as a part of the stopcock variable-frequency ED and equals to M''_{max5} (Figure 3.5). In this case, its value is five times as large as the maximal moment M'_{max5}, occurring on ED shaft with the reduction rate $r_g = 80$ at the stopcock shutting with supply voltage frequency of 5 Hz at the section at $\beta \leq 0.2$. Hence, at the nonlinear law of the stopcock shutting the IM is not used completely by its moment. To provide IM operation with the moment equal to M''_{max5} and reduce the time of the stopcock shutting it is reasonable to reduce the gear reduction rate by $M''_{max5} / M'_{max5} = 5$ times.

Figure 3.7 shows moment M'_{max5} dependence on the gear reduction rate r_g at the supply voltage frequency equal to 5 Hz.

The plot demonstrates that for the considered ED of the stopcock the minimal value of the gear reduction rate is $r_{gmin} = 16$, and the maximal value – $r_{gmax} = 80$. The hatched area represents the area of the possible values of the reduction rate of the gear of the stopcock ED, at which the condition of the invertor overload by current is met.

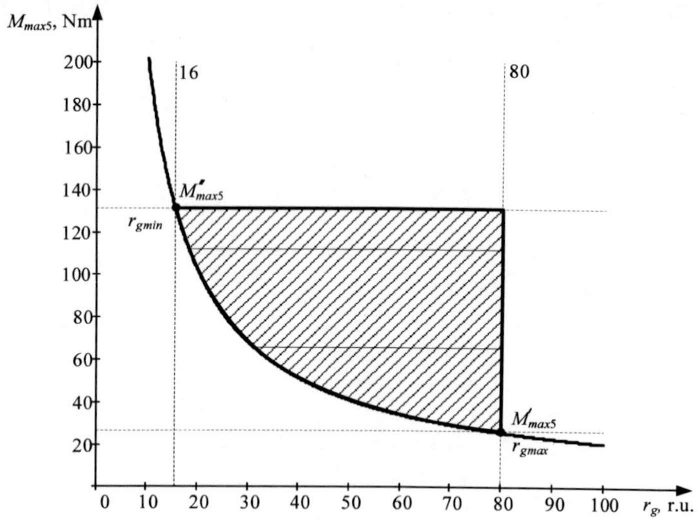

Figure 3.7. Moment M'_{max5} dependence on the gear reduction rate r_g at the supply voltage frequency equal to 5 Hz.

Thus, the IM, as a part of the stopcock variable-frequency ED, used for the reduction of the dynamic loads in PC, is chosen from the condition $r_g < r_{gmax}$.

To determine the gear reduction rate r_g meeting this condition it is necessary to determine the value of IM sliding s''_{max5} corresponding to rotation frequency ω''_{max5}.

Substituting value $1.2I_{1n}$ into the left part of (3.10) and, instead of the current relative frequency, value α_5, corresponding to IM supply voltage frequency, equal to 5 Hz, we will obtain a square equation. The following expression is the solution to this equation:

$$s''_{max5} = \frac{-k_2 - \sqrt{k_2{}^2 - 4k_1k_3}}{k_1}, \tag{3.11}$$

where

$$k_1 = (1.2I_{1n})^2(r^2 + \alpha_5^2) - (U_1^2\alpha_5)(1+\sigma_2)^2;$$
$$k_2 = 2R_1R_2'\alpha_5(1.2I_{1n})^2;$$
$$k_3 = (1.2I_{1n})^2(v^2 + w^2\alpha_5)R_2^2 - \frac{R_2^2(U_1^2\alpha_5)}{X_\mu^2}.$$

By the substitution of values (3.11) and α_5 into (3.9), we will receive the maximal value of IM moment at the supply voltage frequency equal to 5 Hz:

$$M''_{max5} = \frac{3\left(U_1\sqrt{\alpha_5}\right)^2}{\omega_0} \times$$
$$\times \frac{R_2's''_{max5}}{(r^2 + q^2\alpha_5^2)(s''_{max5})^2 + 2R_1R_2'\alpha_5 s''_{max5} + (v^2 + w^2\alpha_5^2)(R_2')^2} \tag{3.12}$$

Then the gear reduction rate will be equal to:

$$r_g = \frac{M'_{max5}}{M''_{max5}} r_{gmax}. \tag{3.13}$$

Table 3.3 contains the parameters of AOC2-41-4 IM, used as a part of an unregulated and a variable-frequency ED for the stopcock with the diameter $d = 1200$ mm and the rated head $H_{pn} = 100$ m, where, in the first case the reduction rate is $r_g = 80$, and in the second case – $r_g = 16$.

Table 3.3. A comparative characteristic of the use of an unregulated and a variable-frequency ED of the stopcock

Motor type	P_{IMn}, kW	r_g	t_{sh}, s	ΔH, m
AOC2-41-4 (unregulated ED)	5.2	80	276	100
AOC2-41-4 (variable-frequency ED)	5.2	16	55	24

3.1.3. The Structure and Algorithm of the Operation of the Electromechanical System of the Dynamic Loads Decrease in a Pumping Complex

An electromechanical system of the decrease of dynamic loads (ESDDL) allows the reduction of the dynamic loads in the hydrosystem in case of an abrupt disappearance of voltage. With this purpose in view, one should search for a nonlinear law of shutting of the shutoff-control valves whereat the head in the pipeline network does not exceed the maximal admissible value. The block diagram of PC with ESDDL is shown in Figure 3.8.

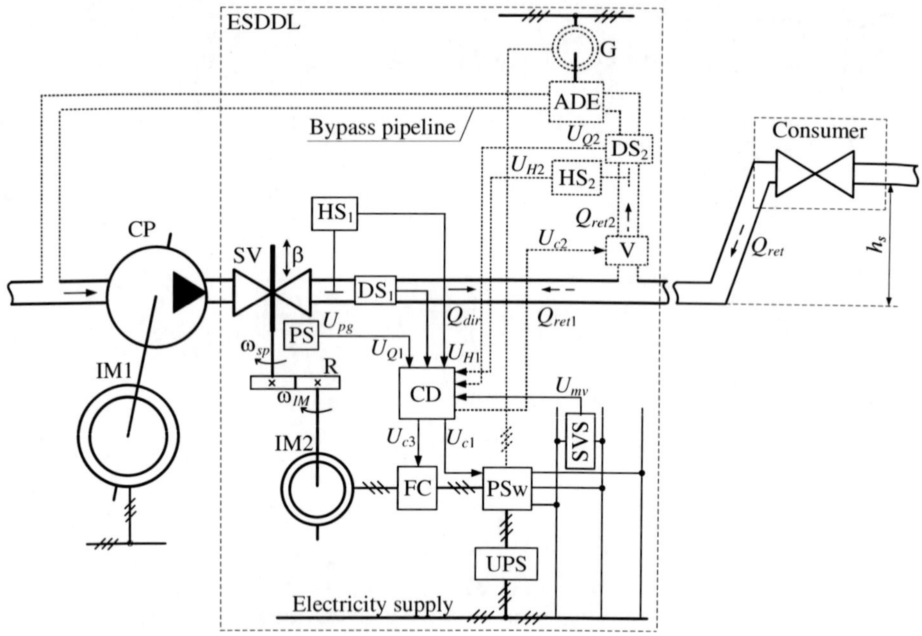

Figure 3.8. A block diagram of PC with ESDDL based on the controlled shutoff-control valves at abrupt disappearance of supply voltage of PU.

It includes [87]: shutoff-control valves (SV) mounted at the output of the centrifugal pump CP with a drive induction motor IM1; a frequency convertor FC connected to the induction drive motor of the valves IM2 whose shaft is connected with its spindle via reducer R; a bypass pipeline including a hydrovalve V, an active dissipator of the energy

ADE of the hydraulic flow with an electric generator G; head sensors HS_1, HS_2 and discharge sensors DS_1, DS_2 mounted in the pump discharge branch and the bypass pipeline before ADE, respectively; a position sensor of the sluice of SV pipeline valves; a supply line voltage sensor SVS; a uninterrupted power supply UPS; a power switch PSw; a control device CD.

The system operates in accordance with the algorithm shown in Figures 3.9–3.11, in which the solid lines refer to the case when an uninterrupted power supply is used, and dashed lines – to the case at additional connection of an active dissipator of the energy of the hydraulic flow with an electric generator in one shaft [87–92].

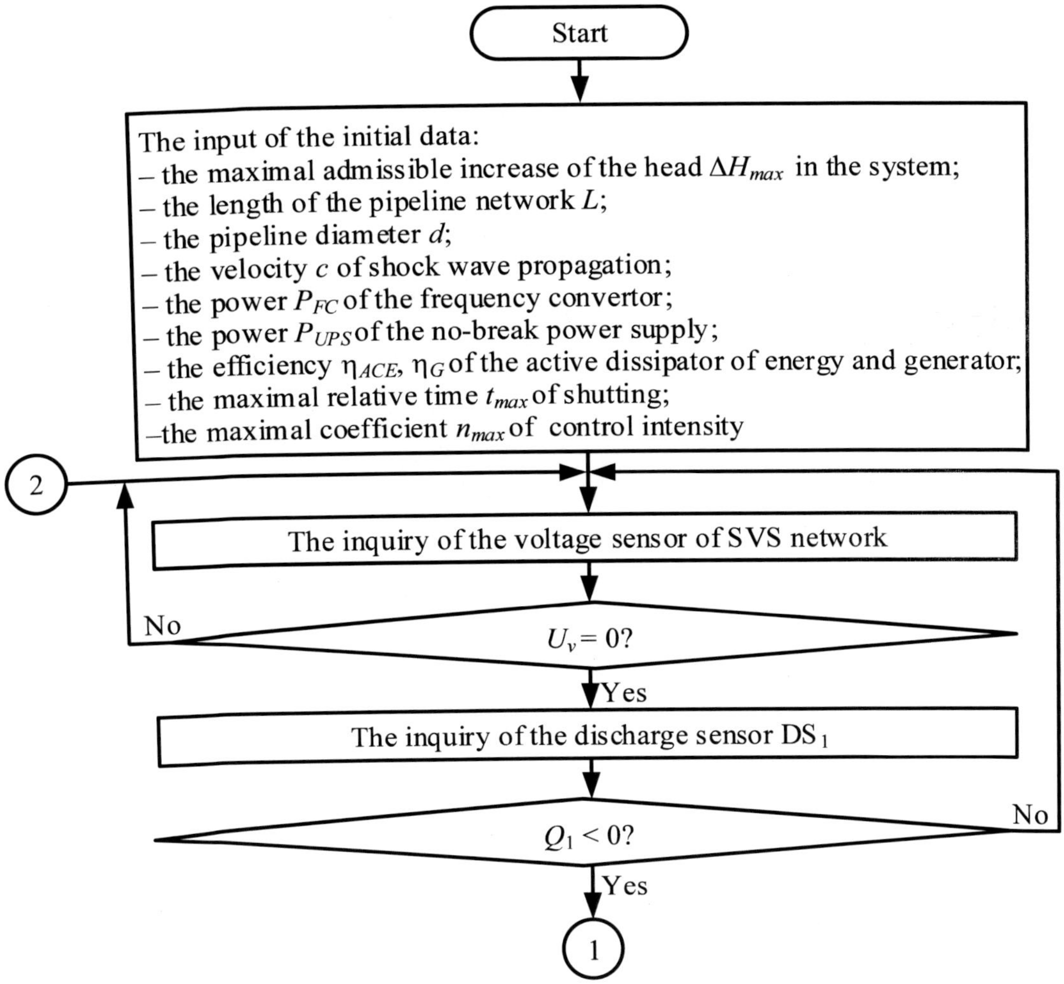

Figure 3.9. The algorithm of ESDDL operation at an abrupt disappearance of voltage in the pumping complex.

At the abrupt stop of electricity supply of induction motors IM1 and IM2 and occurrence of liquid counter flow under the action of static head h_s in the hydrosystem the control device forms control signal U_{c1} to the power switch for the frequency convertor connection to the uninterrupted power supply.

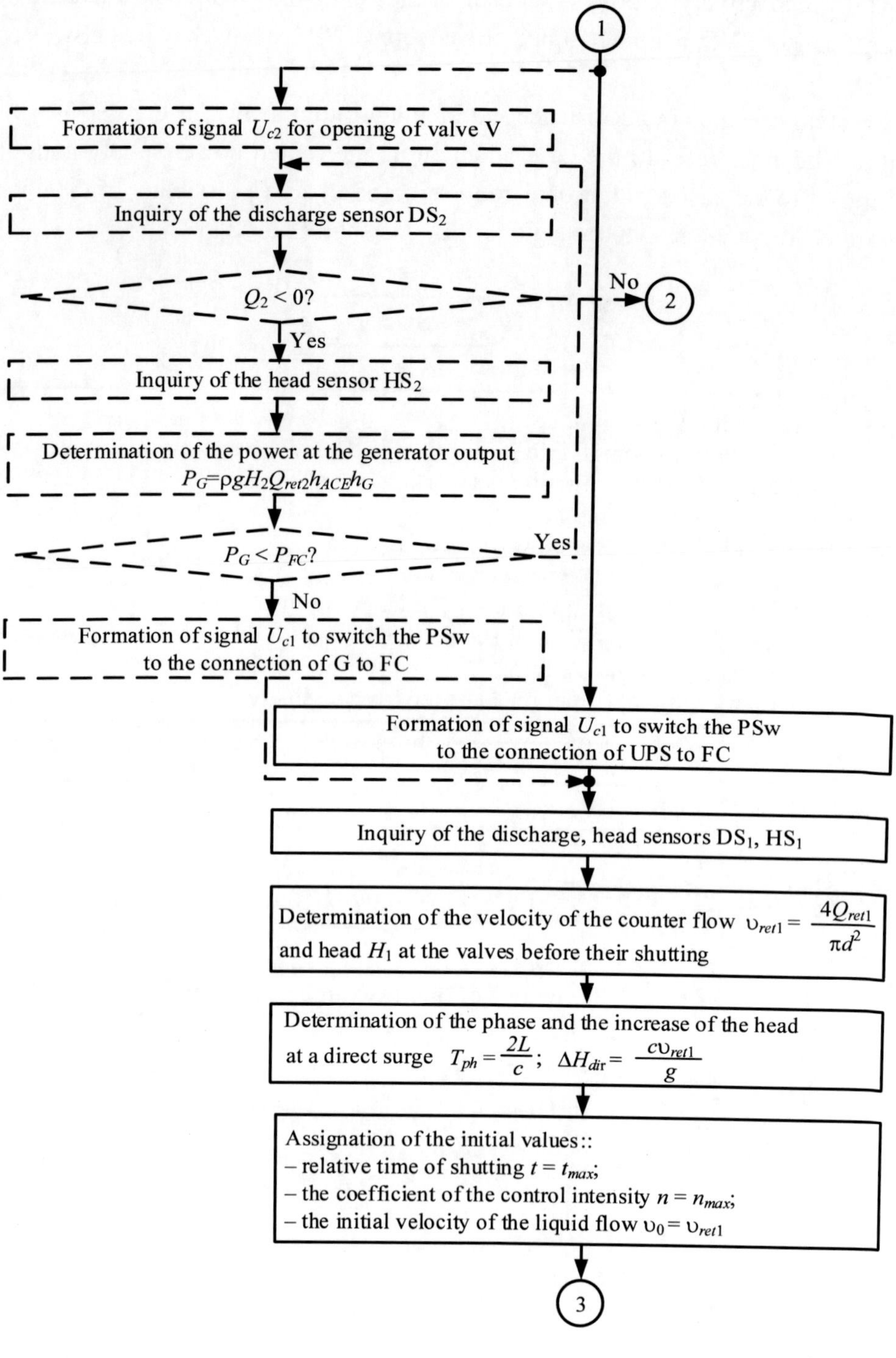

Figure 3.10. The continuation of the operation algorithm.

ESDDL can be additionally equipped with an active dissipator of the energy of the hydraulic flow [91, 92]. In this case, at the abrupt disappearance of voltage in PC and occurrence of liquid counter flow, the control device forms signal U_{c2} for opening of the hydrovalve. A part of the flow is directed via the bypass pipeline to the active dissipator of the energy (hydroturbine) with an electric generator mounted on one shaft.

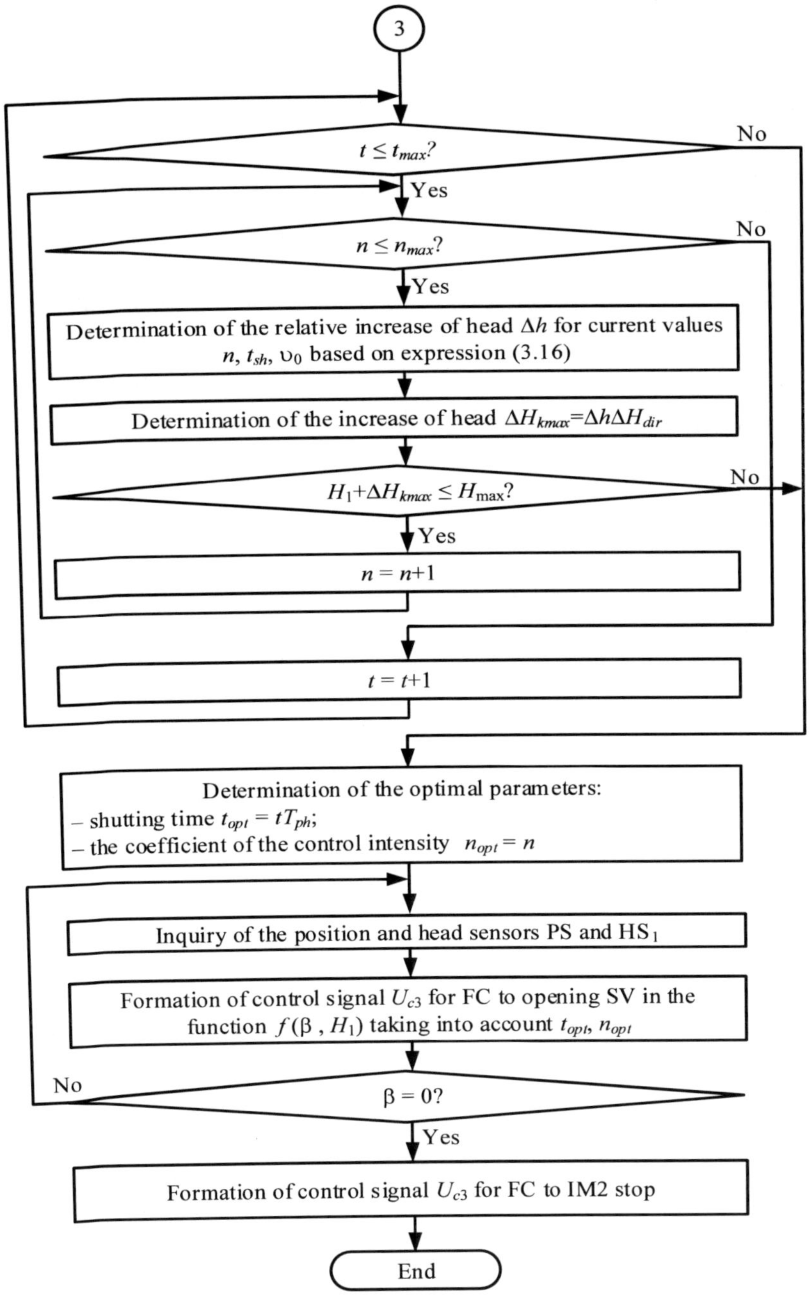

Figure 3.11. The continuation of the operation algorithm.

At the initial time moment after the opening of the hydrovalve, due to the low value of the reverse flow Q_{ret2} in the bypass pipeline, the generator output electric power may be insufficient for the complete shutting of the valves:

$$P_G = \rho g H_2 Q_{ret\,2} \eta_{ACE} \eta_G , \tag{3.14}$$

where H_2 – the head at ADE, m; η_{ACE}, η_G – the efficiencies of the active dissipator of the energy and generator, respectively; ρ – the working medium density, kg/m^3; g – the acceleration of gravity, m/s^2.

When $P_G \geq P_{FC}$, where P_{FC} – the power of the frequency convertor, the control device forms control signal U_{c1} to the power switch for the connection of the generator output terminals to the FC input.

According to the algorithm (Figure 3.11) one searches for the optimal values of intensity coefficient n_{opt} and the pipeline valves control time t_{opt} corresponding to such a control law at which the head in the pipeline at the stopcock $(H_1 + \Delta H_{k\,max})$ will be lower than the maximal admissible value of the head in the hydrosystem $H_{max} = (H_1 + \Delta H_{max})$. Here H_1 – the value of the head at the stopcock in the steady mode, m; $\Delta H_{k\,max}$ – the maximum increase of the head during the time of valves control, chosen from the array of values ΔH_k, corresponding to the assigned values of the control intensity coefficient n, time t_{sh} of the complete shutting of the pipeline valves and the initial velocity υ_0 of liquid motion in the hydronetwork.

To determine the array of values $\Delta H_k = \Delta h \Delta H_{dir}$ we use a multifactor regression model of the form:

$$\Delta h(n,\tau,\upsilon_0) = a_0(\upsilon_0) + a_1(\upsilon_0)n + a_2(\upsilon_0)\tau + a_3(\upsilon_0)n^2 + a_4(\upsilon_0)\tau^2 + \\ + a_5(\upsilon_0)n\tau + a_6(\upsilon_0)n^3 + a_7(\upsilon_0)\tau^3 + a_8(\upsilon_0)n\tau^2 + a_9(\upsilon_0)n^2\tau, \tag{3.15}$$

where $a_k(\upsilon_0) = b_{0k} + b_{1k}n^2(\upsilon_0) + b_{2k}e^{-\upsilon_0}, k = 0-9$; $\Delta h = \Delta H / \Delta H_{dir}$, $\tau = t_{sh}/T_{ph}$ – the relative increase and time of complete shutting of the pipeline valves, respectively; $\Delta H_{dir} = c\upsilon_0/g$ – the value of the direct surge at $t_{sh} \leq T_{ph}$, m.

Parameters n, τ, υ_0 of model (3.15) change in the ranges of 1 to n_{max}, of 1 to τ_{max}, $0.2-5$ m/s, respectively, where $n_{max} = 12$, $\tau_{max} = 10$ – the maximal values of the coefficient of control intensity and the relative time of valves shutting, respectively. The said ranges of the change of values τ and n enable the control of the pressure increase in the hydrosystem within the admissible limits, and the limits of the change of the liquid flow velocity include a wide class of PCs with different parameters of the pipeline network (diameter d, length L, head losses along the length Δh_{fr}).

Figure 3.12 shows the graphs of the change of the relative increase of head Δh depending on relative time τ of shutting and coefficient n of the intensity of the valves control for the chosen range of the change of the initial velocity υ_0 of the flow of the working medium.

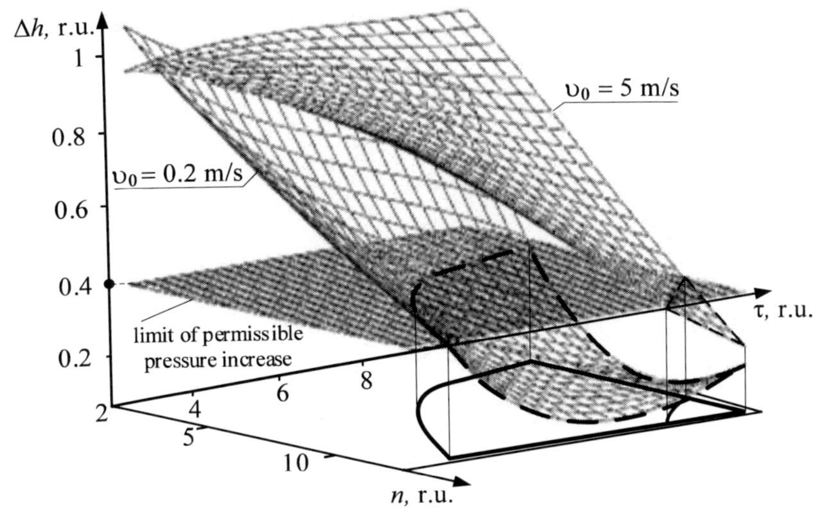

Figure 3.12. PC head relative increase Δh dependence on relative time τ of shutting and the coefficient n of the intensity of the pipeline valves control at different values υ_0 of the initial velocity of the working medium flow in the hydronetwork.

The analysis of the dependence in Figure 3.12 revealed that with the change of parameters n, τ in the ranges 7–12 and 5–10, respectively, the pressure in the communication network does not exceed 40% of the rated value. The figure demonstrates that the reduced dependence has an explicit minimum, which makes it possible to find the optimal values of coefficient n_{opt} of the intensity of the control and time t_{opt} of the stopcock shutting, corresponding to such a control of the pipeline valves, whereat the pressure in the pipeline does not exceed the maximal admissible value.

Table 3.4 contains the values of the coefficients of the regression model (3.15). The coefficient of the determination of the obtained regression model is $R^2 = 0.9$, which allows its use in the carried out calculations.

At the moment of the liquid counter flow ($Q_1 < 0$) under the action of the hydrosystem static head h_s due to an abrupt stop of electric supply ($U_{mv} = 0$) of induction motors IM1 and IM2 by readings H_1, Q_{ret1} of the corresponding head and discharge sensors HS_1 and DS_1 the current parameters of the surge are calculated:

- surge phase T_{ph} determined from expression (2.2);
- the reverse velocity of the flow via the shutoff-control valves:

$$\upsilon_{ret1} = 4Q_{ret1}/(\pi d^2); \tag{3.16}$$

- the pressure increase at the direct surge:

$$\Delta H_{dir} = c\upsilon_{ret1}/g, \tag{3.17}$$

where d – the pipeline diameter, m.

Table 3.4. The values of the coefficients in expressions $a_k(\upsilon_0)$

k	b_{0k}	b_{1k}	b_{2k}
0	0.631222	0.077407	0.669099
1	0.130304	$-2.975237 \cdot 10^{-2}$	-0.112772
2	0.153177	$-3.119610 \cdot 10^{-2}$	-0.255814
3	$-1.053483 \cdot 10^{-2}$	$2.258783 \cdot 10^{-3}$	$1.424503 \cdot 10^{-2}$
4	$-1.741412 \cdot 10^{-2}$	$3.167309 \cdot 10^{-3}$	$3.802698 \cdot 10^{-2}$
5	$-3.300397 \cdot 10^{-2}$	$7.678935 \cdot 10^{-3}$	$-2.526393 \cdot 10^{-2}$
6	$3.288240 \cdot 10^{-4}$	$-6.120325 \cdot 10^{-5}$	$-7.690116 \cdot 10^{-4}$
7	$6.560691 \cdot 10^{-4}$	$-9.776912 \cdot 10^{-5}$	$-1.974709 \cdot 10^{-3}$
8	$1.109118 \cdot 10^{-3}$	$-3.410165 \cdot 10^{-4}$	$1.364034 \cdot 10^{-3}$
9	$7.598641 \cdot 10^{-4}$	$-2.233431 \cdot 10^{-4}$	$1.409865 \cdot 10^{-3}$

Substituting $T_{ph}, \upsilon_{ret1}, \Delta H_{dir}$ into (3.16), we determine optimal values n_{opt} and t_{opt} for the formation of control signal U_{c3} to FC for the provision of such a control of the stopcock ED whereat the head in the pipeline changes within the admissible limits excluding the occurrence of dangerous surges in PC.

The system reserve supply can be provided by industrial uninterrupted power supplies in a wide range of powers (from 100 W and higher), characterized by high values of the efficiency (93-95)%, a long interval of the time of the maintenance of the rated power (from 2.5 min. to 3 h) at the specific cost of 200-250 c. u./kW.

Serial hydroturbine units (hydroturbines with a generator) of low power (up to 100 kW) at the specific cost of 300-350 c. u./kW can be used as active dissipators of energy of the hydraulic flow. Alternating current induction and synchronous generators with a horizontal and vertical shaft, used at small hydroelectric stations with the power within the range from 1 kW to 3 MW, can be used as generators. For the creation of a turbine equipment it is reasonable to use a reaction-type hydraulic turbine (Francis, Kaplan, Pelton turbines), characterized by high values of the efficiency (90-95)% in a wide range of the discharge and power variation. Papers [95, 96] contain a mathematical apparatus for the description of the operating characteristics of the above-mentioned types of the hydroturbines, which are required to determine its parameters (head, discharge, power, etc.) as a part of ESDDL.

3.2. The Method for the Determination of the Limits of the Non-Cavitation Operation of the Pumping Complex When the Process Variables are Regulated

The problem of the cavitation processes influence on the operation characteristics of the pumping equipment is one of the priority issues at the creation of PC parameter control system (CS). In spite of the great amount of research in this field, a clear approach to the determination of the boundaries of the occurrence of cavitation processes in PC has not been formulated up to the present time. Although the joint operating modes of PU with a variable-frequency ED and the pipeline network with time-variable operation mode of the consumer have been taken into account.

Table 3.5. The parameters of the pumping complex

Pump parameters			
Head H_N, m	Discharge Q, m³/h	Power N_N, kW	Approximation coefficients
125	1250	630	$A_2 = 127; B_2 = 124.214; C_2 = -374.325$
Hydronetwork sections parameters			
Hydrodynamic resistance R_s, s²/m⁵	Counter pressure H_{st}, m	Critical number of cavitation χ_{kr}	Liquid temperature T, °C
$R_{s1} = 364.954$	0	250	20
$R_{s2} = 136.858$	0	110	
$R_{s3} = 410.573$	5	150	

The analysis of papers [62-65, 70-73] revealed that every cavitation bubble grows to the final size and then collapses. The resulting shock waves cause considerable pressure difference, breaks of the continuous flow, vibrations, noises, erosive destruction of the material, the increase of the head losses and, consequently, the growth of the power spent on the transportation of the liquid.

The limits of non-cavitation operation were determined for the PC of the water supply system of the city of Komsomolsk. The water supply system parameters are given in Table 3.5.

The operation mode of the pump on the pipeline network is determined by point A of the intersection of their head-discharge characteristics (Figure 3.13) at head $H_A = 122.85$ m, discharge $Q_A = 0.36$ m^3/s and consumed power $N_A = 640$ kW (point B).

The following equation system is true for point A:

$$H = H_0 \nu^2 - R_b Q^2; \quad H = \sum_{i=1}^{n} H_{st\_i} + \sum_{i=1}^{n} R_{s\_i} Q^2, \tag{3.18}$$

where H_0 – the head developed by the pump at the zero discharge, m; $\sum_{i=1}^{n} H_{st\_i} = H_{st1} + H_{st2} + H_{st3}$ – the total static head in PC, m; n – the total number of the researched sections of the hydronetwork into which the pipeline network is divided; i – the number of the researched section; H_{st1}, H_{st2}, H_{st3} – the value of the static head at n sections of the hydrodynamic network, m; $\sum_{i=1}^{n} R_{s\_i} = R_{s1} + R_{s2} + R_{s3}$ – the total hydraulic resistance of the pipeline, s^2/m^5; R_{s1}, R_{s2}, R_{s3} – the hydraulic resistance of the corresponding sections of the pipeline, s^2/m^5.

The head at the pump output at discharge Q_A represents an algebraic sum of the head losses at the pipeline sections (curve $(H-Q)_{s1}$), m:

$$H_A = H_1 = \Delta H_1 + \Delta H_2 + \Delta H_3 + \sum_{i=1}^{n} H_{st\_i}, \tag{3.19}$$

where $\Delta H_1 = Q_A^{\,2} R_{s1}$, $\Delta H_2 = Q_A^{\,2} R_{s2}$, $\Delta H_3 = Q_A^{\,2} R_{s3}$ – the head losses at each section of the pipeline, m.

Thus, the values of the head at the input of the first, second and third sections of the pipeline are determined by curves $(H-Q)_{s1}$, $(H-Q)_{s2}$ and $(H-Q)_{s3}$, obtained according to expressions (3.19) and (3.20).

The analysis of papers [55-66, 78-80] revealed that cavitation develops when the cavitation current number χ is equal or lower than its critical value χ_{kr}, determining the limit of the development of cavitation processes. Then the flow velocity corresponding to the beginning of the cavitation process, m/s:

$$\upsilon_{kr} = \sqrt{\frac{2(p_{pot} - p_{para})}{\rho \chi_{kr}}}. \qquad (3.20)$$

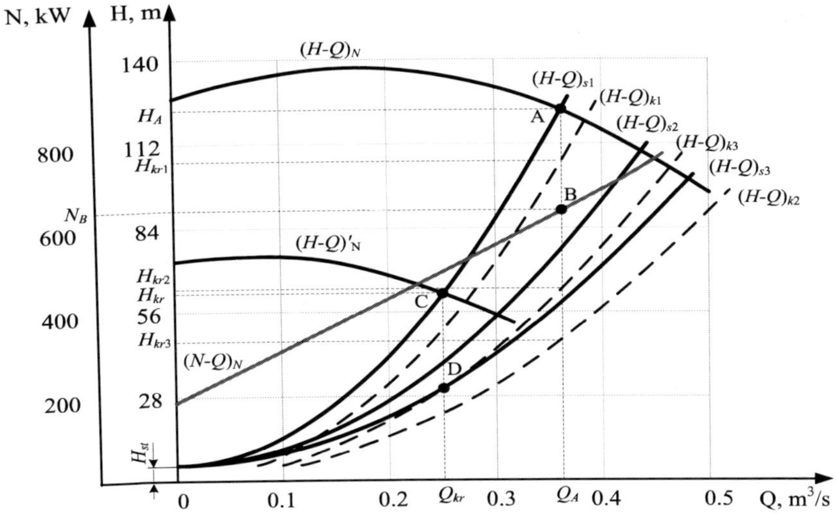

Figure 3.13. The characteristics of the pumping complex and the boundaries of the occurrence of cavitation processes in the pipeline.

Taking into account conventional relations for head $H_{pot} = p_{pot}/\rho g$, discharge $Q = \upsilon S$ and expression (1.14), one can find the head-discharge characteristic, determining the boundary of the cavitation operation mode of any of the sections of the hydrodynamic network, from expression:

$$Q_{kr} = S\sqrt{2(\rho g H_{pot} - p_{para})/\rho \chi_{kr}}. \qquad (3.21)$$

Based on expressions (3.19), (3.20) and (3.22) we obtained curves $(H-Q)_{k1}$, $(H-Q)_{k2}$, $(H-Q)_{k3}$ (Figure 3.13), showing the boundary of the non-cavitation mode of operation of each section of the hydrodynamic network. Their analysis reveals that the pumping station current operation mode corresponding to discharge $Q_A = 0.36$ m³/s, results in meeting the condition $H_i \leq H_{kr}$ and in the development of cavitation processes at the third section of the hydraulic network as head H_{kr3} at this section is lower than the critical value. In this case, the intersection of curve $(H-Q)_{k3}$ with the head-discharge characteristic (curve H_{s3}), showing the change of head at the third section of the pipeline at point C corresponds to the critical value of discharge in PC, determining the beginning of the development of cavitation processes in the hydraulic network.

For point C (Figure 3.13), taking into account expressions (1.1) and (3.20), we have:

$$H_{kr3} = \frac{\rho_3 \chi_{kr3} H_{st3} - 2S_3^2 R_{s3} p_{para3}}{\rho_3 \chi_{kr3} - 2S_3^2 R_{s3} \rho_3 g} \quad ; \tag{3.22}$$

$$Q_{kr3} = \sqrt{\frac{H_{kr3} - H_{st3}}{R_{s3}}}, \tag{3.23}$$

where H_{kr3}, Q_{kr3} – the value of the critical head and discharge at the third section of the pipeline corresponding to the beginning of the development of cavitation processes, respectively, m and m^3/s; H_{st3}, R_{s3}, ρ_3 – the counter pressure, hydraulic resistance and liquid density at the third section of the pipeline, respectively, m, s^2/m^5 and kg/m^3; S_3, χ_{kr3} – the area of the cross section and the critical value of the cavitation number of the third section of the pipeline, m^2.

The joint solution of (3.22), (3.23) and (1.13) enabled obtaining the expression for the determination of the pump ED relative critical rotation frequency whereat cavitation processes begin to develop at the third section of the pipeline network:

$$\nu_{kr3} = \sqrt{\left(\sum_{i=1}^{3} H_{st\_i} + \left(\sum_{i=1}^{3} R_{s\_i} + R_b\right) \frac{H_{kr3} - H_{st3}}{R_{s3}}\right) \Big/ H_0} \, . \tag{3.24}$$

In the general case, the $\nu_{kr3} = 0.711$. Then, to provide the non-cavitation operation of PC with a variable-frequency ED, it is necessary to meet the condition $\nu < \nu_{kr3}$.

In the general case the expression for the determination of the pump ED relative critical rotation frequency providing the non-cavitation operation of the hydrosystem is of the form:

$$\nu_{kr\_i} = \sqrt{\left(\sum_{i=1}^{n} H_{st} + \left(\sum_{i=1}^{n} R_s + R_b\right) \frac{H_{kr\_i} - H_{st\_i}}{R_{s\_i}}\right) \Big/ H_0} \, , \tag{3.25}$$

where $H_{kr\_i} = \left(\rho_i \chi_{kr\_i} H_{st\_i} - 2S_i^2 R_{s\_i} p_{para\_i}\right) / \left(\rho_{\_i} \chi_{kr\_i} - 2S_i^2 R_{s\_i} \rho_i g\right)$ – the value of the critical head at the i-th section of the pipeline corresponding to the beginning of the development of the cavitation processes, m.

Then, to determine the boundaries of the non-cavitation operation of PC with a pump variable-frequency ED the following condition is to be met $\nu_i < \nu_{kr\_i}$.

According to the above stated principles, we propose the following algorithm for the determination of the boundaries of the non-cavitation operation of the hydrosystem (Figure B.1).

It is shown in papers [10–12, 55-66, 78-80, 101-108] that the presence of cavitation processes results in the growth of the pipeline hydraulic resistance and, consequently, in the change of PC operation mode. This is confirmed by the curves (Figure 3.14), obtained in accordance with expression (2.34) showing in the relative units the alteration of the coefficient of resistance ζ' and head losses $\Delta h_m'$ in the PC pipeline depending on volume V_{kav}' of the cavitation cavern for the flow velocity equal to $\upsilon = 2.8$ m/s.

The analysis of the curves (Figure 3.14) revealed that the increase of the volume of the cavitation cavern is accompanied by the intensive growth of the coefficient of resistance and head losses corresponding to the presence of cavitation processes in the pipeline.

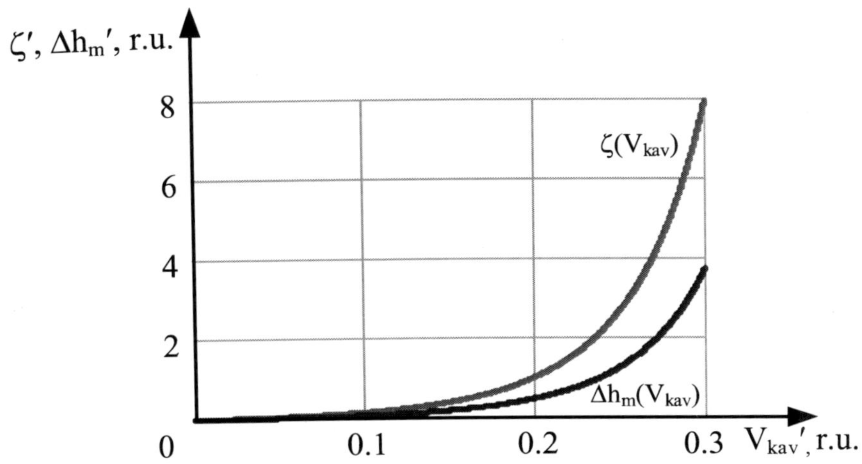

Figure 3.14. The curves representing the alteration of the coefficient of resistance and head losses at the local narrowing depending on the volume of the cavitation cavern.

Then, taking into account the above obtained curves (Figure 3.14) and expressions (2.33), (2.34), (3.18), (3.19); we perform the analysis of the cavitation process influence on the power losses in the PC considered above. The presence of cavitation processes at the third section of the pipeline resulted in the fact that the head-discharge characteristic of the third section $(H-Q)_{s3}$ and the total pipeline characteristic $(H-Q)_{s1}$ shifted upwards along the pump characteristic $(H-Q)_N$ (curves $(H-Q)_{s3}'$ and $(H-Q)_{s1}'$ in Figure 3.15, respectively). As a result, the PC operation mode will be determined by point C, corresponding to head $H_B = 125.6$ m, discharge $Q_B = 0.307$ m³/s and consumed power $N_D = 547.8$ kW (Figure 3.15). In this case, the head losses caused by cavitation $\Delta H_{kav} = H_C - H_E = 125.6 - 97.3 = 28.3$ m.

The pipeline power losses related to the cavitation processes

$$\Delta N_{kav} = 115.2 \text{ kW}.$$

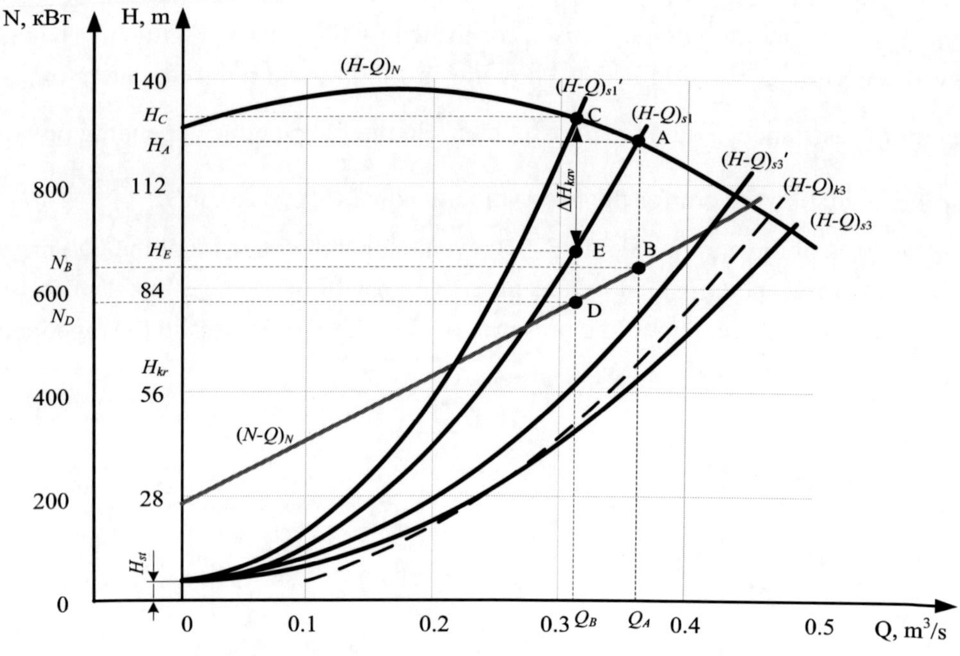

Figure 3.15. The pumping complex characteristics at taking into account the cavitation processes in the pipeline.

Generalizing the above said, it should be noted that the provision of the non-cavitation operation of PC with a variable-frequency ED depends on a number of parameters: the velocity of the flow motion, the cavitation number, the pressure of the saturated steam, the pump ED rotation frequency. The head-discharge characteristic $(H-Q)_k$, representing the boundary of the non-cavitation operation mode, is of a parabolic form, and its intersection with the head-discharge characteristic representing the change of the head at the input into the considered section, enables determining the value of the relative critical rotation frequency of the pump variable-frequency ED.

Thus, one of the possible methods of the reduction of cavitation processes in the pipeline consists in the alteration of the frequency of the pump impeller rotation by means of a variable-frequency ED, which allows meeting the condition of $\nu < \nu_{kr}$ and the change of the technological parameters within the required limits.

For the case when it is impossible to provide PC non-cavitation operation mode by the alteration of the pump ED rotation frequency, the search for the alternative ways of the control of the cavitation processes in the pipeline is topical.

3.3. THE DEVELOPMENT OF THE DEVICE FOR THE CONTROLOF THE CAVITATION PROCESSES IN THE PUMPING COMPLEX BY MEANS OF A VARIABLE-FREQUENCY ELECTRIC DRIVE

3.3.1. The Functional Diagram of the Electromechanical Device for the Control of the Cavitation Processes

Figure 3.16 contains a functional diagram of an electromechanical device for the control of cavitation processes (EDCCP) in PC including a device for the automatic bleeding of the air (DABA) 3, a receiver 4, a pressure sensor 5, a block 6 of control of the regulating valve 7, an electric drive 8 of the regulating valve, an air turbine 9 with an electric generator 10 on one shaft, a block 11 of the generator and the power grid coupling.

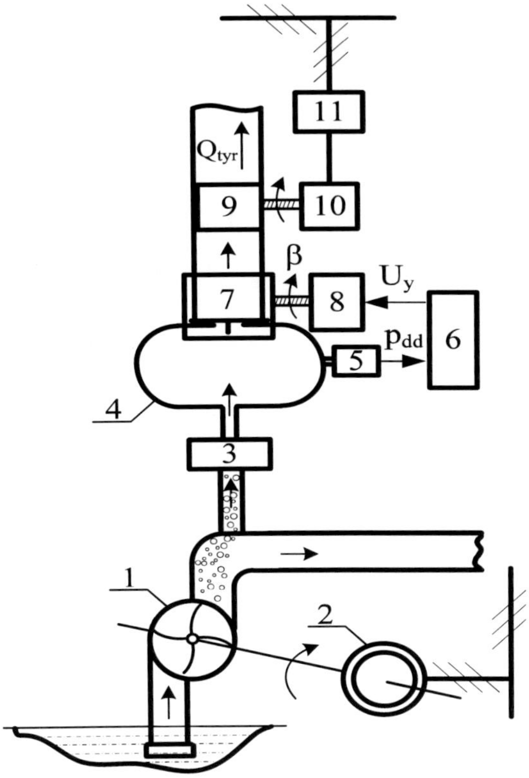

Figure 3.16. The functional diagram of EDCCP in PC: 1 – pump; 2 – induction motor; 3 – automatic air bleeder; 4 – receiver; 5 – pressure sensor; 6 – block of control of regulating valve; 7 – regulating valve; 8 –regulating valve electric drive; 9 – air turbine; 10 – electric generator; 11 – coupling block.

Due to the automatic bleeding of the air from the liquid flow the steam-gas mixture enters the receiver equipped with a pressure sensor and a regulating valve, equipped with ED by means of which the pressure in the receiver and the power at the air turbine input are regulated. The latter converts the energy of the steam-gas mixture into the electric energy by means of an electric generator mounted on one shaft with the turbine and equipped with a device for coupling with the power grid.

When selecting DABA one takes into account the maximal value of the discharge of the steam-gas mixture available in PC liquid flow. Equipment serially manufactured by the industry can be used as a device for air bleeding: an automatic valve for air bleeding with maximal possible discharge of gas 8.1 m^3/hour or a separator with gas discharge exceeding 100 m^3/hour.

An automatic air bleeder (AAB) (Figure 3.17, a) represents a brass body 1, containing a float 2, hinged on a lever 3 with an output valve 4. The operation principle of such an air bleeder consists in the fact that, when there is no air, the float keeps the valve shut; as the float chamber fills with air, it goes down opening the output valve. When the air is bled, the float rises again acting on the lever shutting the output valve. The process continues until all the air gathers in the float chamber.

The amount of the air, that can be removed from the liquid flow, depends on the pressure in the system and is determined by the AAB head-discharge characteristic shown in Figure 3.17, b.

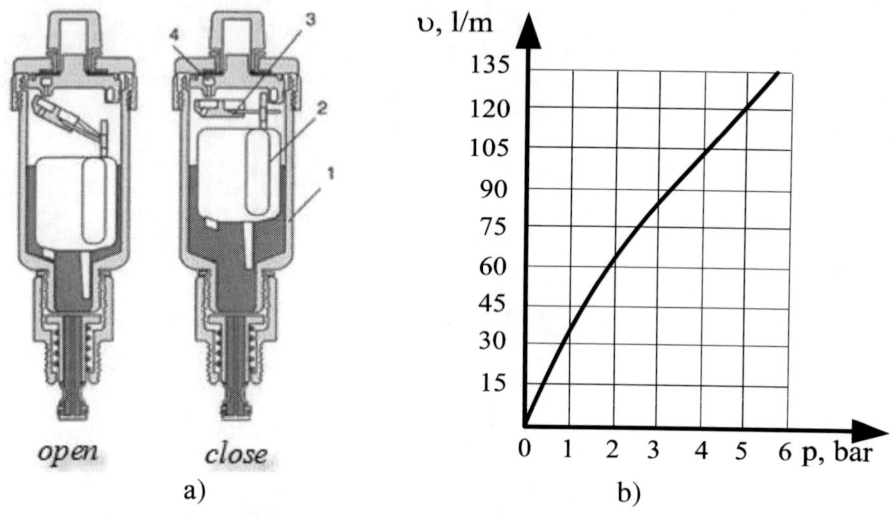

Figure 3.17. Automatic air bleeder: a) design; b) throughput.

DABA should be installed at the point of the biggest accumulation of cavitation bubbles – in the pump delivery branches or the sections of the communication mains located at the highest points of the hydraulic system.

It is reasonable to use in the system the axial ventilators as a turbine device. When selecting an air turbine it is necessary to take into account the following:

- the turbine minimal power $N_{tyr\ min}$ is to be equal to the steam-gas mixture minimal power $N_{kav\ min}$;
- the air turbine maximal power $N_{tyr\ max}$ is to be equal to the steam-gas mixture maximal power $N_{kav\ max}$;
- the air turbine discharge Q_{tyr} is to be within the range of the steam-gas mixture discharge variation Q_{kav}.

The determination of the receiver volume bases on the value of the steam-gas mixture discharge Q_{kav}, maximal $p_{res\ max}$ and minimal $p_{res\ min}$ values of the pressure in the receiver and time t, required to attain the operation pressure p_{dd}. Considering it, the receiver volume is determined according to expression, m^3:

$$V_{res} = \frac{Q_{kav} p_{atm} t}{p_{res\ max} - p_{res\ min}}, \tag{3.26}$$

where p_{atm} – the value of the atmospheric pressure, Pa.

As an air bleeding device for the considered PC it is reasonable to use Flamcovent 400 F separator of the water-gas type with the discharge of 400 m^3/hour. It is possible to use 12B-2 axial-type ventilator with an electric generator, based on a direct- or alternating current machine, as an air turbine – generator system. The volume of the receiver, taking into account time $t = 10$ s of its filling with steam-gas mixture, is 0.47 m^3. Then air collector of V500 11B ZINC type can be used as a receiver. Electromagnetic valve RU16 is used for the remote control of the steam-gas medium at the input of the air turbine.

As mentioned above, the amount of the air that can be removed via DABA is determined by its characteristic (Figure 3.17, b) and depends on the pressure in the system. It should be noted that the pressure created in the receiver in this case, would counteract against the air flow passing via DABA. In this connection, there arises a necessity for the regulation of the value of the pressure in the receiver by means of a regulating valve RV.

Figure B.2 shows the algorithm of EDCCP operation in PC, in accordance with which, at the equality of pressure p_{dd} in the receiver and maximal admissible pressure $p_{res\ max}$, determined by DABA throughput, control signal U_y is formed to the regulating valve ED corresponding to its maximal degree of opening β=1.0 and maximal values of pressure $p_{vx\ tyr} = p_{dd}$, gas discharge $Q_{vx\ tyr}$ at the input of the air turbine and the power of the electric generator:

$$N_g = \frac{p_{vx\,tyr} Q_{vx\,tyr}}{\eta_g}, \tag{3.27}$$

where η_g – the efficiency of the generator.

If pressure p_{dd} in the receiver is less than minimal admissible pressure $p_{res\,min}$, whose value is determined by the required pressure at the input of the air turbine, control signal U_y is formed to full shutting of the regulating valve ($\beta = 0$).

Provided $p_{res\,min} \le p_{dd} < p_{res\,max}$ the current value of the air discharge via the turbine is determined:

$$Q_{tyr_i} = \sqrt{\frac{p_0 - p_{dd}}{R_{tyr}}} \tag{3.28}$$

and signal U_y is formed to the regulating valve ED whose degree of opening is determined according to expression [10–12]:

$$\beta_i = \frac{ab + cC_v^{\,d}}{b + C_v^{\,d}}, \tag{3.29}$$

where p_0 – the pressure of the air turbine idle running; R_{tyr} – the internal resistance of the turbine; a, b, c – the coefficients of approximation of the regulating valve published characteristic whose typical form is shown in Figure 3.18; d – the degree index (determined by the regulating valve published characteristic); $C_v = \frac{Q_{tyr_i}}{Q_{max}}$ – the throughput of the valve; Q_{max} – the maximal discharge of the air via the regulating valve.

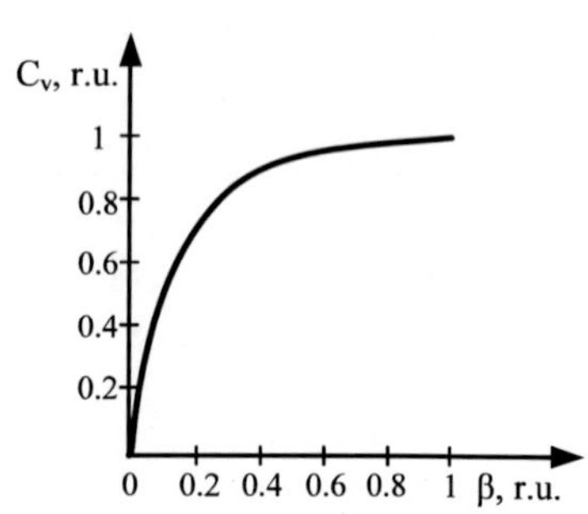

Figure 3.18. A typical published characteristic of the regulating valve.

Then the power of the electric generator:

$$N_{g_i} = \frac{p_{dd} Q_{tyr_i}}{\eta_g}. \qquad (3.30)$$

3.3.2. The Dynamic Modes of the Pumping Complex Operation

The mathematical description of a PC model with a pump variable-frequency ED, a hydrodynamic network represented by three sections of the pipeline network and a cavitation channel is given in the second chapter of the paper.

Let us consider in more detail the mathematical description of EDCCP elements whose input parameters are head H_1 at the output of the first section of the hydronetwork and current volume V_{kav} of the cavitation caverns.

The published characteristic of the automatic air bleeder is represented by an approximation polynomial of the form:

$$Q_{otv} = a_{vo} + b_{vo}\Delta p - c_{vo}\Delta p^2, \qquad (3.31)$$

where a_{vo}, b_{vo}, c_{vo} – the coefficients of approximation of the published characteristic of the automatic air bleeder (in the considered case the approximation coefficients are equal to $a_{vo} = 2.21\cdot 10^{-5}$, $b_{vo} = 5.49\cdot 10^{-9}$, $c_{vo} = 3.12\cdot 10^{-15}$); Q_{otv} – the discharge of the automatic air bleeder, m^3/s; $\Delta p = p_{pot} - p_{res}$ – the difference of the pressures at the input p_{pot} and output p_{res} of the automatic air bleeder, Pa; p_{pot} – the pressure in the liquid flow, Pa; p_{res} – the pressure in the receiver, Pa.

We have obtained that the determination coefficient at the use of the approximation polynomial (3.13) is 0.986.

The value of the discharge of the steam-gas mixture entering the receiver, m^3/s:

$$\Delta Q_{otv} = Q_{kav} - |Q_{kav} - Q_{otv}|, \qquad (3.32)$$

where Q_{kav} – the discharge of the steam-gas mixture in the liquid flow, m^3/s.

The receiver accumulating the steam-gas mixture is represented by a transfer function:

$$W_{res}(p) = \frac{p_{res}(p)}{\Delta Q_{otv}(p) - Q_{tyr}(p)} = \frac{k_{res}}{T_{res}p + 1}, \tag{3.33}$$

where $T_{res} = V_{res} / \Delta Q_{otv}$ – the constant of the receiver time, s; V_{res} – the volume of the receiver, m^3; $k_{res} = \frac{c\rho}{S_{res}}$ – the coefficient of the receiver proportionality, kg/s·m^4; c – the velocity of the sound in the medium at rest, m/s; S_{res} – the area of the receiver, m^2; ρ – the density of the substance, kg/m^3.

The amount of the steam-gas mixture flowing via the regulating valve depends on the degree of its opening and is determined by the expression of the form:

$$Q_{kl} = \beta Q_{tyr}, \tag{3.34}$$

where Q_{kl} – the discharge of the steam-gas mixture coming to the turbine input, m^3/s; β – the degree of opening of the regulating valve; Q_{tyr} – the discharge of the turbine, m^3/s.

The pressure at the turbine input and the losses in it, respectively, Pa:

$$p_{tyr} = p_{res} - \Delta p_{set}; \tag{3.35}$$

$$\Delta p_{tyr} = R_{tyr} Q_{tyr}^{\ 2}, \tag{3.36}$$

where $\Delta p_{set} = R_{set} Q_{kl}^{\ 2}$ – the losses of the pressure in the pipeline, Pa; R_{set} – the resistance of the air duct, kg/m^7; R_{tyr} – the aerodynamic resistance of the turbine, kg/m^7.

The torque developed by the turbine is determined by the expression of the form, N·m:

$$M_{tyr} = \frac{p_{tyr} Q_{kl} \eta_{tyr}}{\omega_{tyr}}, \tag{3.37}$$

where η_{tyr} – the efficiency of the air turbine; ω_{tyr} – the angular rotation frequency of the air turbine impeller, s^{-1}.

The moment of resistance and EMF developed by the direct current generator are determined by the expression of the form, M_{cg} and E_g, respectively:

$$M_{cg} = k\Phi_g I_g; \tag{3.38}$$

$$E_g = k\Phi_g \omega_{tyr}, \tag{3.39}$$

where k – the design constant; Φ_g – the magnetic flux of the direct current generator, Wb; I_g – the current flowing along the generator armature circuit, A.

The operation of the air turbine – generator system is described by the transfer functions of the form:

$$W_{tyr}(p) = \frac{Q_{tyr}(p)}{p_{tyr}(p) - \Delta p_{tyr}(p)} = \frac{1/R_{tyr}}{T_{tyr}p + 1}; \tag{3.40}$$

$$W_{t\omega}(p) = \frac{\omega_{tyr}(p)}{M_{tyr}(p) - M_{cg}(p)} = \frac{1}{J_{tg}p}; \tag{3.41}$$

$$W_g(p) = \frac{I_g(p)}{E_g(p) - U_n(p)} = \frac{1/R_g}{T_g p + 1}, \tag{3.42}$$

where J_{tg} – the inertia moment of the air turbine – generator system, kgm^2; U_n – the load voltage, V; R_g – the resistance of the generator armature circuit, Ohm; T_g – the electromagnetic constant of the generator time, s; T_{tyr} – the constant of the turbine time, s.

The output parameters of EDCCP are: discharge ΔQ_{otv} of the steam-gas mixture removed from the liquid flow; recuperated power maximal $N_{g\max} = p_{tyr}Q_{kl}$ and current $N_g = U_n I_g$ values depending on PC operation mode (the electric generator power is regulated by means of the alteration of the degree of opening β of the regulating valve).

Researching the developed PC mathematical model with EDCCP, we obtained curves (Figures 3.19-3.22), representing the alteration of the technological, power and mechanical parameters of the electromechanical system.

The parameters of PC with EDCCP adopted for modeling: D1250-62.5 pump rated head and discharge – $H_{nom} = 62.5$ m and $Q_{nom} = 1250$ m^3/h respectively; the maximal volume of the caverns – $V_{kr} = 0.05128Q$ m^3; the power of the pump electric drive – $N_{nom} = 250$ kW; the supply voltage – $U = 380$ V; the pump efficiency – $\eta_{nom} = 0.86$; the rated angular rotation frequency – $\omega_{nom} = 154.17$ s^{-1}; the number of the hydronetwork sections – $n = 3$; the pipeline diameter – $d = 0.4$ m; the length of the first, second and third sections of the pipeline – $l_1 = 1000$ m, $l_2 = 500$ m and $l_3 = 800$ m, respectively; the critical number of cavitation –

$\chi_{kr} = 350$; the velocity of sound in the liquid – $c = 1450$ m/s; the specific weight of the liquid – $\gamma = 9792.9$ N/m^3; the temperature of the pumped liquid – $T_V = 25$ °C; the radius of the bubbles of the air undissolved in the liquid – $R_0 = 10^{-4}$ m; the initial volume of the cavitation caverns – $V_0 = 10^{-7}$ m^3; the pressure in the cavern – $p_{kav} = 10000$ Pa; the pump time constant – $T_N = 0.005$ s; the inertia time constants of the hydronetwork sections – $T_{u1} = 4.05$ s, $T_{u2} = 6.75$ s, $T_{u3} = 10.35$ s; the capacitive time constants of the hydronetwork sections – $T_{e1} = 5.86 \cdot 10^{-4}$ s, $T_{e2} = 8.79 \cdot 10^{-4}$ s, $T_{e3} = 0.00135$ s.

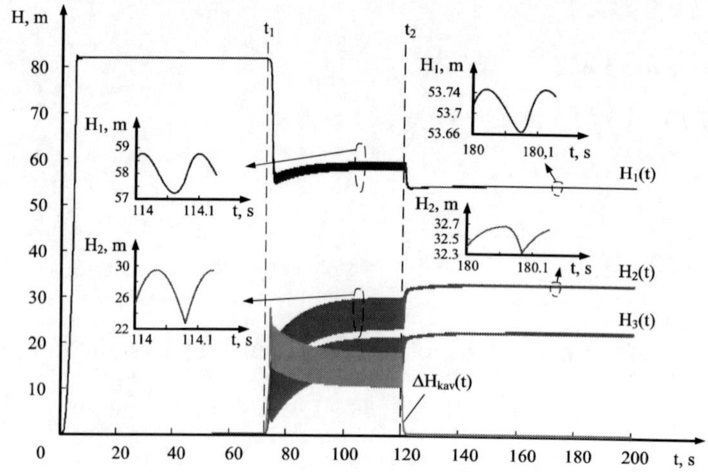

Figure 3.19. The pumping complex head alteration curves: $H_1(t)$ – $H_3(t)$ – the head at the corresponding sections of the pipeline; $\Delta H_{kav}(t)$ – the head losses caused by the presence of the cavitation processes in the pipeline.

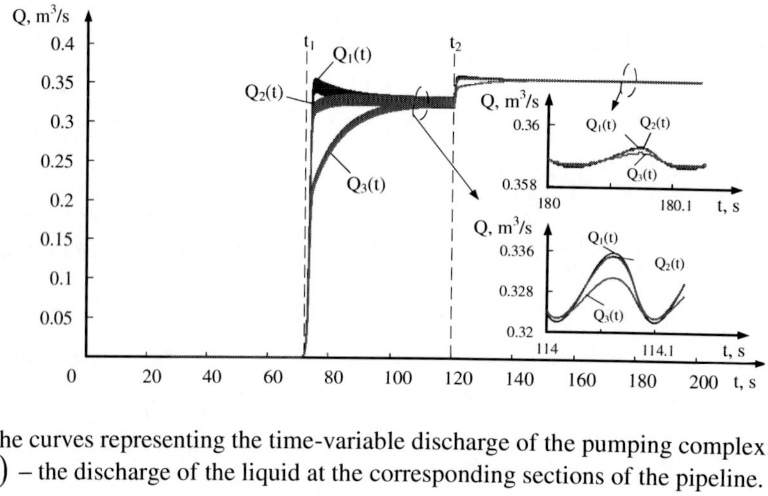

Figure 3.20. The curves representing the time-variable discharge of the pumping complex: $Q_1(t)$ – $Q_3(t)$ – the discharge of the liquid at the corresponding sections of the pipeline.

Figures 3.19–3.22 contain curves representing time-variable head, discharge, hydraulic power in PC, the pump ED rotation frequency, the moment developed by PU, respectively. The analysis revealed that in the time interval $0-70$ s the pump unit starts on the shut stopcock. In this case, head H_1, developed by the pump and power N_N consumed by it correspond to the published values of PU at the zero discharge.

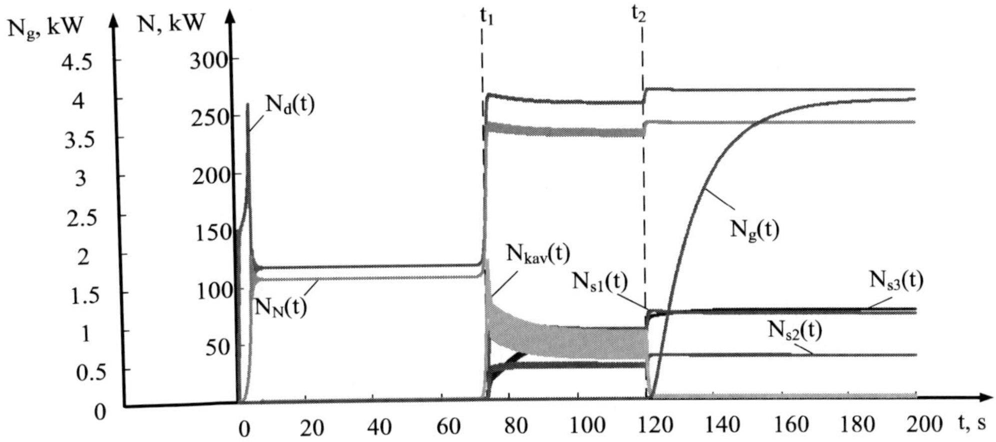

Figure 3.21. PC power variation curves: $N_d(t)$ – the power entering the induction motor on the stator side; $N_N(t)$ – the power of the pump; $N_{s1}(t)$ – $N_{s3}(t)$ – the power of the liquid flow at corresponding sections of the pipeline; $N_{kav}(t)$ – the power losses caused by the presence of cavitation processes in the pipeline; $N_g(t)$ – the power of the recuperation of the steam-gas mixture energy.

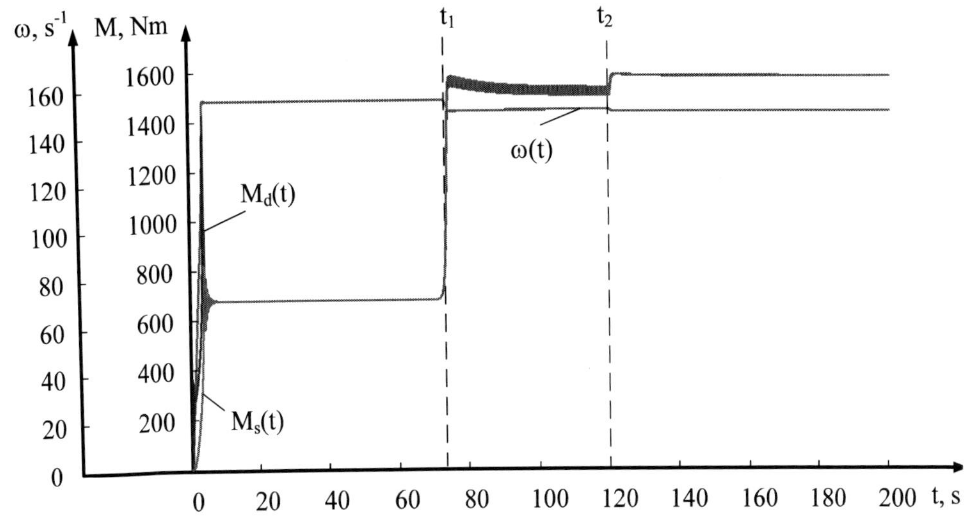

Figure 3.22. The curves representing the time-variable mechanical parameters of the pump unit: $M_d(t)$ – the moment developed by IM; $M_s(t)$ – the moment of resistance created by the turbomechanism; $\omega(t)$ – the angular rotation frequency of the pump ED.

Time moment $t_1 = 70$ s is characterized by shutting the stopcock in the pressure pipeline, which is accompanied by the decrease of head H_1 at the first section and the increase of head H_2 at the second and head H_3 at the third section of the hydronetwork located after the stopcock. It results in meeting the condition $\nu_i > \nu_{kr\_1}$ and is accompanied by the development of cavitation processes at the first section and occurrence of the head pulsations with the frequency of 10 Hz. In this case, losses of head ΔH_{kav} and power N_{kav}, caused by cavitation, make about 30% and 24% of the corresponding PC rated parameters. Besides, the increase of the pump load (opening of the stopcock) results in the growth of resistance moment M_s created by the turbomechanism and, consequently, in the increase of power N_d consumed by ED and power N_N consumed by the pump, moment M_d, developed on the pump ED shaft and the decrease of PU angular rotation frequency ω.

Time moment $t_2 = 120$ s is characterized by the switch of EDCCP, which results in the reduction of head losses ΔH_{kav} and power N_{kav} on cavitation. In this case, the values of discharge Q_1 – Q_3 and head H_2, H_3 at the second and the third sections of the hydronetwork grow, which results in the shift of the point characterizing PC operating mode. It is accompanied by inessential increase of the resistance moment M_s created by the turbomechanism and the reduction of angular rotation frequency ω of the pump ED.

The analysis of the obtained curves (Figure 3.21) reveals that power N_g of the recuperation of the steam-gas mixture energy changes from zero to the maximal value, corresponding to the full opening of the regulating valve in EDCCP. In the considered case $N_g = 3.9$ kW, which makes 1.56% of the pump rated power N_{nom}.

CONCLUSION

1. The analysis of curves $M_v(\beta, \omega_{sp})$, representing the alteration of the stopcock resistance moment, revealed that at the rated velocity of the stopcock spindle rotation at the section where $\beta \le 0.2$ there is a significant growth of moment M_v at the decrease of the relative degree of its opening β. A tenfold reduction of the spindle rotation frequency results in the decrease of the maximal value of the stopcock resistance moment by 1.8 times at its full shutting. In this case, the head in the pipeline network considerably reduces. Thereby, an efficient way of the control of the dynamic loads in the pumping complex consists in the formation of the law of control of the pipeline valves electric drive by means of the decrease of the supply voltage frequency.

2. The decrease of the supply voltage frequency of the stopcock ED, chosen by the catalog, results in the reduction of the dynamic loads in PC. However, in this case, the time of shutting of the pipeline valves considerably increases, which is caused by a significant reduction rate of the stopcock ED gear. This problem is solved by the use of the law of frequency control with the maintenance of the constant power and specification of the reduction rate of the stopcock ED gear, whereat the condition of the invertor overload by current is met. In this case, such an approach makes it possible to provide the minimal values of the pipeline valves control time and the increase of the head in the hydronetwork whereat a surge in the system is impossible.
3. The developed structure and the algorithms of functioning of the electromechanical system of the reduction of the dynamic loads in the pumping complex, bases on the variable-frequency electric drive of the shutoff-control stopcock. It uses the reserve electric supply from an uninterrupted power supply or an active dissipator of the hydraulic flow energy. It enables eliminating the inadmissible increases of the pressure in the pipeline, the efficient use of the energy of the liquid reverse flow, the prolongation of the technological equipment life span.
4. We have proposed a method for the determination of the boundaries of the non-cavitation operation of the pumping complex with a variable-control electric drive. The method bases on the analysis of the joint operation modes of the pumping facility on the pipeline network and allows the determination of the range of the admissible values of the alteration of the pump electric drive rotation frequency at the regulation of the discharge within the required limits. So, for the considered pumping facility, at the alteration of the pump discharge within (0-1250) m^3/h, the non-cavitation operation of the electrohydraulic system will be provided in the range of the alteration of the pump electric drive rotation frequency of (0-0.7)n_n.
5. We have developed a method for the control of cavitation processes in the pumping complex, making it possible, by means of the removal of the steam-gas mixture released from the liquid, to reduce the unproductive power losses in the pipeline, caused by the presence of cavitation with a simultaneous possibility of its energy recuperation.
6. We have substantiated the structure of the electromechanical system of the control of the cavitation processes in the pumping complex based on an automatic air-bleeding device, a receiver and an air turbine with an electric generator on one shaft. We have proposed recommendations as to the selection of the elements of the electromechanical device for the control of the cavitation processes, taking into account the operation characteristics of the used equipment, the mode of the pumping complex operation and the degree of the development of the cavitation processes.

7. We have developed a mathematical model of a pumping complex with an electromechanical device for the control of the cavitation processes in the pipeline. The obtained dynamic curves representing the time-variable head, discharge, and hydraulic power in PC proved the efficiency of the operation of this device: the possibility of practically complete elimination of the pipeline network power losses caused by the accumulation of cavitation bubbles in the liquid flow. In the considered case at the pump power of 250 kW, the flow hydraulic power of 217 kW, the recuperation power is 3.9 kW.

Chapter 4

The Identification of the Pumping Complex Parameters in Non-Steady Operation States

4.1. The Pumping Complex Equivalent Electric Circuits Based on the Electrohydraulic Analogy

Pump complexes (PC) of the systems of urban and industrial water supply and drainage are complicated electrotechnological objects. They include pumping units (PU), a pipeline network, a consumer (a settlement, a reservoir, a water tower). The changeable character of water consumption, sewage inflow results in the necessity for variation of PU functioning conditions in time. The most efficient method of PC parameters regulation consists in the variation of the pump impeller rotational speed by means of the variable-frequency electric drive (ED). It provides the most accurate correspondence to the required water consumption schedule with minimum power losses [51, 109–111].

When PC operation conditions change, complex hydrodynamic processes occur in the pipeline network, they are caused by PU activation/halt, failure to comply with the tempo of pipeline valves control, sudden disconnection of PU supply, etc. [49, 73, 112]. Such processes result in variation of pressure, flow rate, hydraulic resistance and power in time.

The analysis [52, 113] revealed that to research dynamic processes in PC it is expedient to use the method of electrohydraulic analogy (MEHA). The essence of MEHA consists in the similarity of processes taking place in electric circuits and pipeline networks represented by long lines with distributed parameters. In this case, the main electrical equations are transformed into corresponding hydraulic relations and a PC equivalent electric circuit is created on their basis.

The basic similar electrical and hydraulic values are given in Table 4.1.

Table 4.1 demonstrates that the basic variables, characterizing PC operation conditions, are:

- discharge Q – the volume of liquid fed by the pump through the nozzle in a unit of time;
- head H – the difference of full specific energies of the liquid flow in the pump output and input nozzles.

The mentioned parameters are analogs of corresponding electrical values: current I and voltage U. In this case, the head at zero discharge, developed by the pump with shut stopcock on the head pipeline, corresponds to electromotive force E of the electric circuit.

Hydraulic resistance R – is a value related to the forces of the resilient friction between the liquid layers, the liquid and the channel walls. The presence of hydraulic resistance results in energy dissipation in the form of heat in the pump and the pipeline.

Inertial resistance L is reactance taking into account the inertial head counteracting the discharge change in the pipeline.

Capacitance C takes into account the properties of liquid compressibility under the influence of external actions. A reservoir with liquid fed and drained through the pipeline is analogous to the capacitance in a hydraulic system.

Table 4.1. Basic electrical and hydraulic relations

Electric circuit	Hydraulic circuit
Current I, A	Discharge Q, m³/s
Voltage U, V	Head H, m
Electromotive force E, V	Head H_0 at zero discharge (head of idling), m
Resistance R, Ohm	Hydraulic resistance R, s/m²
Inductance L, Hn	Inertial resistance (hydraulic inductance) L, s²/m²
Capacitance C, F	Capacitance (hydraulic capacitance) C, m²

It should be noted that the major difference between hydraulic circuits and electric ones consists in the fact that head drop ΔH is proportional to squared discharge Q^2.

To describe PC operation conditions it is expedient to use the known electrotechnical laws of the theory of branched electrical circuits that are true for hydrodynamic processes research:

the sum of liquid discharges in the i-th unit is equal to zero:

$$\sum_{i=1}^{n} Q_i = 0 , \tag{4.1}$$

where n – general number of the units;

the sum of head losses for any closed circuit is equal to the sum of the heads developed by the sources (pumps):

$$\sum_{i=1}^{n} R_i Q_i^2 = \sum_{i=1}^{n} \Delta H_i \ . \qquad (4.2)$$

Head losses at every hydraulic element are determined according to the expressions:

for hydraulic resistance

$$\Delta H_R = RQ^2 \ ; \qquad (4.3)$$

hydraulic inductance

$$\Delta H_L = L\frac{dQ}{dt} \ ; \qquad (4.4)$$

hydraulic capacitance

$$\Delta H_C = \frac{1}{C}\int_0^T Qdt \ , \qquad (4.5)$$

where T – period of repetition of the analyzed signal.

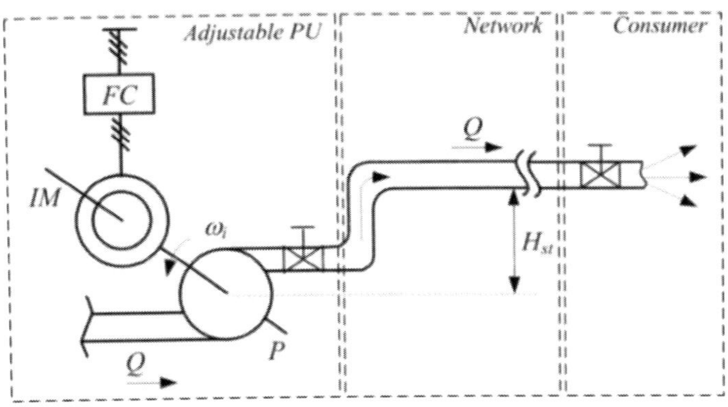

Figure 4.1. Technological scheme of a simplest PC.

Thus, the use of the known laws of the electric circuit theory (Ohm's and Kirghoff's laws) in accordance with MEHA makes it possible to create the equations of head losses at PC elements for every hydraulic section. The obtained equation system is to be the basis for a mathematical model of a hydraulic system in steady and unsteady modes.

Let us consider a simplest PC technological scheme (Figure 4.1), including a frequency converter (FC), an induction motor (IM), a pump (P), shut-off-and-regulating fittings, a pipeline and a consumer. Figure 4.2 contains its equivalent electric circuit, wherein PU with controlled rotational speed is presented in the form of an idealized hydromachine with constant structural parameters. The scheme in Figure 4.2 contains the following designations: $H_0\nu^2$ – hydraulic source of energy; R_p – hydraulic resistance of the pump, s/m²; $\nu = \omega_i / \omega_n$ – relative rotational speed of the pump impeller; ω_i, ω_n – current and rated values of PU angular rotational speed, respectively, s^{-1}.

Thus, the PU head-discharge characteristic can be described by an expression of the form [114]:

$$H = H_0\nu^2 - R_p Q^2 . \tag{4.6}$$

The pipeline (Figure 4.2) is represented by a source of static backpressure H_{st}, m, related to the necessity for raising the liquid at certain height and active-inductive $R_{net} - L_{net}$ load taking into account friction head losses and inertial losses of the liquid flow. Pipeline hydraulic resistance R_{net} depends on the length of communication network, the pipeline cross section, walls roughness, presence of local resistances (elbows, check valves, stopcocks). The consumer is represented by hydraulic resistive impedance R_{con}, located at the end of the pipeline network.

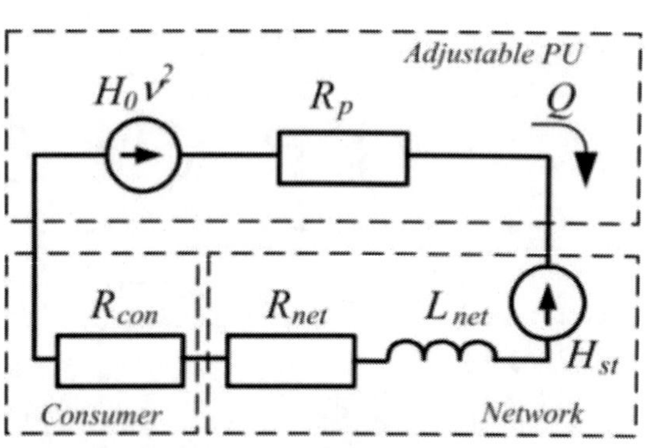

Figure 4.2. PC equivalent electric circuit.

Then the pipeline characteristic is described by equality [114]:

$$H = H_{st} + (R_{net} + R_{con})Q^2 . \tag{4.7}$$

Taking (4.3) – (4.5) into consideration, the general equation of heads balance for the given equivalent circuit is of the form:

$$H_0 \nu^2 = \Delta H_{Rp} + \Delta H_{Rnet} + \Delta H_{Lnet} + \Delta H_{st} + H_{Rcon} \,, \tag{4.8}$$

where ΔH_{Rp} – head losses at the pump hydraulic resistance; ΔH_{Rnet}, ΔH_{Lnet} – head losses at the pipeline network hydraulic resistance and inductance, respectively; ΔH_{st} – head losses at overcoming the backpressure; H_{Rcon} – head at the consumer.

In reality PUs consist of several turbomachines connected in parallel (Figure 4.3 a) or in series (Figure 4.3, b) to the pipeline network, which is caused by the necessity for plant operation aiming at meeting the schedule of changeable discharge or increasing the head in the hydrosystem. Hydraulic machines, operating simultaneously, mutually influence each other: the discharge, head, power and efficiency of each of them essentially depend on the load modes of the other operating machines.

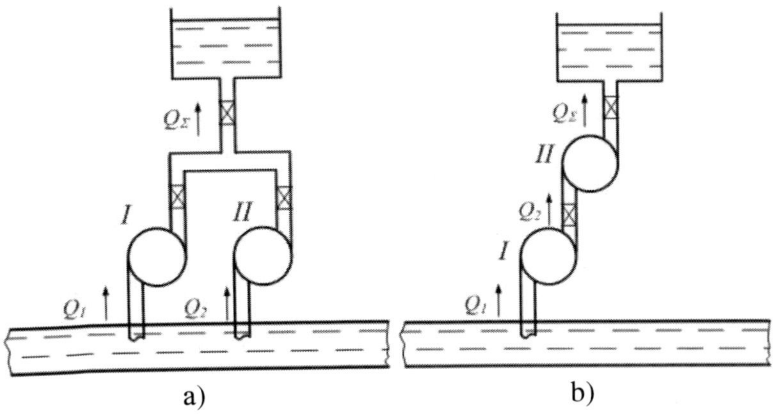

Figure 4.3. Technological scheme of pumping units connection in parallel a) and in series b).

PC equivalent electric circuits taking into account group operation of PU with pumps adjustable and unregulated electric drives are shown in Figures 4.4, 4.5.

The following equation system can be written down based on equivalent electric circuit of PC with PU operating in parallel (Figure 4.4) [109, 114]:

$$\left.\begin{aligned} &H_{01}\nu^2 - H_{02} = R_{p1}Q_1^2 - R_{p2}Q_2^2; \\ &H_{01}\nu^2 - H_{st} = R_{p1}Q_1^2 + R_{net}Q_3^2; \\ &Q_1 + Q_2 = Q_\Sigma \,, \end{aligned}\right\} \tag{4.9}$$

where H_{01}, H_{02} – head developed by pumps at zero discharge of an adjustable and an unregulated pump, respectively; R_{p1}, R_{p2} – hydraulic resistances of an adjustable and an unregulated pump, respectively; Q_1, Q_2, Q_Σ – discharges of an adjustable and an unregulated pump, as well as PU total discharge, respectively.

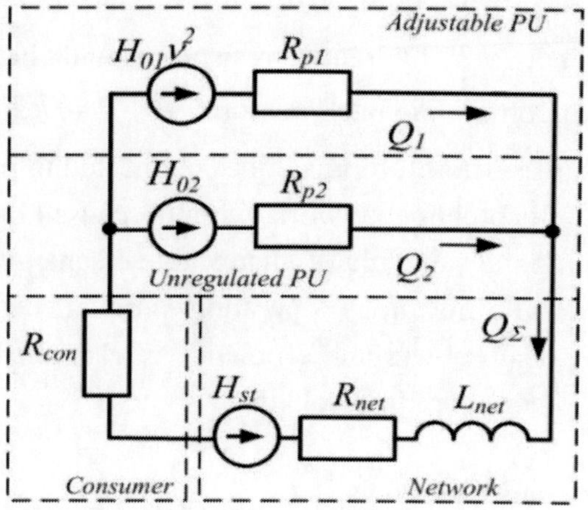

Figure 4.4. PC equivalent electric circuit with PU operating in parallel.

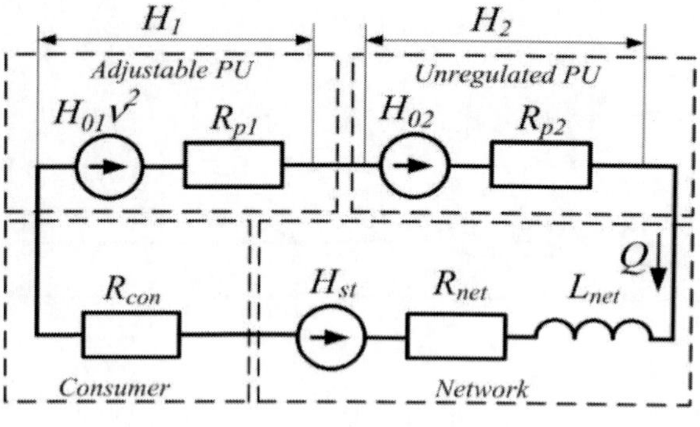

Figure 4.5. PC equivalent electric circuit with PU operating in series.

For PU operating in series (Figure 4.5), provided the discharge is regulated by variation of rotational speed of one of the pumps, the following equation system is true:

$$\left.\begin{aligned} &H_{01}v^2 + H_{02} - H_{st} = R_{b1}Q_\Sigma^2 + R_{b2}Q_\Sigma^2 + R_{net}Q_\Sigma^2;\\ &H_1 = H_{01}v^2 - R_{b1}Q_\Sigma^2;\\ &H_2 = H_{02} - R_{b2}Q_\Sigma^2;\\ &H_\Sigma = H_1 + H_2, \end{aligned}\right\} \tag{4.10}$$

where $Q_\Sigma = Q_1 = Q_2$ – discharge created by PP; H_1, H_2, H_Σ – heads of the adjustable and unregulated pumps and PP total head, respectively.

Operation of a real unidealized pump is characterized by final number K_L of impeller paddles, presence of volume losses Q_∞ of working fluid in gaskets and by-passes, hydraulic ΔW_{hydr} and mechanical ΔW_{mech} losses of energy [113]. It should be taken into consideration by introducing corresponding hydraulic resistances, in particular:

$R_t = (H_0 - 1/\eta_g \mu_H)\eta_0 \mu_Q$ – the hydraulic resistance, caused by PU structural parameters; μ_H – the coefficient of head reduction under the influence of the number of paddles; μ_Q – the coefficient of working flow volume compression by the paddles (coefficient of filling in the operating zone cross section with active flow); η_g, η_o – pump hydraulic and volumetric efficiencies, respectively;

$R_{\mu H} = (H_0/Q - R_t)(1-\mu_H)$ – the unit hydraulic resistance taking into account the final number of paddles;

$R_{\Delta H} = C_2 / Q(Q - C_1 Q_n)^2 + C_0 Q$ – the hydraulic resistance reflecting pump hydraulic losses consisting of eddy (impact, diffuser) losses and losses along the length; C_1, C_2, C_3 – constant coefficients; Q, Q_n – current and rated values of discharge;

$R_{\mu Q} = (H_0/Q - R_t - R_{\mu H})(1/(1-\mu_Q))$ – the pump hydraulic resistance characterizing the variation of liquid discharge (the volumetric compression of the working flow) depending on the number of paddles; k – the current value of the coefficient of reduction of the idling head.

PC equivalent electric circuit with unidealized PU is given in Figure 4.6, where Q_μ, Q_Δ – the discharge losses depending on the number of paddles and hydraulic losses of energy, respectively; Q_p, Q_{net}, Q_{con} – the discharge at the pump output, in the pipeline network and at the consumer, respectively.

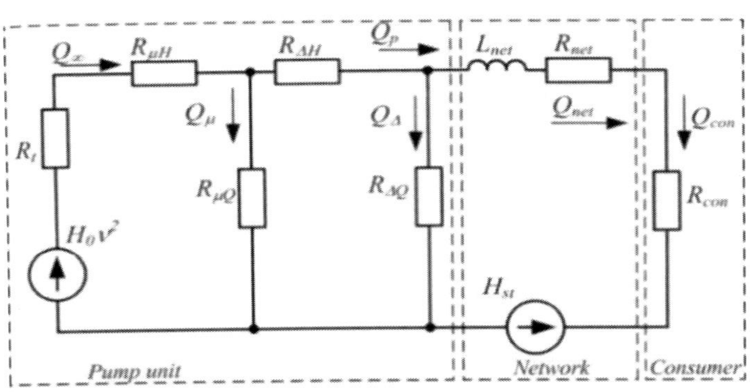

Figure 4.6. PC equivalent electric circuit with an unidealized PU.

When PC operates for a long pipeline, the hydraulic network can be represented by j number of sections, each of which is characterized by certain parameters: length l, m, diameter d, m, roughness coefficient k_e etc. For simplicity, we assume that the parameters of the pipeline network section are invariable in time.

During pipeline networks operation there occur wave processes of different nature, caused by water hammers, forced pressure fluctuations, self-excited fluctuations, resonance phenomena, etc. They are accompanied by the occurrence of vibrations spreading along the pipeline. These vibrations result from interaction of the liquid molecules between the main flow and the flow appearing because of depressurization.

The form of the equations of the distribution of pressure waves in the pipes resembles the laws of the distribution of electromagnetic waves that can be described by wave (telegrapher's) equations known in electrical engineering [52, 104]. When MEHA for pipeline networks is used, the following equation system is true:

$$\left.\begin{aligned} &\frac{\partial H}{\partial x}+\frac{1}{gS}\frac{\partial Q}{\partial t}+\frac{\lambda}{d}\frac{1}{S^2}\frac{Q|Q|}{2g}=0; \\ &\frac{\partial H}{\partial t}+\frac{c^2}{gS}\frac{\partial Q}{\partial x}=0, \end{aligned}\right\} \tag{4.11}$$

where H, Q – the head and discharge of liquid in the current section of the pipeline, respectively, m, m³/s; $S=\pi d^2/4$ – the area of the pipeline cross section, m²; λ – the dimensionless coefficient of resistance of the pipeline; c – the sound velocity in the medium, m/s (for water c = 1450 m/s); $g=9.81$ m/s² – the acceleration of gravity.

The final element method [52, 104] is used to solve telegrapher's equations; it enables presentation of a pipeline as a final number of elements of equal length. In this case, the head and discharge at every interval are assumed constant. Such transformation corresponds to the change of a long line for a circuit scheme of j number of the pipeline network sections of the final length and the transfer from partial derivatives, included in system (4.11), to differential analogs:

$$\left.\begin{aligned} &H_{net\,j}-H_{net\,j-1}+L_{net\,j}l_{sect}\frac{dQ_{net\,j-1}}{dt}+R_{net\,j}l_{sect}Q_{net\,j-1}\left|Q_{net\,j-1}\right|=0; \\ &\frac{dH_{net\,j}}{dt}+C_{net\,j}\frac{1}{l_{sect}}\left(Q_{net\,j}-Q_{net\,j-1}\right)=0, \end{aligned}\right\} \tag{4.12}$$

where $l_{sect} = L/n$ – the length of the pipeline section; $R_{net\,j} = \lambda/2dgS^2$; $C_{net\,j} = c^2/Sg$; $L_{net\,j} = 1/Sg$ – the hydraulic resistance, inductance and capacitance of the j-th section of the pipeline, s/m^2, m^2, s^2/m^2, respectively; $H_{net\,j}, Q_{net\,j}, H_{net\,j-1}, Q_{net\,j-1}$ – the head and discharge at the output and input of the j-th section of the network, respectively.

Electric equivalent circuit of PC operating for a long pipeline is shown in Figure 4.7 where Q'_j – the discharge losses at the j-th section of the pipeline.

Thus, application of the described method to PC modeling on MEHA basis makes it possible to determine head and discharge at any point of the hydrosystem, head losses at PC elements at different consumer's parameters and discharge regulation by variation of the pump impeller rotational speed.

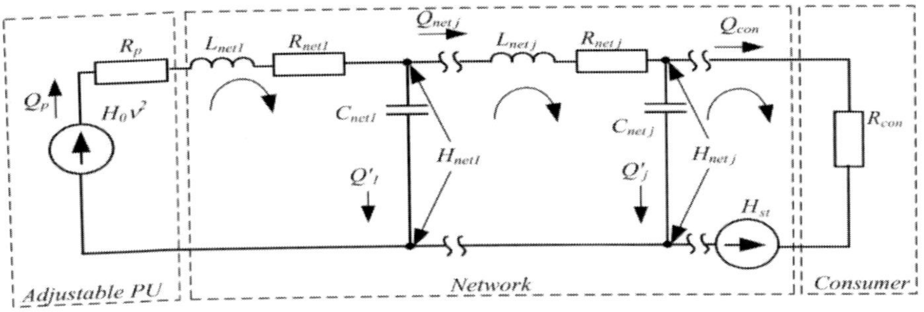

Figure 4.7. Electric equivalent circuit of PC operating for a long pipeline.

4.2. WORKING OUT THE PUMPING COMPLEX POWER MODEL TAKING INTO ACCOUNT THE NONLINEAR PROCESSES IN THE PIPELINE NETWORK

In practical PC operation, there often occur conditions resulting in the appearance of hydrodynamic processes, at which pressure, flow rate, hydraulic resistance, power, etc. are time functions [115–120].

The described processes may appear in different ways when liquid interacts with some elements or devices, hydraulic machines operating units, etc.

So, forced fluctuations caused by the irregularity of PU supply may be generated in a number of hydrosystems. In this case, the pressure in PC changes, which is accompanied by additional dynamic loads on the elements of the pumps and the pipelines [121].

When the discharge is regulated, PP is often equipped with different (as to their operation principle and design) distributing devices (valves, stopcocks), whose flow elements represent short channels or openings with constant or time-variable flow areas. In the latter case, rate profiles are greatly distorted. It results in the occurrence of vortexes due to which the medium non-stationary motion influence on the processes of liquid outflow via distribution devices becomes essential [98].

The above said explains the topicality of the problems of PC transient processes research and the determination of the parameters of hydraulic equipment with periodic variation of technological parameters.

Let us represent discharge at PU output in the considered case as a periodic time function:

$$Q(t)=Q_0+Q_m\sin(\Omega t-\varphi)=Q_0+Q_a\cos(\Omega t)+Q_b\sin(\Omega t), \tag{4.13}$$

where Q_0, Q_m – the amplitude values of the discharge signal constant and variable components, respectively, m³/s; $Q_a=Q_m\cos(\Omega t)$, $Q_b=Q_m\sin(\Omega t)$ – the orthogonal cosine and sine components of the discharge signal, respectively; Ω – the angular frequency of the discharge signal, s^{-1}; φ – the shift angle of the discharge signal, degs.

Time variation of discharge $Q(t)$ and head $H(t)$ results in the generation of hydraulic power periodic signal at the corresponding i-th elements of PC:

$$p_i(t)=\rho g H_i(t)Q_i(t), \tag{4.14}$$

where ρ – the density of liquid; $H_i(t)$, $Q_i(t)$ – the head and discharge at the i-th element of PC, respectively.

Using the method of harmonic analysis based on the Fourier series [122], the harmonic power at PC elements can be presented as a sum of constant and harmonic (cosine and sine) components [123–126].

Therefore, the hydraulic power at the output of the hydraulic power source (pump):

$$\begin{aligned} p_s(t)&=\rho g H_0 \nu^2 Q_p(t)= \\ &=\rho g H_0 \nu^2\left(Q_{p0}+Q_{pa}\cos(\Omega t)+Q_{pb}\sin(\Omega t)\right)= \\ &=P_{s0}+P_{s1a}\cos(\Omega_k t)+P_{s1b}\sin(\Omega_k t), \end{aligned} \tag{4.15}$$

where $P_{s0}=\rho g H_0 Q_{p0}$ – the constant component of the power at the output of the hydraulic power source; $P_{s1a}=\rho g H_0 Q_{pa}$, $P_{s1b}=\rho g H_0 Q_{pb}$ – the amplitude values of the cosine and sine components of the first power harmonic of the hydraulic power source, respectively; $\Omega_k=2\Omega$ – the angular frequency of the power of the hydraulic power source.

The power losses at hydraulic resistance of the pump:

$$\begin{aligned}\Delta p_p(t) &= \rho g R_p Q_p^3(t) = \rho g R_p\left(Q_0 + Q_a\cos(\Omega t) + Q_b\sin(\Omega t)\right)^3 = \\ &= P_{Rp\,0} + P_{Rp\,1a}\cos(\Omega_k t) + P_{Rp\,1b}\sin(\Omega_k t) + P_{Rp\,2a}\cos(2\Omega_k t) + \\ &+ P_{Rp\,2b}\sin(2\Omega_k t) + P_{Rp\,3a}\cos(3\Omega_k t) + P_{Rp\,3b}\sin(3\Omega_k t).\end{aligned} \tag{4.16}$$

The hydraulic power losses at the j-th section of the pipeline:

$$\Delta p_{net\,j}(t) = \Delta p_{R\,net\,j}(t) + \Delta p_{L\,net\,j}(t) + \Delta p_{C\,net\,j}(t); \tag{4.17}$$

where the power losses at the hydraulic resistance

$$\begin{aligned}\Delta p_{R\,net\,j}(t) &= \rho g R_{net\,j} Q_{net\,j}^3(t) = \\ &= \rho g R_{net\,j}\left(Q_0 + Q_a\,cos(\Omega t) + Q_b\,sin(\Omega t)\right)^3 = \\ &= P_{Rnetj\,0} + P_{Rnetj\,1a}\,cos(\Omega_k t) + P_{Rnetj\,1b}\,sin(\Omega_k t) + \\ &+ P_{Rnetj\,2a}\,cos(2\Omega_k t) + P_{Rnetj\,2b}\,sin(2\Omega_k t) + \\ &+ P_{Rnetj\,3a}\,cos(3\Omega_k t) + P_{Rnetj\,3b}\,sin(3\Omega_k t);\end{aligned} \tag{4.18}$$

the power losses at the inductance

$$\begin{aligned}\Delta p_{L\,net\,j}(t) &= \rho g L_{net\,j} Q_{net\,j}(t)\, d\left(Q_{net\,j}(t)\right)/dt = \\ &= \rho g L_{net\,j}\left(Q_0 + Q_a\cos(\Omega t) + Q_b\sin(\Omega t)\right)\times \\ &\times\left(-Q_a\Omega_{kav}\sin(\Omega t) + Q_b\Omega_{kav}\cos(\Omega t)\right) = \\ &= P_{Lnetj\,1a}\cos(\Omega_k t) + P_{Lnetj\,1b}\sin(\Omega_k t) + \\ &+ P_{Lnetj\,2a}\cos(2\Omega_k t) + P_{Lnetj\,2b}\sin(2\Omega_k t);\end{aligned} \tag{4.19}$$

the power losses at the capacitance

$$\begin{aligned}\Delta p_{C\,net\,j}(t) &= \rho g Q_{net\,j}(t)/C_{net\,j}\int_0^T Q_{net\,j}(t)dt = \\ &= \rho g\left(Q_0 + Q_a\cos(\Omega t) + Q_b\sin(\Omega t)\right)\times \\ &\times\left(Q_0 + (Q_a/\omega)\sin(\Omega t) + (Q_b/\omega)\cos(\Omega t)\right) = \\ &= P_{Cnetj\,1a}\cos(\Omega_k t) + P_{Cnetj\,1b}\sin(\Omega_k t) + \\ &+ P_{Cnetj\,2a}\cos(2\Omega_k t) + P_{Cnetj\,2b}\sin(2\Omega_k t).\end{aligned} \tag{4.20}$$

The hydraulic power losses for overcoming the backpressure in the pipeline network:

$$
\begin{aligned}
&\Delta p_{st}(t) = \rho g H_{st} Q_{con}(t) = \\
&= \rho g H_{st}\left(Q_{p0} + Q_{pa}\cos(\Omega t) + Q_{pb}\sin(\Omega t)\right) = \\
&= P_{st\,0} + P_{st\,1a}\cos(\Omega_k t) + P_{st\,1b}\sin(\Omega_k t).
\end{aligned}
\tag{4.21}
$$

The hydraulic power at the consumer:

$$
\begin{aligned}
&p_{con}(t) = \rho g R_{con} Q_{con}^3(t) = \\
&= \rho g R_{con}\left(Q_0 + Q_a\cos(\Omega t) + Q_b\sin(\Omega t)\right)^3 = \\
&= P_{con\,0} + P_{con\,1a}\cos(\Omega_k t) + P_{con\,1b}\sin(\Omega_k t) + \\
&+ P_{con\,2a}\cos(2\Omega_k t) + P_{con\,2b}\sin(2\Omega_k t) + \\
&+ P_{con\,3a}\cos(3\Omega_k t) + P_{con\,3b}\sin(3\Omega_k t).
\end{aligned}
\tag{4.22}
$$

In this case, in expressions (4.16) – (4.22) the following designations are adopted: $P_{R_{net\,j}\,0}$, $P_{st\,0}$, $P_{con\,0}$ – the constant components of the power losses at the pump hydraulic resistance, for overcoming the static backpressure and the power of the consumer, respectively; P_{Rpka}, P_{Rpkb}, $P_{Rnetj\,ka}$, $P_{Rnetj\,kb}$, $P_{Lnetj\,ka}$, $P_{Lnetj\,kb}$, $P_{Cnetj\,ka}$, $P_{Cnetj\,kb}$, $P_{st\,ka}$, $P_{st\,kb}$, $P_{st\,kb}$, $P_{con\,ka}$, $P_{con\,kb}$ – the amplitude values of cosine and sine components of the k-th harmonic of the power losses at the pump hydraulic resistance, the pipeline section, at the pipeline inductance and capacitance, for overcoming the static backpressure and the power of the consumer, respectively.

The parts of constant and variable components in expressions (4.16) – (4.22) are obtained by the multiplication of the discharge initial signal by the values of the corresponding parameters of PC equivalent circuit elements with their following transformation to the form of Fourier's trigonometric series.

Thus, a scheme of power losses distribution in PC containing a hydraulic power source, a pipeline network and a hydraulic power consumer is shown in Figure 4.8.

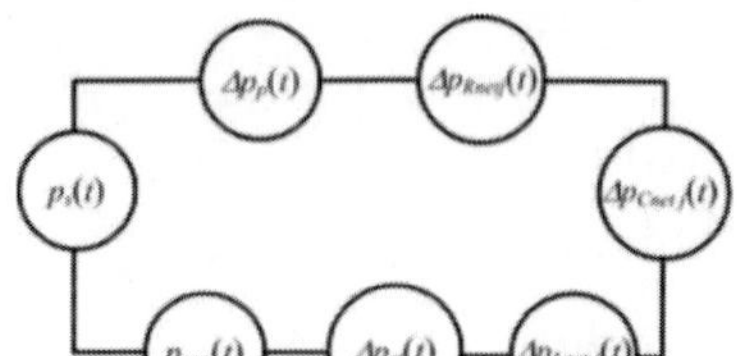

Figure 4.8. The scheme of the power losses distribution in PC power channel.

The energy balance equation of the following form is true for the shown scheme of the distribution of power losses at PC elements:

$$p_s(t) = \Delta p_p(t) + \Delta p_{Rnet\,j}(t) + \Delta p_{Lnet\,j}(t) + \\ + \Delta p_{Cnet\,j}(t) + \Delta p_{st}(t) + p_{con}(t). \tag{4.23}$$

Expression (4.23) is a PC energy model, allowing the research of energy processes flowing in the hydrosystem and the determination of the hydraulic power losses at any point of PC.

The development of a PC energy model taking into acount the cavitation processes in the pipeline network.

The regulation of PC operating modes, switching on/off of PU or separate pipeline sections, the actuation of shut-off-and-regulating fittings are accompanied by the development of non-stationary hydrodynamic processes, in particular, head pulsations, water hammers, cavitational fluctuations, etc. [51, 73, 112].

Cavitational phenomena are one of the most negative non-stationary processes as to their consequences. They are characterized by the periodic generation and collapse of cavities filled with steam or gas [104, 117–119]. Researchers' interest to this problem is caused by the diversity of steam- and gas-liquid mixtures in nature and their intensive use in heat engineering, hydrodynamics, etc. [117–119].

Cavitation bubbles appear in the liquid flow because of the local drop of the pressure in the liquid (below the pressure of saturated steams), the growth of the temperature of the pumped operating medium, the availability of the local resistances in the pipeline network, the insufficient support at PU suck [51, 73, 112, 117–119].

Cavitation processes may develop in impellers, pump intake and head nozzles, at the points of narrowing and turn of the pipeline, in siphon pipelines [104, 117–119].

The dissipation of the cavitation wave energy due to the uneven heat exchange between the bubbled gas and liquid presents the determining mechanism in the pressure wave distribution in gas- and steam-liquid mixtures [117–119]. Nonlinear hydrodynamic waves may both damp and intensify, because of heat energy release and dispersion, the change of the liquid phase state. It is essential that cavitation caverns destruction is accompanied by the release of big amount of liquid energy, resulting in the growth of temperature of the pump impeller or the liquid flow, the periodic self-fluctuations of the head and discharge (efficiency, hydraulic resistances) of the pump and hydronetwork [73, 112].

As stated above, the occurrence of cavitation processes results in the appearance of cavitation caverns – cavities filled with steam or gas. They are characterized by volume $V_{kav} = 1/T_{kav} \int_0^T (Q_{net}(t) - Q_p(t))dt$, certain length l_{kav}, and period of pulsations

$T_{kav} = V_{kav}\gamma/G$ (Figure 4.9), where G – the weight discharge of liquid; γ – the specific weight of liquid [73, 112, 104].

The cavitation cavern volume variation curve (Figure 4.9) is characterized by the gradual growth (time moment t_1) and sharp collapse (time moment t_2) of gas bubbles when the volume reaches the critical value V_{kr}.

In this case, the variation of the hydraulic resistance at the pipeline network section, when cavitation is developed, is described by the introduction of a cavitation channel into the equivalent circuit (Figure 4.10). The cavitation channel includes key K and nonlinear resistance $R_{kav}(t)$. The condition for actuation of key K consists in drop of pressure in the pipeline below the saturated steam pressure caused by the variation of the consumer operation condition R_{con}.

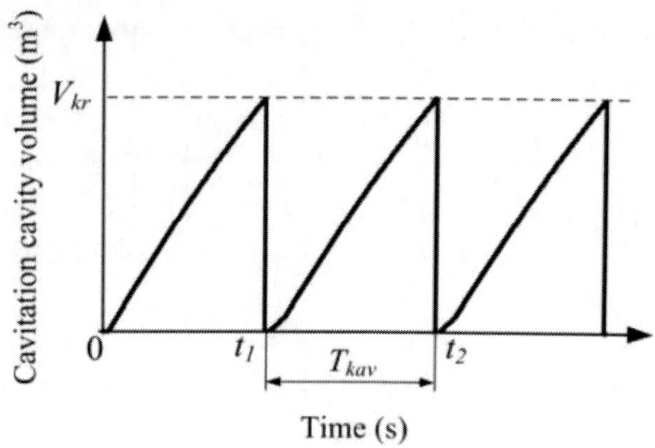

Figure 4.9. The curve of cavitation cavern volume variation.

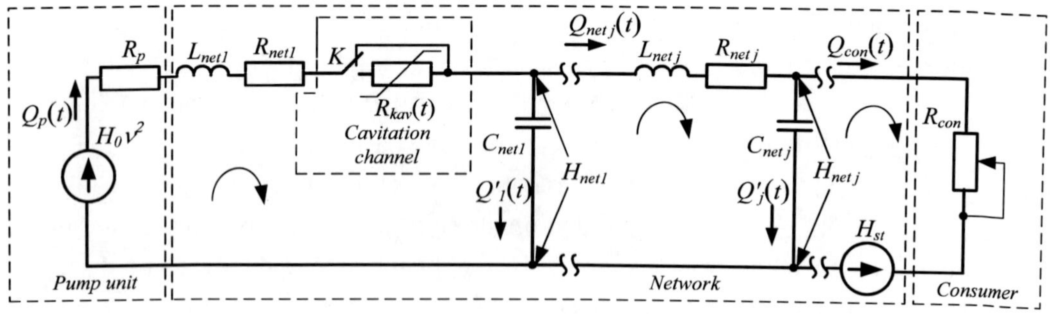

Figure 4.10. PC equivalent electric circuit taking into account the development of the cavitation processes in the pipeline network.

To research PC characteristics taking into account the cavitation processes in the pipeline network consisting of *j*-th number of sections the system of head and discharge balance equations is true:

$$\left.\begin{aligned}
&H_0 \nu^2 = R_p Q_p^2(t) + R_{net1} Q_p^2(t) + R_{kav}(t) Q_p^2(t) + \\
&+ L_{net1} d(Q_p(t))/dt + 1/C_{net1} \int_0^T Q_1'(t) dt; \\
&0 = R_{net2} Q_2^2(t) + L_{net2}\, d(Q_2(t))/dt + \\
&+ 1/C_{net2} \int_0^T Q_2'(t) dt - 1/C_{net1} \int_0^T Q_1'(t) dt; \\
&0 = R_{net3} Q_3^2(t) + L_{net3}\, d(Q_3(t))/dt + \\
&+ 1/C_{net3} \int_0^T Q_3'(t) dt - 1/C_{net2} \int_0^T Q_2'(t) dt; \\
&\cdots\cdots\cdots\cdots\cdots\cdots\cdots\cdots\cdots\cdots \\
&0 = R_{net\,j} Q_j^2(t) + L_{net\,j}\, d(Q_j(t))/dt + \\
&+ 1/C_{net\,j} \int_0^T Q_j'(t) dt - 1/C_{net\,j-1} \int_0^T Q_{j-1}'(t) dt; \\
&0 = H_{st} + R_{con} Q_{con}(t) - 1/C_{net\,j} \int_0^T Q_j'(t) dt; \\
&Q_p(t) = Q_{con}(t) + Q_1'(t) + Q_2'(t) + \ldots + Q_j'(t).
\end{aligned}\right\} \tag{4.24}$$

The pipeline section time variable hydraulic resistance related to periodic cavitation fluctuations can be described by a trigonometric series of the following form with sufficient accuracy [122]:

$$R_{kav}(t) = a_0 + \sum_{n=1}^{N} a_n \sin(n\Omega_{kav} t), \tag{4.25}$$

where a_0, a_n – the coefficients of trigonometric series; $\Omega_{kav} = 2\pi f_{kav}$ – the angular frequency of the cavitation fluctuations; $f_{kav} = 1/T_{kav}$ – the frequency of the cavitation fluctuations; n, N – the number and quantity of the harmonics of the cavitation channel hydraulic resistance signal.

The occurrence of the cavitation processes in the pipeline results in the periodic nonsinusoidal variation of the discharge at PU output:

$$\begin{aligned}
Q_p(t) &= Q_0 + \sum_{l=1}^{L} Q_l \cos(\Omega_l t - \gamma) = \\
&= Q_0 + \sum_{l=1}^{L} Q_{la} \cos(\Omega_l t) + \sum_{l=1}^{L} Q_{lb} \sin(\Omega_l t),
\end{aligned} \tag{4.26}$$

where l, L – the number and quantity of the discharge signal harmonics, respectively; Q_0, Q_l – the amplitude values of the constant and variable components of the discharge signal, respectively; $Q_{la} = Q_l \cos\gamma$, $Q_{lb} = Q_l \sin\gamma$ – the orthogonal cosine and sine components of the discharge signal, respectively; Ω_l, γ – the angular frequency and shift angle of the discharge signal phase in relation to the origin of coordinates, respectively.

Then the hydraulic power at the output of the hydraulic supply source (pump):

$$
\begin{aligned}
&p_s(t) = \rho g H_0 \nu^2 Q_p(t) = \\
&= \rho g H_0 \nu^2 \left(Q_0 + \sum_{l=1}^{L} Q_{la} \cos(\Omega_l t) + \sum_{l=1}^{L} Q_{lb} \sin(\Omega_l t) \right) = \\
&= \sum_{r=1}^{R} P_{rs0} + \sum_{r=1}^{R} P_{rsa} \cos(\Omega_r t) + \sum_{r=1}^{R} P_{rsb} \sin(\Omega_r t),
\end{aligned}
\tag{4.27}
$$

where r, R – the number and quantity of the harmonics of the hydraulic power signal at the output of the hydraulic supply source, respectively; $P_{rs0}, P_{rsa}, P_{rsb}$ – the amplitude values of the constant and orthogonal cosine and sine components of the signal of the hydraulic power of the hydraulic supply source, respectively; Ω_r – the angular frequency of the pump hydraulic power signal.

The power losses at the pump hydraulic resistance:

$$
\begin{aligned}
&\Delta p_p(t) = \rho g R_p Q_p^3(t) = \\
&= \rho g R_p \left(Q_0 + \sum_{l=1}^{L} Q_{la} \cos(\Omega_l t) + \sum_{l=1}^{L} Q_{lb} \sin(\Omega_l t) \right)^3 = \\
&= \sum_{r=1}^{R} P_{rRp0} + \sum_{r=1}^{R} P_{rRpa} \cos(\Omega_r t) + \sum_{r=1}^{R} P_{rRpb} \sin(\Omega_r t).
\end{aligned}
\tag{4.28}
$$

The hydraulic power losses at the j- th section of the pipeline:

$$
\Delta p_{net\,j} = \Delta p_{Rnet\,j}(t) + \Delta p_{Lnet\,j}(t) + \Delta p_{Cnet\,j}(t);
\tag{4.29}
$$

where the power losses at the hydraulic resistance

$$\Delta p_{Rnet\,j}(t) = \rho g R_{net\,j} Q^3_{net\,j}(t) =$$
$$= \rho g R_{net\,j} \left(Q_0 + \sum_{l=1}^{L} Q_{la} \cos(\Omega_l t) + \sum_{l=1}^{L} Q_{lb} \sin(\Omega_l t) \right)^3 = \tag{4.30}$$
$$= \sum_{r=1}^{R} P_{rRnetj0} + \sum_{r=1}^{R} P_{rRnetja} \cos(\Omega_r t) + \sum_{r=1}^{R} P_{rRnetjb} \sin(\Omega_r t);$$

the power losses at the inductance

$$\Delta p_{Lnet\,j}(t) = \rho g L_{net\,j} Q_{net\,j}(t) d(Q_{neti} j(t))/dt =$$
$$= \rho g L_{net\,j} \left(Q_0 + \sum_{l=1}^{L} Q_{la} \cos(\Omega_l t) + \sum_{l=1}^{L} Q_{lb} \sin(\Omega_l t) \right) \times$$
$$\times \left(-\sum_{l=1}^{L} Q_{la} \Omega_l \sin(\Omega_l t) + \sum_{l=1}^{L} Q_{lb} \Omega_l \cos(\Omega_l t) \right) = \tag{4.31}$$
$$= \sum_{r=1}^{R} P_{rLnetj0} + \sum_{r=1}^{R} P_{rLnetja} \cos(\Omega_r t) + \sum_{r=1}^{R} P_{rLnetjb} \sin(\Omega_r t);$$

the power losses at the capacitance

$$\Delta p_{Cnet\,j}(t) = \frac{\rho g Q_{net\,j} j(t)}{C_{net\,j}} \int_0^T Q_{net\,j}(t) dt =$$
$$\frac{\rho g}{C_{net\,j}} \left(Q_0 + \sum_{l=1}^{L} Q_{la} \cos(\Omega_l t) + \sum_{l=1}^{L} Q_{lb} \sin(\Omega_l t) \right) \times$$
$$\times \left(\sum_{l=1}^{L} Q_{l0} + \sum_{l=1}^{L} \frac{Q_{la}}{\Omega_l} \sin(\Omega_l t) - \sum_{l=1}^{L} \frac{Q_{lb}}{\Omega_l} \cos(\Omega_l t) \right) = \tag{4.32}$$
$$= \sum_{r=1}^{R} P_{rCnetj0} + \sum_{r=1}^{R} P_{rCnetja} \cos(\Omega_r t) + \sum_{r=1}^{R} P_{rCnetjb} \sin(\Omega_r t).$$

The hydraulic power losses for overcoming the backpressure in the pipeline network:

$$\Delta p_{st}(t) = \rho g H_{st} Q_{con}(t) =$$
$$\rho g H_{st} \left(Q_0 + \sum_{l=1}^{L} Q_{la} \cos(\Omega_l t) + \sum_{l=1}^{L} Q_{lb} \sin(\Omega_l t) \right) = \tag{4.33}$$
$$= \sum_{r=1}^{R} P_{rst0} + \sum_{r=1}^{R} P_{rsta} \cos(\Omega_r t) + \sum_{r=1}^{R} P_{rstb} \sin(\Omega_r t).$$

The hydraulic power at the consumer:

$$
\begin{aligned}
& p_{con}(t)=\rho g R_{con} Q_{con}^3(t)= \\
& =\rho g R_{con}\left(Q_0+\sum_{l=1}^{L} Q_{la}\cos(\Omega_l t)++\sum_{l=1}^{L} Q_{lb}\sin(\Omega_l t)\right)^3= \\
& =\sum_{r=1}^{R} P_{rcon0}+\sum_{r=1}^{R} P_{rcona}\cos(\Omega_r t)+\sum_{r=1}^{R} P_{rconb}\sin(\Omega_r t).
\end{aligned}
\tag{4.34}
$$

In this case, in expressions (4.28)–(4.34) the following designations are adopted: P_{rRp0}, $P_{rRnetj0}$, $P_{rLnetj0}$, $P_{rCnetj0}$, $P_{rst\,0}$, $P_{rcon\,0}$ – the constant components of the power losses at the pump hydraulic resistance, hydraulic resistance, inductance and capacitance at the *j*-th section of the pipeline, for overcoming static backpressure and power of the consumer, respectively; P_{rRpa}, $P_{rRnetja}$, $P_{rLnetja}$, $P_{rCnetja}$, $P_{rst\,a}$, $P_{rcon\,a}$, P_{rRpb}, $P_{rRnetjb}$, $P_{rLnetjb}$, $P_{rCnetjb}$, $P_{rst\,b}$, $P_{rcon\,b}$ – the amplitude values of the cosine and sine components of the r-th harmonic of the power losses at the pump hydraulic resistance, hydraulic resistance, inductance and capacitance of the *j*-th pipeline section, for overcoming the static backpressure and the power of the consumer, respectively.

When cavitation self-fluctuations occur in the pipeline network, it is necessary to take into account the power losses for the cavitation:

$$
\begin{aligned}
& \Delta p_{kav}(t)=\rho g R_{kav}(t) Q^3(t)= \\
& =\rho g\left(a_0+\sum_{n=1}^{N} a_n\sin(n\Omega_{kav}t)\right)\times \\
& \times\left(Q_0+\sum_{l=1}^{L} Q_{la}\cos(\Omega_l t)+\sum_{l=1}^{L} Q_{lb}\sin(\Omega_l t)\right)^3= \\
& =\sum_{r=1}^{R} P_{rkav0}+\sum_{r=1}^{R} P_{rkava}\cos(r\Omega_r t)+\sum_{r=1}^{R} P_{rkavb}\sin(r\Omega_r t),
\end{aligned}
\tag{4.35}
$$

where P_{rkav0} – the constant component of the power losses for the cavitation; P_{rkava}, P_{rkavb} – the amplitude values of the cosine and sine components of the j-th harmonic of the power losses for the cavitation, respectively; Ω_r – the resultant circular frequency of the cavitation fluctuations power.

PC energy balance equation during the development of cavitation processes in the pipeline network is of the form:

$$\begin{aligned} p_s(t) &= \Delta p_p(t) + \Delta p_{Rnet\,j}(t) + \Delta p_{Lnet\,j}(t) + \\ &+ \Delta p_{Cnet\,j}(t) + \Delta p_{st}(t) + \Delta p_{kav}(t) + p_{con}(t). \end{aligned} \tag{4.36}$$

The obtained energy model enables the analysis of any PC element technological and energy parameter variation when non-stationary (cavitation) processes develop in the pipeline network.

4.3. The Power Processes in the Pumping Complex at the Development of Cavitation in the Pipeline Network

A mathematical model proposed for the research of the transient processes in PC includes FC, a block generating the law of frequency converter control, PU, a pipeline network with a consumer at the end.

FC generating the quadratic law of variation of voltage U_{phn} depending on frequency f_{net} is described by the first order aperiodic link:

$$W_{fc}(s) = \frac{U_m(s)}{f(s)} = \frac{k_{fc}}{T_{fc}s + 1}, \tag{4.37}$$

where $U_m(s)$ – the voltage at FC output, V; $f(s)$ – the frequency of the supply voltage, Hz; k_{fc} – the coefficient of proportionality and FC time constant, respectively, s.

To describe the operation of the pump induction ED with frequency control a known model in $u, v, 0$-coordinates is used; the differential equations in it are represented in the form of flux linkages for a synchronous coordinate system [99]:

$$\left.\begin{aligned}
&\frac{d\psi_{1u}}{dt}=-\frac{R_s\psi_{1u}}{L_r\sigma}+\frac{R_sL_\mu\psi_{2u}}{L_rL_s\sigma}+\omega_0\psi_{1v}+U_{1m};\\
&\frac{d\psi_{1v}}{dt}=-\frac{R_s\psi_{1v}}{L_r\sigma}+\frac{R_sL_\mu\psi_{2v}}{L_rL_s\sigma}-\omega_0\psi_{1u};\\
&\frac{d\psi_{2u}}{dt}=-\frac{R'_r\psi_{2u}}{L_s\sigma}+\frac{R'_rL_\mu\psi_{1u}}{L_rL_s\sigma}+\omega_{2e}\psi_{2\upsilon};\\
&\frac{d\psi_{2v}}{dt}=-\frac{R'_r\psi_{2v}}{L_s\sigma}+\frac{R'_rL_\mu\psi_{1v}}{L_rL_s\sigma}-\omega_{2e}\psi_{2u};\\
&i_{1v}=\frac{\left(\psi_{1v}L_s-\psi_{2v}L_\mu\right)}{\left(L_rL_s-L_\mu^2\right)};\\
&i_{1u}=\frac{\left(\psi_{1u}L_s-\psi_{2u}L_\mu\right)}{\left(L_rL_s-L_rL_\mu\right)};\\
&u_{1v}=R_si_{1v}+\frac{d\psi_{1v}}{dt}+\omega_0\psi_{1u};\\
&u_{1u}=R_si_{1u}+\frac{d\psi_{1u}}{dt}+\omega_0\psi_{1v};\\
&M=\left(\frac{3}{2}\right)\frac{p_wL_\mu}{\left(L_rL_s\sigma\right)}\left(\psi_{1v}\psi_{2u}-\psi_{1u}\psi_{2v}\right);\\
&\omega_{2e}=\omega_0-p_w\omega,
\end{aligned}\right\}\qquad(4.38)$$

where ψ_{1u}, ψ_{1v}, ψ_{2u}, ψ_{2v} – the flux linkages of the electric machine stator and rotor by coordinates u and v, respectively, Wb; U_{1m} – the amplitude value of the supply voltage of the stator winding, V; R_s, R_r – the resistances of the stator and rotor windings, respectively, Ohm; L_s, L_r – the inductances of the stator and rotor phases, respectively, H; L_μ – the mutual inductance of the stator and rotor windings, H; σ – the coefficient of dispersion; ω_0, ω_{2e}, ω – the synchronous angular rotation frequency of the magnetic field, the slip frequency and IM rotor current rotation frequency, respectively, s^{-1}; M, M_s – the electromagnetic and static torques at the shaft of electric machine, respectively, N·m; p_w – the number of pole pairs; J_Σ – the total reduced moment of inertia of IM and pump, kgm^2.

The moment of resistance created by the pump, N·m [114]:

$$M_s=\frac{P_s}{\omega}=\frac{\rho gH_pQ_p}{\eta_p\omega},\qquad(4.39)$$

where P_s – the power supplied to the pump shaft, W; η_p – the pump efficiency.

PU operation is described by a head-discharge characteristic, its equation is given in expression (6), and the persistence of its operation is taken into account by the introduction of the first order aperiodic link:

$$W_p(s)=\frac{Q_p(s)}{H_p(s)}=\frac{k_p}{T_p s+1}, \qquad (4.40)$$

where T_p – the pump time constant; k_p – the pump coefficient of transmission.

The physical processes in the pipeline are described with the use of MEHA and are represented by equation system (4.12).

The hydraulic characteristic of the consumer is of the form:

$$H_{con}-H_{st}=R_{con}Q_{con}^2. \qquad (4.41)$$

The head critical value corresponding to the beginning of the development of the cavitation processes at the *j*-th section of the pipeline is determined by expression [104, 127, 128]:

$$H_{kj}=\frac{Q_j^2\chi_{kr\,j}}{S_j^2 2g}+\frac{h_{kav\,j}}{g\rho_j}, \qquad (4.42)$$

where $\chi_{kr\,j}$ – the critical value of the number of cavitation at the *j*-th section of the pipeline; $h_{kav\,j}$ – the pressure in the cavern, Pa; ρ_j, S_j – the liquid density and the area of cross section at the *j*-th section of the pipeline, kg/m^3, m^2, respectively.

In the case when the value of head $H_{net\,j}$ at the *j*-th section of the pipeline is lower or equal to the value of head H_{kj} in the flow of liquid there is growth of cavitation caverns of the volume of:

$$V_{kav}=V_0+V'_{kav}, \qquad (4.43)$$

where V_0, V'_{kav} – the initial and current values of the volume of the cavitation cavern, respectively.

The critical value of the cavitation cavern, at which it collapses, is determined by expression of the form [104, 127, 128]:

$$V_{kr\,j} = f\left(h_{kav\,j}\right) = \frac{4\pi}{3}\left(\frac{2\sigma}{h_{kav\,j} - h_{net\,j}}\right)^3, \tag{4.44}$$

where $h_{net\,j}$ – the pressure at the j-th section of the pipeline, Pa.

The block diagram of cavitation channel is shown in Figure 4.11. Such representation of the cavitation channel allows its connection to any section of the pipeline where cavitation development is possible.

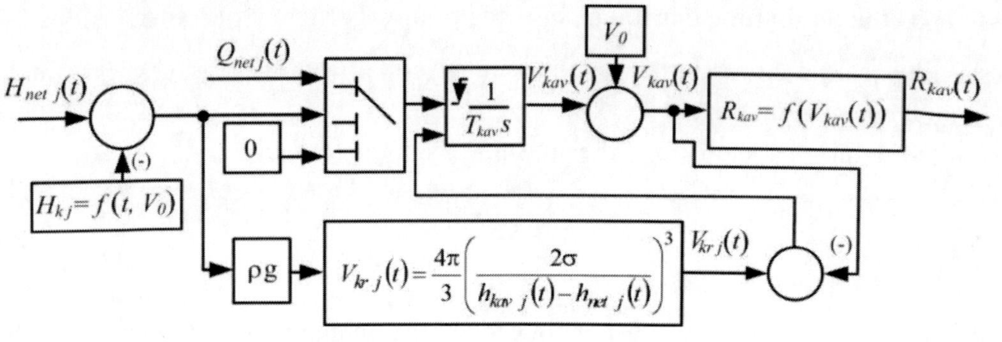

Figure 4.11. The block diagram of the cavitation channel.

The electric power supplied to the motor stator windings is determined by an expression of the form:

$$p_{el}(t) = u(t)i(t), \tag{4.45}$$

where $u(t)$, $i(t)$ – the signals of the electric motor voltage and current, V, A, respectively.

The mechanical power at the induction motor shaft:

$$p_{mech}(t) = M(t)\omega(t), \tag{4.46}$$

where $M(t)$, $\omega(t)$ – the signals of the torque and angular rotation frequency of the electric motor, N·m, s^{-1}, respectively.

The hydraulic power at PU output, at the j-th section of the pipeline and at the consumer is generated by the multiplication of the corresponding signals of the head and discharge according to expression (4.14).

The block diagram of PC model taking into account the development of the cavitation processes in the pipeline network is shown in Figure 4.12.

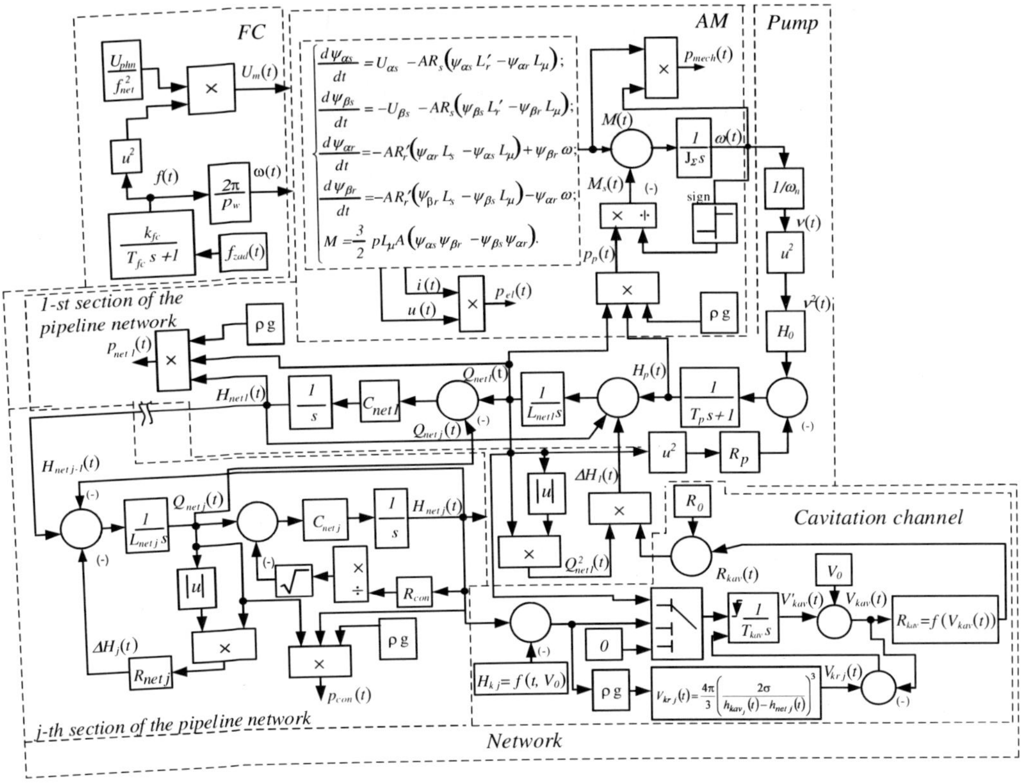

Figure 4.12. Mathematical model of an electrohydraulic complex.

The head and discharge variation curves for PP with the rated parameters $P_n = 200$ kW, $\omega_n = 154.9$ s^{-1}, $Q_p = 0.002$ m^3/s, $H_0 = 18$ m, $H_p = 4.7$ m and a pipeline network with parameters $d = 0.04$ m, $l = 6$ m are shown in Figures 4.13, 4.14, respectively.

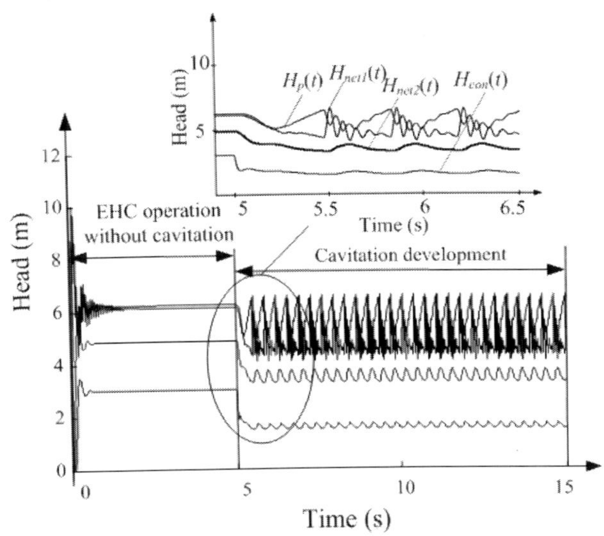

Figure 4.13. The variation curves of the head at the pipeline network sections.

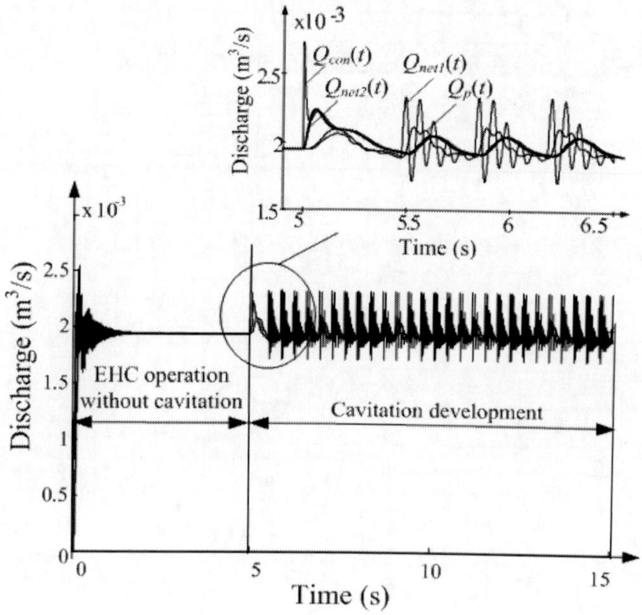

Figure 4.14. The variation curves of the discharge at the pipeline network sections.

At the time moment $t = 2$ s PU reaches operating values of the head and discharge; at $t = 5$ s the head at the pump output reduces, which is accompanied by the occurrence of cavitation processes in the hydronetwork. It is obtained that development of the pipeline network cavitation fluctuations of the frequency of $f_{kav} = 3.2$ Hz results in periodic variation of the head and discharge with the fluctuation amplitude of (10-15)% of the average value of these parameters.

The variation curves of the hydraulic power at PC elements and the cavitation power losses are shown in Figures 4.15, 4.16, respectively.

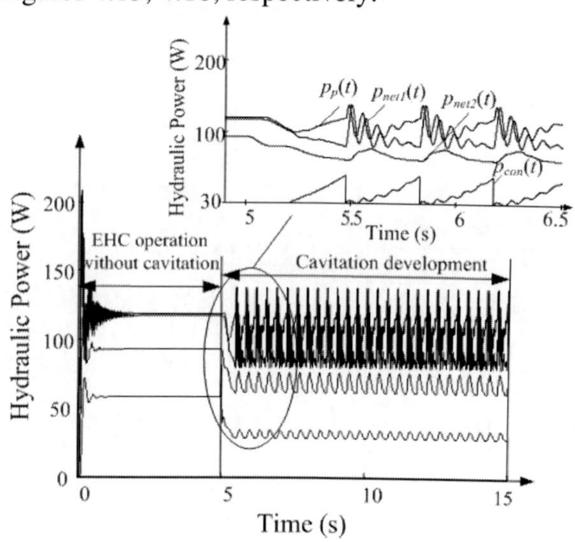

Figure 4.15. The variation curves of the power and power losses at PC elements.

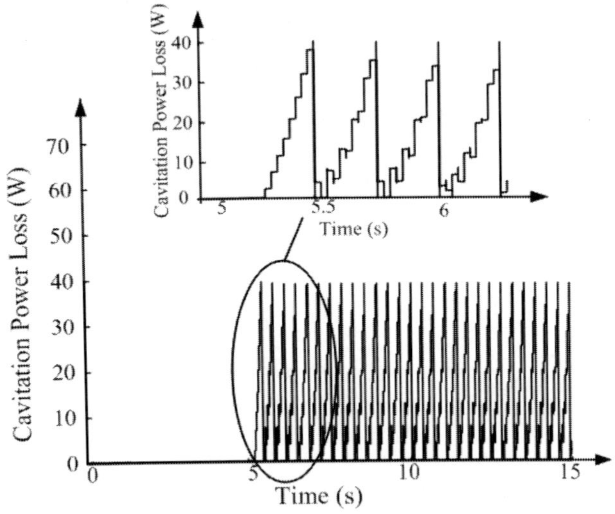

Figure 4.16. The variation curves of the cavitation power losses.

The analysis of the obtained curves (Figure 4.15) reveals that the development of the cavitation processes at the output of the first section of the pipeline is accompanied by the power losses that make 17% of the power consumed by the pump. The cavitation self-fluctuations result in the appearance of a variable component in the power signals at PC elements. The amplitude of their fluctuation goes up to 15% of the rated value.

The cavitation processes accompanied by the change of the hydrosystem load condition are directly related to the energy parameters of the electrohydraulic equipment operation: the electric power supplied to the motor stator windings, the mechanical power at PU shaft, the hydraulic powers at the pump output, as well as at the sections of the pipeline and at the consumer [123–126].

The application of the method of the power signal harmonic analysis based on the Fourier series enabled the assessment of the energy processes in PC at the periodic head and discharge variations caused by the cavitation processes in the hydrosystem. Figures 4.17–4.19 contain the amplitude spectra of the head, discharge and hydraulic power when the cavitation processes occur in the pipeline network.

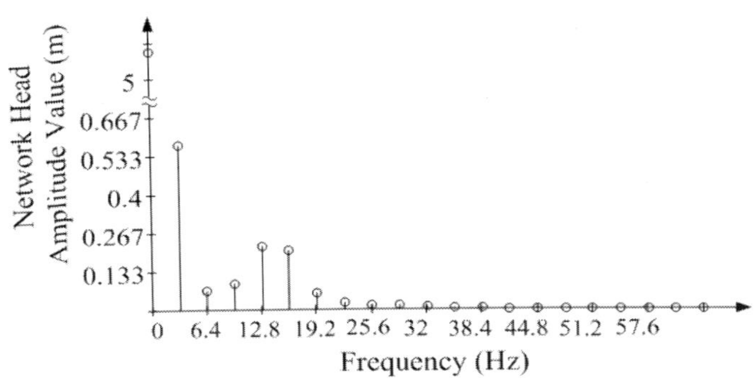

Figure 4.17. The amplitude spectrum of the head in the pipeline network in the presence of cavitation.

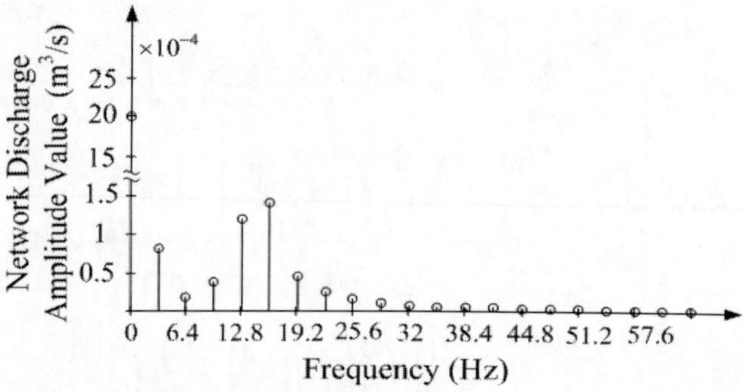

Figure 4.18. The amplitude spectrum of the discharge in the pipeline network in the presence of cavitation.

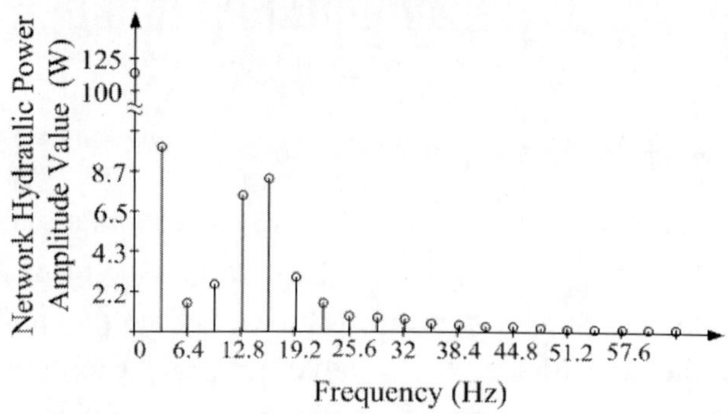

Figure 4.19. The amplitude spectrum of the hydraulic power in the pipeline network in the presence of cavitation.

The frequency analysis of the signals of the head, discharge and hydraulic power at the output of the first section (Figures 4.17–4.19) revealed that the development of the cavitation processes results in the appearance of the harmonic components in the amplitude spectrum of the order of 1 to 20 Hz, whose share makes 40% of signals constant component.

4.4. The Method for the Identification of the Pumping Complex Parameters in Non-Steady Operation States

Energy processes taking place in PC during the development of nonlinear hydrodynamic processes in the pipeline network have the character of periodic steady fluctuations of power at the analyzed time interval. Due to this, it is admissible to synthesize a hydraulic power time function by a harmonic series generated based on the orthogonal components of the heads and discharges.

The proposed energy model (4.23) can be the basis for the solution to the problems of PC parameter identification: the resistances of the pump and the hydronetwork sections when nonlinear hydrodynamic processes develop in the pipeline [129, 130].

The system of energy balance equations for the hydraulic power components between the source and PC elements at the periodic change of the discharge is of the form:

$$\left.\begin{array}{l} P_{s0} = \Delta p_{st0} + \Delta p_{R\Sigma 0} + \Delta p_{Lnet\,j0} + \Delta p_{Cnet\,j0} + p_{Rcon0}; \\ P_{s1a} = \Delta p_{st1a} + \Delta p_{R\Sigma 1a} + \Delta p_{Lnet\,j1a} + \Delta p_{Cnet\,j1a} + p_{Rcon1a}; \\ P_{s1b} = \Delta p_{st1b} + \Delta p_{R\Sigma 1b} + \Delta p_{Lnet\,j1b} + \Delta p_{Cnet\,j1b} + p_{Rcon1b}; \\ \cdots\cdots\cdots\cdots\cdots\cdots\cdots\cdots\cdots\cdots\cdots\cdots \\ P_{ska} = \Delta p_{stka} + \Delta p_{R\Sigma ka} + \Delta p_{Lnet\,jka} + \Delta p_{Cnet\,jka} + p_{Rconka}; \\ P_{skb} = \Delta p_{stkb} + \Delta p_{R\Sigma kb} + \Delta p_{Lnet\,jkb} + \Delta p_{Cnet\,jkb} + p_{Rconkb}, \end{array}\right\} \quad (4.47)$$

where the power constant components correspond to index “0”, the cosine and sine power components correspond to indices “*a*”, “*b*”, respectively; $\Delta p_{R\Sigma}$ – the amplitude value of the hydraulic power losses at total hydraulic resistance $R_\Sigma = R_p + \sum_{j=1}^{J} R_{net\,j}$; k – the number of power signal harmonic.

To determine n of PC unknown parameters (the hydraulic resistances of the pump, the pipeline and the consumer, the inductances and capacitances of the hydronetwork sections) it is necessary to use as many as n identification equations of the system (4.47), reflecting the pattern of the energy balance between the most important components of power in the PC power channel.

So, for the simplest PC equivalent circuit (Figure 4.2), where the discharge signal is of the form $Q(t) = Q_0 + Q_a \cos(\Omega_{kav} t) + Q_b \sin(\Omega_{kav} t)$, the following expressions are obtained to determine the hydrosystem parameters:

PC total hydraulic resistance consisting of a pump resistance, a pipeline section and a consumer

$$R_\Sigma = R_p + R_{net} = [2H_0 - 2H_{st} - 2Q_0/C]/[2Q_0^2 + 3Q_a + 3Q_b]; \quad (4.48)$$

the pipeline inductance

$$L_{net} = [H_0 Q_a - H_{st} Q_a - R_\Sigma (3Q_b Q_0 + 3Q_a Q_b^3/4 + 3Q_a^3/4) + \\ + Q_0 Q_b / C\omega - Q_0 Q_a / C]/\omega Q_0 Q_b; \quad (4.49)$$

the pipeline capacitance

$$C_{net} = \left[H_0 Q_b - H_{st} Q_b - R_\Sigma \left(3Q_b Q_0^2 + 3Q_a^2 Q_b/4 + 3Q_b^3/4\right) - \\ - Q_0 Q_a / C\omega - Q_0 Q_b / C\right] / \omega Q_0 Q_a . \quad (4.50)$$

Under real conditions the resistances of the pump, the pipeline and the consumer depend on the change of the discharge in the system [113]. The determination of the resistances at every PC element results in the increase of the number of the unknown parameters of the equivalent circuit and components making the system (4.47).

When PC operates for a long pipeline network including *j* number of pipeline sections, it is necessary to use a great number of energy balance equations for the hydraulic power components, which is possible by generation of a harmonic master control of a certain amplitude and frequency at the input of PP variable-frequency ED [124, 129].

A generalized algorithm of the procedure of the identification of PC parameters is shown in Figure 4.20.

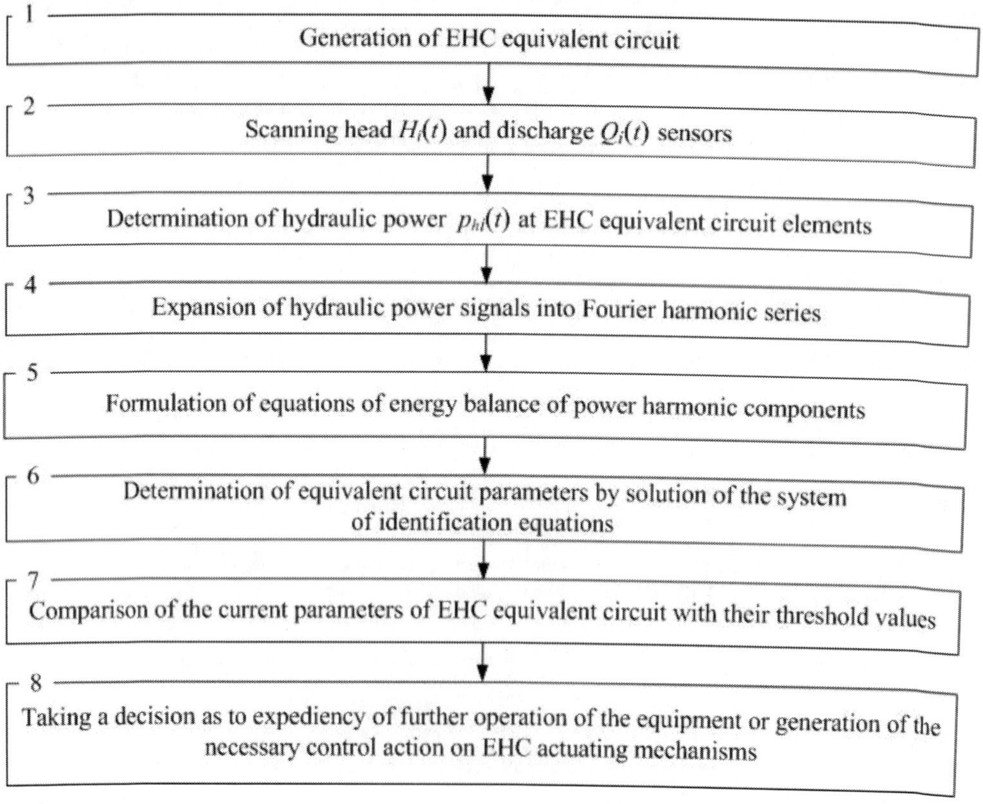

Figure 4.20. The algorithm of the procedure of PC parameters identification.

It consists of the following basic stages:

- the creation of PC equivalent circuit taking into account PU connection, the pipeline network configuration, the occurrence of various nonlinear processes in the hydrosystem;
- the pickup of the data from the head and discharge sensors;
- the determination of the hydraulic power at the elements of the equivalent circuit;
- the approximation of the hydraulic power by the Fourier trigonometric series;
- the formulation and solution of energy balance equations;
- the comparison of the obtained values of equivalent circuit parameters with their threshold (critical) values and taking decisions as to further operation of the electric equipment.

Let us consider a particular case of the application of the described method for the determination of PC equivalent circuit parameters (Figure 4.7) when nonlinear processes develop in the pipeline network.

The variation curve of the cavitation cavern volume (Figure 4.9) can be described by the following expression with a high degree of accuracy:

$$\begin{aligned} V_{kav}(t) &= 1/T_{kav}\int_0^{T_{kav}}\left(Q_{neti}(t)-Q_p(t)\right)dt = K_{kav}R_{kav}(t) = \\ &= K_{kav}\left(a_0 + a_1\sin(\Omega_{kav}t) + a_2\sin(2\Omega_{kav}t) + a_3\sin(3\Omega_{kav}t)\right), \end{aligned} \tag{4.51}$$

where K_{kav} – the coefficient of proportionality.

The periodic variation of cavitation cavern volume results in the formation of discharge at the *j*-th section of the pipeline network as a time function:

$$\begin{aligned} Q_j(t) &= \frac{d\left(V_{kav}(t)\right)}{dt} = Q_{j0} + Q_1\cos(\Omega_{kav}t) + \\ &+ Q_{j2}\cos(2\Omega_{kav}t) + Q_{j3}\cos(3\Omega_{kav}t). \end{aligned} \tag{4.52}$$

A system of identification equations was obtained for PC in the presence of nonlinear processes in the pipeline network. This system of equations is based on the equations of energy balance between the sources of PC elements:

$$\left.\begin{aligned}
&P_{s0} = \Delta p_{st0} + \Delta p_{Rp0} + \Delta p_{Rnet0} + p_{Rcon0};\\
&P_{s1a} = \Delta p_{st1a} + \Delta p_{Rp1a} + \Delta p_{Rnet\,j1a} + p_{Rcon1a};\\
&0 = \Delta p_{Lnet\,j1b} + \Delta p_{Cnet\,j1b};\\
&P_{s2a} = \Delta p_{st2a} + \Delta p_{Rp2a} + \Delta p_{Rnet\,j2a} + p_{Rcon2a};\\
&0 = \Delta p_{Lnet\,j2b} + \Delta p_{Cnet\,j2b};\\
&P_{s3a} = \Delta p_{st3a} + \Delta p_{Rp3a} + \Delta p_{Rnet\,j3a} + p_{Rcon3a};\\
&0 = \Delta p_{Lnet\,j3b} + \Delta p_{Cnet\,j3b};\\
&0 = \Delta p_{Rp3a} + \Delta p_{Rnet\,j3a} + p_{Rcon3a};\\
&0 = \Delta p_{Lnet\,j4b} + \Delta p_{Cnet\,j4b};\\
&\cdots\cdots\cdots\cdots\cdots\cdots\cdots\cdots\cdots\cdots\\
&0 = \Delta p_{Rp9a} + \Delta p_{Rnet9a} + p_{Rcon9a};\\
&0 = \Delta p_{Lnet\,j9b} + \Delta p_{Cnet\,j9b},
\end{aligned}\right\} \tag{4.53}$$

where:

the amplitude values of the constant component

$$P_{s0} = H_0 Q_0;\ \Delta P_{st0} = H_{st} Q_0;$$

$$\Delta P_{R_\Sigma 0} = {}^3\!/_2 R_\Sigma Q_0\left(Q_1^2 + Q_2^2 + Q_3^2\right) + R_\Sigma Q_0^3 + {}^3\!/_4 R_\Sigma Q_1^2 Q_2 + {}^3\!/_2 R_\Sigma Q_1 Q_2 Q_3;$$

the amplitude values of the variable cosine component of the first harmonic

$$P_{s1a} = H_0 Q_{1a};\ \Delta P_{st1a} = H_{st} Q_{1a};$$

$$\begin{aligned}\Delta P_{R_\Sigma 1a} = {}&{}^3\!/_4 R_\Sigma Q_1^2 Q_3 + {}^3\!/_4 R_\Sigma Q_2^2 Q_3 + {}^3\!/_4 R_\Sigma Q_1^3 + 3R_\Sigma Q_0^2 Q_1 +\\ &+ 3R_\Sigma Q_0 Q_1 Q_2 + {}^3\!/_2 R_\Sigma Q_1 Q_2^2 + {}^3\!/_2 R_\Sigma Q_1 Q_3^2 + 3R_\Sigma Q_0 Q_2 Q_3;\end{aligned}$$

the amplitude values of the variable sine component of the first harmonic

$$\Delta P_{L_{net\,i}1b} = -L_{net\,i} Q_0 Q_1 \Omega;$$

$$\Delta P_{C_{net\,i}1b} = Q_0 Q_1 / \Omega C - 3Q_1 Q_2 / 4\Omega C + Q_2 Q_3 / 12\Omega C;$$

the amplitude values of the variable cosine component of the second harmonic

$$P_{s2a} = H_0 Q_{2a};\ \Delta P_{st2a} = H_{st} Q_{2a};$$

$$\Delta P_{R_\Sigma 2a} = \tfrac{3}{4} R_\Sigma Q_2^3 + 3R_\Sigma Q_0^2 Q_2 + \tfrac{3}{2} R_\Sigma Q_1^2 Q_2 + \tfrac{3}{2} R_\Sigma Q_0 Q_1^2 +$$
$$+ \tfrac{3}{2} R_\Sigma Q_2 Q_3^2 + 3R_\Sigma Q_0 Q_1 Q_3 + \tfrac{3}{2} R_\Sigma Q_1 Q_2 Q_3;$$

the amplitude values of the variable sine component of the second harmonic

$$\Delta P_{L_{neti}\,2b} = -L_{net\,i} Q_0 Q_2 \Omega - L_{net\,i} Q_1^2 / 2;$$

$$\Delta P_{C_{neti}\,2b} = -Q_0 Q_2 / 2\Omega C + Q_1^2 / 2\Omega C - 2 Q_1 Q_3 / 3\Omega C.$$

Therefore, for PP with the rated data given in Table 4.2, the calculation of PC equivalent circuit parameters was performed based on the equations of energy balance of the hydraulic power components for a system without cavitation and with cavitation in the pipeline network (Table 4.3).

Table 4.2. PC rated parameters

Q_n, m³/s	H_0, m	H_n, m	d, m	l, m
0.0014	20.8	18	0.04	6

Table 4.3. Calculated parameters of PC equivalent circuit on the basis of an energy model

Parameter	System without cavitation	System with cavitation
R_Σ, s/m^2	$9.6 \cdot 10^5$	$2.1 \cdot 10^6$
L_{net}, s^2/m^2	10.41	14.41
C_{net}, m^2	$1.3 \cdot 10^{-4}$	$1.17 \cdot 10^{-4}$

During the analysis of the pipeline network cavitation fluctuations influence on PC parameters special attention should be paid to the variation of total hydraulic resistance R_Σ. The above said is caused by the fact that the growth of cavitation caverns volume in the liquid flow is accompanied by a considerable increase of hydraulic resistance R_Σ. When

R_{Σ} approaches the hydraulic resistance threshold value R_{lim}, at which the pipeline is plugged, it is necessary to use the pipeline air bleeding technology.

Thus, it is demonstrated that for PC parameters identification it is expedient to apply the energy model based on the harmonic analysis of the hydraulic power signals. Such an approach makes it possible to single out the power variable component caused by the nonlinear periodic processes in the hydronetwork to obtain the frequency characteristic of the energy processes in the hydrosystem.

CONCLUSION

1. It has been shown that the presentation of an electrohydraulic complex by an electric equivalent circuit based on the method of electrohydraulic analogy makes it possible to determine the hydraulic power and the power losses at any section of the hydrosystem when nonlinear processes develop in the pipeline network.
2. It has been determined that the energy processes taking place in the electrohydraulic complex are of the character of periodic stationary fluctuations of the power at the analyzed time interval. Thereby, it is admissible to perform the synthesis of the time function of power (electric, hydraulic) by a harmonic series generated based on the orthogonal components of the voltages and currents, heads and discharges.
3. An electrohydraulic complex energy model in the form of the equations of the hydraulic power harmonic components balance between the power supply and the hydrosystem elements at the periodic variation of the technological parameters has been proposed. This approach is the basis for the solution to the problems of the determination of the parameters of the electrohydraulic complexes, whose technology systems take into account a group operation of the pump units with variable-frequency electric drives in the pumps and uncontrolled electric drives, the occurrence of the periodic processes both in the turbomechanisms and at the pipeline network sections. The analyzed method allows singling out the power variable component, caused by the nonlinear periodic processes in the hydrosystem, obtaining the frequency characteristic of the energy processes in the hydrosystem.

Chapter 5

Working out the Systems of the Automatic Control of the Pumping Complex Electric Drive at the Development of the Non-Steady Processes in the Hydronetwork

5.1. The Development and Research of the Closed Loop Electromechanical System for the Decrease of the Dynamic Loads in the Pumping Complex

5.1.1. The Requirements to Making a Closed Loop System

The performed analysis of the methods for the decrease of the dynamic loads in PC with shutoff-control valves revealed that the conventional approaches base on the following:

- the increase of the stopcock shutting time with the observance of the linear law;
- the formation of the nonlinear law of the shutoff-control valves shutting with the velocity reducing in time, which is realized discretely in several stages due to the use of unregulated induction ED having several rotation frequencies.

The drawbacks of these methods include: the absence of the possibility for the control of the shutoff-control valves at an abrupt disappearance of PC power supply; the absence of the pipeline valves variable-frequency ED enabling an uninterrupted alteration of the velocity and formation of any trajectory of the valves shutting; in the valves control the nonlinear character of hydraulic resistance coefficient ξ dependence on the relative degree β of its opening is not taken into account.

The liquid flow initial velocity, the pipeline network parameters, the hydraulic characteristics and the pipeline valves control character influence on the values of the dynamic loads in the hydronetwork is described in the second chapter, based on a model of PC with controlled pipeline valves. It allowed the substantiation of the ways of the reduction of the dynamic loads, namely:

- the formulation of the nonlinear law of the pipeline valves shutting by means of the use of the stopcock variable-frequency electric drive;
- the reduction of the time of the pipeline valves control due to the use of a gear with a lower reduction rate.

ESDDL can be used for the valves control both in normal (operation) modes and at the alteration of the PC head or discharge on the side of the consumer, and in PC emergency operation related to an abrupt disconnection of PU power, occurrence of surges.

At the emergency disconnection of PU from the power network and the occurrence of the liquid counter flow in the pipeline network in case of the damage of the check valve (the disk break off the spindle; the spindle damage) or its absence there occurs a surge, the liquid overflow from the pressure pipeline through the pump and the rotor rotation in the reverse direction, a sharp increase of the dynamic loads in the unit. In this case, the pump may break, because of the impeller jamming, which, in its turn, results in the breakage of its body.

At the emergency disconnection of PU from the power supply network and the presence of a check valve a more dangerous situation may occur when the check valve goes off at the wrong time, which is caused by its incorrect regulation, jamming or sticking to the valve disk seat, getting of solid particles between the disk and the seat, etc. In this case, the check valve shuts at the time moment when the liquid flows through the pump in the reverse direction at a high velocity. In this case, there occurs a surge in the hydraulic network, due to which the several-fold pressure increase exceeds the rated value of the pressure in PC. Such a surge is able to cause cracks in the pipes, destroy pipe joints, break off the check valve and damage the pump and/or its electric drive.

Taking into account the above said, the closed loop ESDDL in PC is to provide:

- the minimization of the loads in the hydrosystem in the normal (operation) and emergency modes;
- the operative control of the pipeline valves in the emergency (non-steady) situations caused by an abrupt disconnection of the electricity.

5.1.2. The Substantiation of the Choice of the Criterion of the Quality of the System for the Decrease of the Dynamic Loads in the Pumping Complex

At the development of the closed loop systems of the automatic control of the dynamic loads in PC the problem of the choice and the substantiation of the components of the system quality criterion is important. To meet the above-mentioned requirements to the creation of the closed loop ESDDL in PC the quality criterion is to include a number of components:

$$I = \int_0^{t_{sh}} \left(\nu_v(t)\Delta h(t) + k_t \Delta t \right) dt \rightarrow \min , \qquad (5.1)$$

where $\nu_v = \omega_{IM}/\omega_0$ – the weighting coefficient equal to the relative frequency of the stopcock ED rotation; ω_{IM}, ω_0 – the current and synchronous rotation frequencies of the stopcock IM, respectively, s^{-1}; Δh – the relative increase of the head at the stopcock; Δt – the relative time of the stopcock shutting; k_t – the weighting coefficient at component Δt, taking into account the stopcock shutting time, s; t_{sh} – the time of the stopcock complete shutting.

The first component Δh in (5.1) is determined by the necessity for the minimization of the increase of the head in the hydrosystem at the control of the stopcock ED:

$$\Delta h = \left(H - H_{pl} \right) / \Delta H_{dir} , \qquad (5.2)$$

where $H_{pl} = H_0$, $H_{pl} = h_s$ – the head in the pipeline in the steady mode after shutting the stopcock in PC normal (operation) and the emergency mode, respectively, m; H, H_0 – the current head in the pipeline network and the head of the pump at the zero discharge and the rated rotation frequency, respectively, m; h_s – the static head in the pipeline systems, m; $\Delta H_{dir} = \dfrac{\upsilon_0 c}{g}$ – the increase of the head at the direct surge (the maximal increase of the head at the stopcock shutting), m; $\upsilon_{\upsilon 0} = \sqrt{H_0 / \left(R_p + R_{pl} + R_{cn} \right)} / S$, $\upsilon_{\upsilon 0} = \sqrt{h_s / \left(R_p + R_{pl} \right)} / S$ – the velocity of the liquid flow in the pipeline before the stopcock shutting in the steady condition in the normal (operation) and emergency mode of PC operation, respectively, m/s;

$R_p, R_{pl} = \frac{\lambda_{pl}}{S^2 d}\frac{L}{2g}, R_{cn}$ – the pump internal resistance determined by the published data, the network hydrodynamic resistance, taking into account the head losses on the friction along the length of the pipeline network and the rated resistance of the consumer, respectively s²/m⁵; λ_{pl} – a dimensionless coefficient of the pipeline resistance; d, L, S – the diameter, length and cross section area of the pipeline network, respectively, m, m, m²; c – the velocity of the shock wave propagation, m/s.

The second component in the criterion (5.1) refers to the provision of the minimal time of the stopcock ED control:

$$\Delta t = (t - t_0)/t_{max}, \tag{5.3}$$

where t, t_0 – the current time corresponding to the beginning of the stopcock shutting, s; t_{max} – the maximal time of shutting at the minimal value of the stopcock ED supply voltage frequency, s.

Weighting coefficients $\nu_v(t)$ and k_t at the summands in integrand (5.1) are assumed based on the analysis of the priority of the problems solved by ESDDL in PC.

Weighting coefficient $\nu_v(t)$ takes into account the value of component $\Delta h(t)$ of the quality criterion. The upper boundary of integration in (5.1) t_{sh} depends on current rotation frequency ω_{IM} of the stopcock ED. To provide the values of the same order of criterion I at different time t_{sh} weighting coefficient $\nu_v(t)$ must also depend on ω_{IM} and change in the range of 0-1. That is why in expression (5.1) coefficient $\nu_v(t)$ is equal to the stopcock ED relative rotation frequency.

Weighting coefficient k_t determines the value of component Δt of the quality criterion. It is assumed constant in the range of 0-1 within the framework of the solution of a particular problem. Therefore, in normal (operation) modes the closed loop ESDDL is to provide the minimization of the dynamic loads in PC pipeline network, in this case, component Δt insignificantly influences the value of criterion I. That is why in PC operation modes the weighting coefficient at component Δt is assumed equal to $k_t = 0.005$. In emergency (non-steady) situations, caused by an abrupt disconnection of the power supply, the closed loop ESDDL is to provide not only the minimization of the pressure increase in PC pipeline network, but also the decrease of the time of pipeline valve shutting to prevent PU continuous rotation in the reverse direction at the liquid counter flow. That is why, in emergency (non-steady) PC operation modes the weighting coefficient at component Δt is assumed higher by one order $k_t = 0.05$.

5.1.3. The Optimization of the Closed Loop Electromechanical System for the Decrease of the Dynamic Loads in the Pumping Complex

The choice of the quality criterion of the form (5.1) results in the necessity for the use of not analytical methods but search algorithms for its finding. The search for the optimal law of the pipeline valves control corresponds to such ED supply voltage frequency f_0 dependence on the relative degree β of its opening, at which the minimal value of the quality criterion is provided (5.1). For its determination, we used the dynamic programming method allowing the optimization of the law of control at every stage of the variation of the relative degree β of opening the pipeline valves thereby eliminating the complete examination of the variants of control [131, 132].

PC with the following parameters was taken for the research: the pump rated head $H_{pn} = 100$ m, the velocity of the operation medium flow $\upsilon_n = 1.4$ m/s, the hydraulic resistance of the consumer $R_{cn} = 52$ s²/m⁵; the length $L = 5000$ m and the diameter $d = 1.2$ m of the pipeline; the number of the pipeline sections $N = 20$; the velocity of the shock wave propagation $c = 1000$ m/s; the number of turns of the spindle of the regulating stopcock for its full shutting $z = 28$. The specific parameters of every section of the pipeline: $r_0 = 0.000162$ s^2/m^6, $l_0 = 0.0901$ s^2/m^3, $c_0 = 90132$ m^{-1}. Static head h_s in the hydrosystem was assumed equal to 40 and 60 m.

With the reference to PC analyzed in the monograph, we propose the procedures of the search for the optimal law of control of the regulating stopcock ED in the operation (Appendix I, Figure I1, I2) and the emergency (Appendix I, Figure I3, I4) PC modes with different static head where at every step of β variation either the supply voltage frequency f_0 reduces (the diagonal sections) or it remains unchanged (the vertical sections) [133, 134]. The increment of the quality criterion is given for each section:

$$\Delta I = I_m - I_{m-1}, \tag{5.4}$$

where I_m, I_{m-1} – the values of the quality criterion at the following m and the previous $m-1$ step, respectively.

To determine values I_m, I_{m-1} in normal (operation) and emergency modes of the technological equipment we used a PC imitation model with a closed loop ESDDL, whose block diagram is shown in Figure I5 of Appendix I and includes:

- a PC model with controlled pipeline valves and a check valve;
- an ESDDL model with an induction variable-frequency ED of the shutoff-control stopcock;
- a block of the search for the optimal law of control of the pipeline valves;
- a block of the determination of the quality criterion.

To calculate values ΔI we used the script whose initial code is given in Appendix I.

To find the optimal law of control of the regulating stopcock ED the plane $\beta f_0 0$ (Figure 5.1) is divided into m_1 and m_2 values by axes β and f_0, respectively.

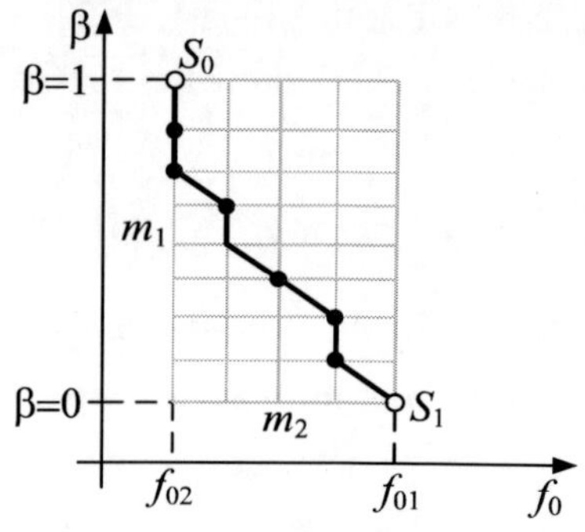

Figure 5.1. The trajectory of the movement of the stopcock shutting operating element from the initial state S_0 to the final state S_1.

Values m_1 and m_2 are chosen based on the requirements to the accuracy of the solution of the problem. In this case, the trajectory of the movement of the stopcock shutting operating element from the initial state S_0 to the final state S_1 on plane $\beta f_0 0$ represents a certain broken line.

Referring to the considered problem, axis β is divided into $m_1 = 18$ values – [0; 0.0075; 0.01; 0.02; 0.025; 0.05; 0.075; 0.1; 0.125; 0.15; 0.175;0.2; 0.225; 0.25; 0.5; 0.75; 0.85; 1]; axis f_0 is divided into $m_2 = 10$ values – [5; 10; 15; 20; 25; 30; 35; 40; 45; 50], Hz (Figure I1–I4).

The optimal law of the stopcock ED control was searched for in the reverse-time direction – from point S_1 at the stopcock full shutting ($\beta = 0$) to point S_0, corresponding to its full opening ($\beta = 1$). According to the principle of dynamic programming, at every step at a certain value of β the quality criterion minimal values achieved at the optimal control starting from the current point S_0 and ending with the final point S_1 were pointed out.

Below there is a fragment of the procedure of the search for the optimal law of control of the regulating stopcock ED for the case of PC emergency operation mode at the static head of $h_s = 40$ m at the final step (Figure 5.2) and at the last step but one (Figure 5.3).

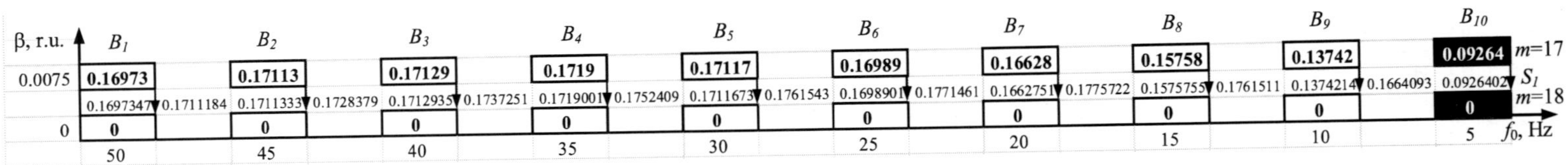

Figure 5.2. The fragment of the procedure of the search for the optimal law of control of the regulating stopcock ED at the last step.

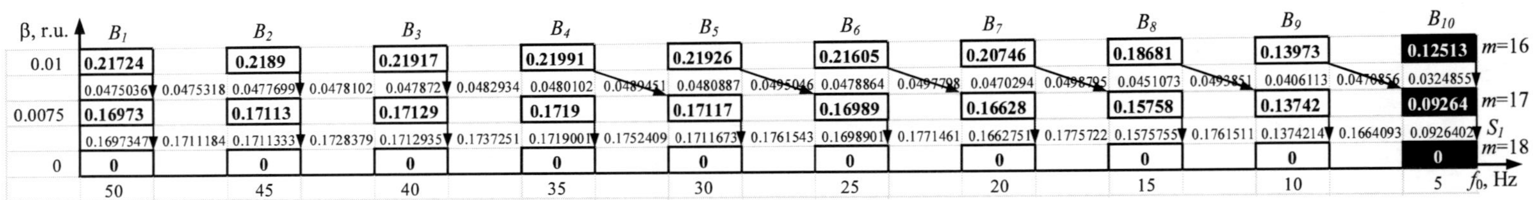

Figure 5.3. The fragment of the procedure of the search for the optimal law of control of the regulating stopcock ED at the last but one-step.

The figures demonstrate that at the last step it is possible to move to point S_1 from ten neighboring points $B_1, B_2, \ldots, B_{10}$ at different values of the stopcock ED supply voltage frequency f_0.

Transition $B_{10} \to S_1$ takes place at the supply voltage frequency $f_0 = 5$ Hz, in this case, the increment of the quality criterion equal to $\Delta I = 0.0926402$ corresponds to it. Transition $B_9 \to S_1$ may take place at supply voltage frequency $r_g = 80$ equal to 5 and 10 Hz, respectively. In the first case, the quality criterion increment ΔI is equal to 0.0926402, in the second case – 0.1664093.

The optimal control at the transition from the point at the first step to the point at the second step corresponds to the minimal increment ΔI of the quality criterion and is indicated by an arrow. The values of the quality criterion equal to $I_{m-1} = I_m + \Delta I$ are shown in the rectangles. In this case, the value of the quality criterion at point S_1 ($m = 18$), corresponding to the start of the movement in the reverse-time direction, is equal to $I_{18} = 0$. The calculations at the following steps are performed analogously.

Thus, passing from point to point vertically or diagonally top-down, starting from the final point S_1, for every intermediate point the optimal equation is chosen at the following step. In this case, we consider two possible transition variants corresponding to different values of the supply voltage frequency: the vertical and the diagonal ones. For each variant the value of the quality criterion is determined $I_{m-1} = I_m + \Delta I$. The transition, for which this sum is lower, is chosen out of two variants. The optimal control is marked by an arrow and the value of the quality criterion at the optimal control, starting from every intermediate point to the final point S_1, is indicated in the corresponding black rectangle. The procedure of finding the optimal control ends at the initial point S_0. The transition by the marked arrows from point S_0 at full opening of stopcock ($\beta = 1$) to point S_1, at its full shutting ($\beta = 0$) corresponds to the optimal law of control of the regulating stopcock ED.

The arrows in Figures I1–I4 in Appendix I show the optimal law of control of the stopcock ED of the considered PC.

5.1.4. The Research of the Dynamic Processes in the Pumping Complex

Two cases are considered for the research of the dynamic processes in PC with closed loop ESDDL and optimal control:

- PC operation mode when stopcock SV2 on the side of the consumer, equipped with a variable-frequency ED, shuts (Figure 5.4, a);

- PC emergency mode when PU abruptly disconnects from the electrical grid and stopcock SV1, equipped with a variable-frequency ED, shuts in the pump delivery branch at the liquid counter flow (Figure 5.4, b).

In the closed loop ESDDL for the research we analyzed AOC2-41-4U34 induction motor with the rated power of $P_{IMn} = 5.2$ kW and a gear with the reduction rate of $r_g = 16$, chosen in the third chapter.

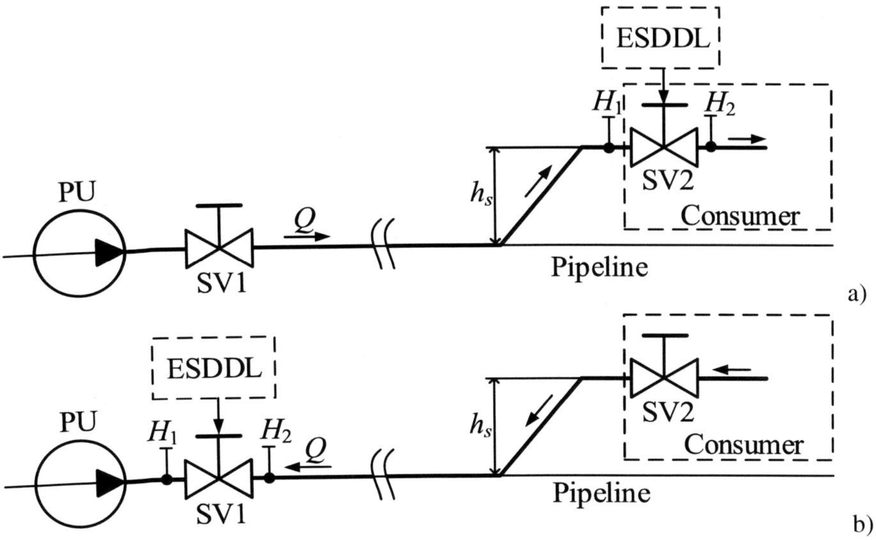

Figure 5.4. The flow sheets for the research of the dynamic processes in PC in operation modes.

For the comparative analysis, we also considered the dynamic processes in the system without ESDDL:

- in the emergency mode of operation of PC (Figure 5.5, a), with a pressure header equipped with a check valve (CV). At the abrupt CV disconnection from the grid CV goes off with a lag when the liquid counter flow starts;
- in the above mentioned PC operation (Figure 5.5, b) and emergency (Figure 5.5, c) modes, wherein AOC2-41-4U3 induction motor with the rated power of $P_{IMn} = 5.2$ kW and a gear with the reduction rate of $r_g = 80$ is chosen as the ED for the regulating stopcock SV1.

The following designations are adopted in Figures 5.4, 5.5: H_1, H_2 – the head in the pipeline network, respectively, before and after shutting of stopcocks SV1 and SV2 and the check valve CV; Q – the discharge in the pipeline network.

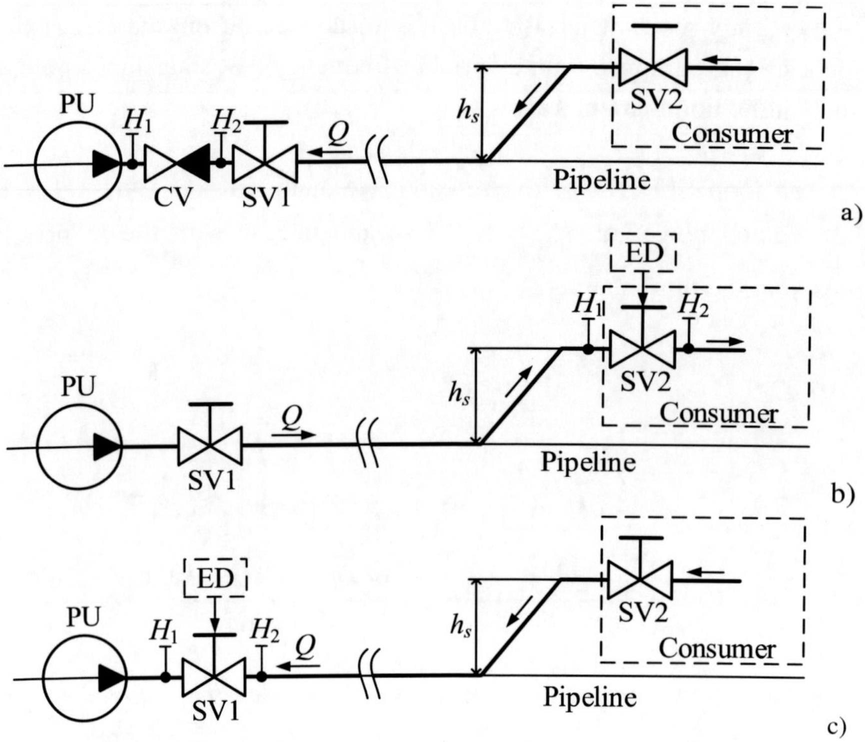

Figure 5.5. The flow sheets for the research of the dynamic processes in PC in the emergency modes.

Figures I6–I17 in Appendix I contain head variation curves before $H_1(t)$ and after $H_2(t)$ the stopcock, discharge $Q(t)$ in the pipeline network, the relative degree $\beta(t)$ of the stopcock opening, rotation frequency $\omega_{IM}(t)$ and moment $M(t)$ of the stopcock IM for the operation and emergency modes of PC with a closed loop ESDDL and without it at different values of the static head h_s.

Figures 5.6 and 5.7 contain curves representing the variation of head $H_2(t)$ and discharge $Q(t)$ in the pipeline network for the case of PU abrupt disconnection (time moment t_1) and a sharp actuation of the check valve CV in the PC pressure header.

The graphs demonstrate that after the appearance of the liquid counter flow (Figure 5.7, time moment t_2) in PC pipeline network a sharp actuation of the check valve results in the occurrence of a surge in the pipeline system in the form of high- and low-pressure waves. In this case, the value of the head in the pipeline at the check valve at the static head in the hydrosystem of 40-60 m is, respectively, by 1.75-2.6 times higher than PU rates head.

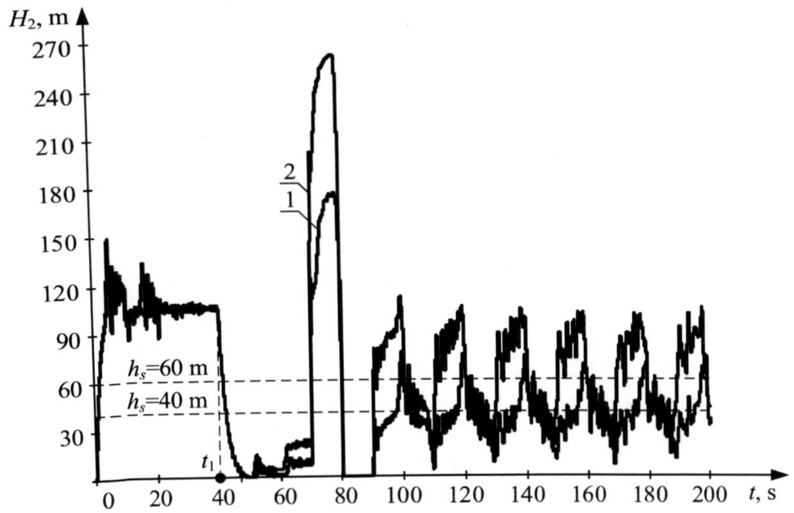

Figure 5.6. The curves representing the variation of the head $H_2(t)$ after the check valve in PC emergency mode with the static head h_s equal to 40 m (1) and 60 m (2).

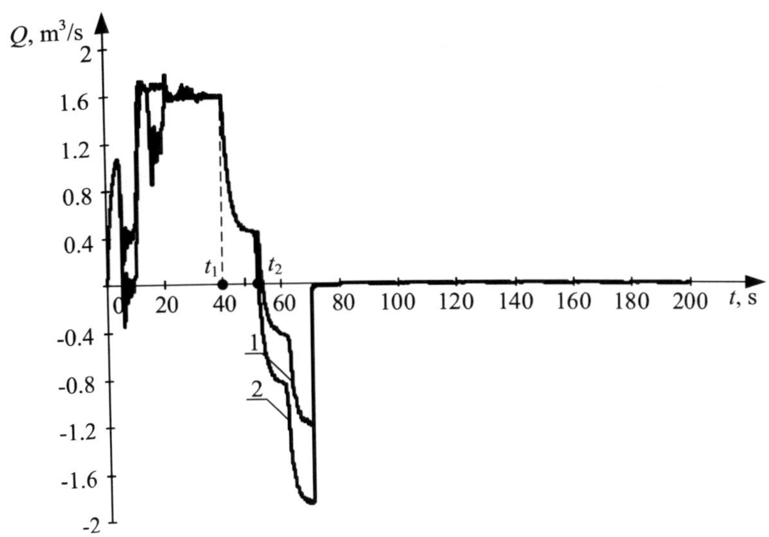

Figure 5.7. The curves representing the variation of discharge $Q(t)$ at the actuation of the check valve in PC emergency mode with the static head h_s equal to 40 m (1) and 60 m (2).

In Figures I6–I17 of Appendix I numbers 1, 2 correspond to the graphs at the optimal law of the stopcock ED control in PC with ESDDL, in which the static head h_s is equal to 40 and 60 m, respectively; 3, 4 – to the graphs at the linear law of shutting of the stopcock with ED with the supply voltage frequency equal to 50 Hz in PC without ESDDL wherein the static head h_s is equal to 40 and 60 m, respectively.

The following designations are adopted in Figures I6–I17 in Appendix I and in Figure 5.6, 5.7: t_1 – the moment of beginning of the stopcock shutting corresponding to the relative degree β=1 of its opening, and PU disconnection from the electric grid; t_2 – the moment of the start of the liquid counter flow, $Q(t)<0$; t_3 – the moment of the highest increase of pressure ΔH in the pipeline network in relation to the pump rated head H_{pn} at which β→0.

The graphs in Figure I6 demonstrate that in the operation mode at the optimal law of control of the stopcock ED the head in the pipeline network exceeds the pump rated head by 33.8–36.8%. During the use of the pipeline valves, unregulated ED of the power of 5.2 kW, at the linear law of control with the supply voltage frequency of $f_0 = 50$ Hz, in comparison with the optimal law, the head at the stopcock is practically twice as high as the rated value. In this case, when the optimal law of ED control is observed, the time of the stopcock shutting is half as big.

The stopcock shutting in PC emergency mode at the appearance of the counter flow by means of the pipeline valves unregulated ED results in the increase of the head by (2.6–30)% in comparison with the rated values (Figure I9). At the optimal law of control (Figure I8) and $h_s = 40$ m the head in the pipeline network does not exceed PU rated head, at $h_s = 60$ m the head increase makes 22% of value H_{pn}. The analysis of the stopcock shutting time t_{sh} in the emergency mode revealed that at the optimal law of control its value decreases by 3.4 times in comparison with the linear law of ED control, which prevents PU long rotation in the reverse direction under the impact of the liquid counter flow.

The comparative analysis of the curves representing the time variation of PC head (Figures I6–I9) and discharge (Figures I10–I11) revealed that functioning of closed loop ESDDL in PC at the optimal law of control makes it possible to reduce the amplitude of the head pulsations and the discharge in the pipeline network by one order.

The analysis of the curves representing the variation of rotation frequency $\omega_{IM}(t)$ and moment $M(t)$ of the stopcock IM for PC operation and emergency modes (Figures I14–I17) revealed that at the law of the frequency regulation with the constant power in both cases the values of the moment and frequency of the stopcock ED shaft rotation vary within the admissible limits, and do not result in the overload by current of the frequency converter invertor.

The transient processes indices obtained by the results of the analysis of the curves in Figures I6–I9 in Appendix I and in Figures 5.6, 5.7, are given in Tables 5.1, 5.2.

The head increase ΔH in the hydronetwork is shown in the Tables for the above stated PC modes in absolute units and the percent relation between the relative and rated head H_{pn} of the pump.

Table 5.1. The indices of the transient processes in PC operation mode

The law of control	h_s, m	t_{sh}, s	ΔH, m	$\Delta H/H_{pn}$, %
Optimal	40	137	36.8	36.8
Linear		277	96.8	96.8
Optimal	60	137	33.8	33.8
Linear		277	96.8	96.8

Table 5.2. The indices of the transient processes in PC emergency modes

The law of control	h_s, m	t_{sh}, s	ΔH, m	$\Delta H/H_{pn}$, %
Optimal	40	80	0	0
Linear		274	2.6	2.6
Optimal	60	80	22	22
Linear		274	30	30
CV sharp actuation	40	-	75	75
	60	-	160	160

5.2. The Development and Research of the Closed Loop System for the Automatic Control of the Pumping Complex Electric Drive with the Minimization of the Power Losses in the Pipeline Network

5.2.1. The Requirements to Making the System

The problems of the increase of PC power efficiency are among the basic directions of the energy and resource saving [1–5, 135–138]. The operation mode of water- and heat-supply systems PS is determined by the schedule of water consumption continuously changing in time and depending on the number of population, the operation schedule of enterprises, climatic conditions, the season of the year, the day of the week, the time of the day, etc. [1–9, 135–138]. The variable character of water consumption results in the necessity for the constant regulation of the output data of PC technological parameters.

The analysis of the methods of PC parameter regulation, performed in the first chapter, revealed that the conventional approaches base on the action on the pump electric drive, the turbomechanism or the pipeline network. In this case, the physical processes of the liquid flow in the pipeline are not taken into account. So, the decrease of the pressure in the liquid flow below the certain value or the increase of the flow velocity above the critical value results in meeting the conditions for the creation and development of cavitation processes. In their turn, the latter are characterized by the increase of cavitation power losses in the pipeline network up to 30% of the pump rated power.

The cavitation self-oscillations influence on PC technological, energy and mechanical characteristics is shown by means of the cavitation channel model proposed in the second chapter of the monograph. It enables the substantiation of the ways of the improvement of PC efficiency taking into account the cavitation processes in the pipeline network:

- the decrease of the flow motion velocity below the critical value determined by expression $\upsilon_{kr} = \sqrt{\dfrac{2(p - p_{para})}{\rho\chi_{kr}}}$;
- the removal of the steam-gas mixture discharged from the liquid with simultaneous possibility for its energy recuperation.

The former variant can be realized by means of the pump variable-frequency ED, the latter variant implies the installation of EDCCP at the points of the highest accumulation of air in the pipeline.

During the operation of water supply systems PS, it is often necessary to maintain the pressure at the consumer at the required level. Taking into account the above said, the closed loop automatic control system (ACS) of PC ED is to meet the following requirements:

- providing the stabilization of the pressure at the consumer where the deviation by the head Δh should not exceed 1%;
- minimizing the pipeline power losses N_{kav} caused by the cavitation processes;
- providing the maximal value of the recuperated power N_g of the steam-gas mixture.

5.2.2. The Substantiation of the Choice of the Criterion of the Quality of the System for the Automatic Control of the Pumping Complex Electric Drive

To meet the above stated requirements to the creation of a PC electric drive closed loop ACS the quality criterion is to include a number of components:

$$I = \int_0^t \left(K_h \Delta h(t) + K_n \Delta N_{kav}(t) + K_g \Delta N_g(t)\right) dt \rightarrow \min , \tag{5.4a}$$

where K_h, K_n, K_g – the weighting coefficients; $\Delta h(t)$ – the disparity by the head, r. u.; $\Delta N_{kav}(t)$ – the relative value of the power losses caused by the presence of cavitation processes, r. u. $\Delta N_g(t)$ – the disparity by the steam-gas mixture recuperation power, r. u.

The first component in (5.4) is determined by the necessity for the minimization of the deviation of the head at the consumer from the assigned value, r. u.:

$$\Delta h(t) = \frac{H_{zad}(t) - H_{tek}(t)}{H_{zad}(t)},$$

where $H_{zad}(t)$, $H_{tek}(t)$ – the assigned and the current values of the heads, respectively, m.

The second component refers to the provision of the minimal relative value of the power losses caused by the presence of cavitation processes in the hydrosystem, r. u.:

$$\Delta N_{kav}(t) = \frac{N_{kav}(t)}{N_{NK}(t)},$$

where $N_{NK}(t)$ – the current value of the power consumed by PC without cavitation processes in the pipeline kW; $N_{kav}(t)$ – the hydraulic power losses caused by the presence of cavitation processes in the pipeline, kW.

The third component characterizes the deviation of the steam-gas mixture recuperation power from the maximal possible value, r. u.:

$$\Delta N_g(t) = \frac{\left| N_{g\max}(t) - N_g(t) \right|}{N_{g\max}(t)},$$

where $N_{g\max}(t)$, $N_g(t)$ – the maximal and the current values of the recuperation power, kW.

The weighting coefficients at the summands in the integrand (5.4) are taken based on the analysis of the priority of the problems solved in the system of PC control [135, 136, 139–142]. As the main purpose of PP consists in the stabilization of the head at the consumer, it has the maximal weight equal to one. The solution to the problem of the energy and resource saving in PC is provided by the maintenance of the minimal value of power losses $\Delta N_{kav}(t)$, caused by the presence of cavitation processes in the pipeline and the maximal value of the recuperation power $\Delta N_g(t)$ of the steam-gas mixture energy. These are the attendant problems at the operation of the required technological mode in accordance with the assigned accuracy of the maintenance of the controlled parameter. That is why the choice of the values of the weighting coefficients K_n and K_g, taking part in expression (5.4), is to base on the analysis of the curves representing the variation of the criterion of the quality of the researched system $I = f(K_n, K_g)$ at $K_n = \text{var}$ and $K_g = \text{var}$.

For the maximal value of the water supply of PC whose parameters are given in Table 2.1 we obtained a surface (Figure 5.8) representing the variation of the quality criterion depending on the weighting coefficients $I = f(K_n, K_g)$. The analysis of the surface (Figure 5.8) enabled the determination of the minimum of the quality criterion $I = 0.11$ for the following values of the weighting coefficients $K_n = 0.75$ and $K_g = 0.75$.

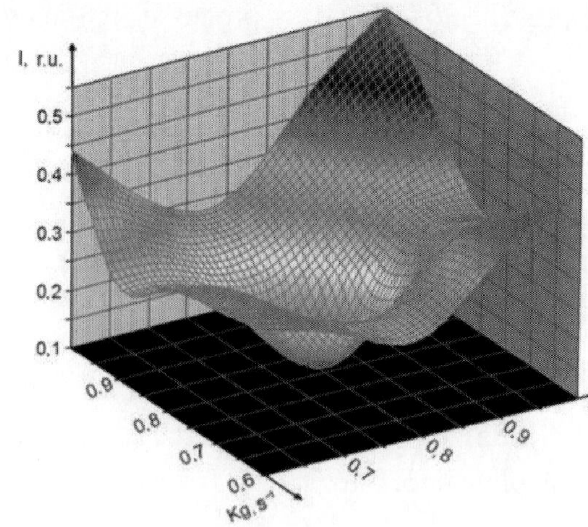

Figure 5.8. The quality criterion dependence on weighting coefficients K_n and K_g.

The choice of PC ED operation mode corresponding to the minimum of the quality criterion I enables obtaining the required pressure in the pipeline network at the minimal possible losses of power N_{kav}, caused by the presence of cavitation processes in the pipeline network and the maximum of the generated power N_g.

The choice of the quality criterion of the form (5.4) results in the necessity for the use of not the analytical methods for its finding but search algorithms. They make it possible to determine the dependence of the values of the control actions – supply voltage frequency f at FC output and degree β of opening of EDCCP regulating valve – on the pressure at the checkpoint of the hydronetwork, the volume of the cavitation caverns and the power of the steam-gas mixture recuperation. Thus, the values of the control actions at the variation of PC ED operation mode should provide the minimum of the quality criterion I. The character of the variation of the latter is determined for various modes of PP water supply based on the imitation model of the electrohydraulic system whose block diagram includes:

- a PC model with a cavitation channel in the pipeline whose parameters are given in Table 2.2;
- EDCCP model;
- a block of the determination of the quality criterion.

Figure D.1 contains an algorithm allowing the determination of the character of the variation of the quality criterion I depending on the control actions at the changing mode of water consumption.

Using algorithm shown in (Figure D.1), based on the enumerative technique for PC whose parameters are given in Table 2.2, we obtained surfaces (Figure 5.9), representing the variation of the quality criterion depending on the control actions $I = f(f, \beta)$ for different modes of PC water consumption: $Q = Q_{max}$, $Q = 0.3Q_{max}$. Their analysis revealed that for the mode of maximal water consumption of the considered PC the minimum of the quality criterion $I = 0.11$ is provided at $f = 49.7$ Hz and $\beta = 0.9$. In this case, it is obvious that the reduction of the value of water consumption results in the shift of the minimum of the quality criterion into the domain of the low frequencies of the mains.

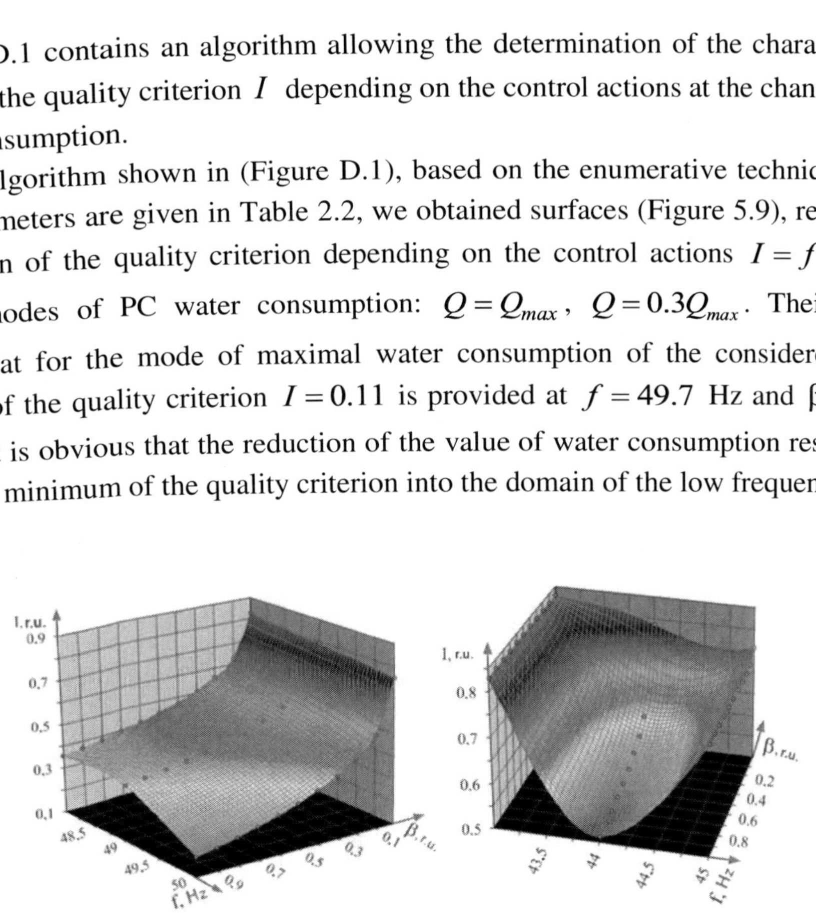

a) b)

Figure 5.9. The quality criterion dependence on the control actions for different modes of water consumption: a) $Q = Q_{max}$; b) $Q = 0.3Q_{max}$.

5.2.3. Working out the Functional Diagram of the Closed Loop System for the Automatic Control of the Pumping Complex Electric Drive

Figure 5.10 shows a functional diagram of the closed loop ACS of PC ED, taking into account the cavitation processes in the pipeline, including: a variable-frequency electric drive (FC-IM) of the pump P, a device for automatic bleeding of the air DABA, a receiver R, pressure sensors PS_1 and PS_2, a regulating valve S_3, an air turbine AT, an electric energy generator EG, a control system based on a logic controller LC, a device for coupling CD of the generator with the power network.

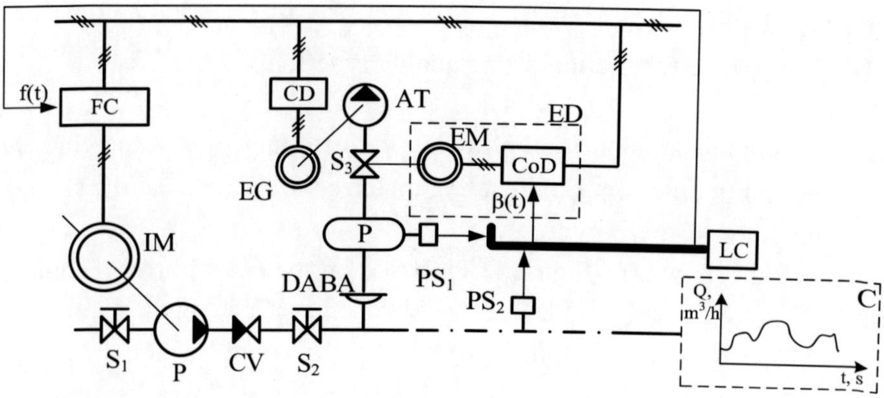

Figure 5.10. A functional diagram of the closed loop ACS of PC ED taking into account the cavitation processes in the pipeline: FC – frequency convertor; IM – induction motor; P – pump; CV – check valve; S_1, S_2 – stopcocks at the suction pipe and the delivery branch of the pump, respectively; DABA – a device for automatic bleeding the air; R – receiver; PS_1, PS_2 – pressure sensors at the receiver and the check point of the pipeline, respectively; S_3 – regulating valve; EM – electric machine; CoD – EM control device; ED – electric drive of the regulating valve; AT – air turbine EG – electric energy generator; CD – device for coupling the generator with the power network; LC – logic controller; $\beta(t)$, $f(t)$ – signals of assigning the degree of opening the regulating valve and the supply voltage frequency, respectively; C – consumer.

The basic problem of ACS of PC ED consists in the stabilization of the pressure at the checkpoint of the pipeline network and the accumulation of energy of the steam-gas mixture with its further recuperation into electric energy [139, 141, 142]. The former is achieved by the regulation of the supply voltage frequency $f(t)$ by means of a variable-frequency ED of the pump P (FC-IM system) with simultaneous minimization of head losses in the pipeline, caused by the presence of cavitation caverns in the liquid flow. The energy of cavitation bubbles accumulates in the receiver R, its value depends on PC ED operation mode varying in accordance with the water consumption curve, the intensity of cavitation processes and the degree of opening $\beta(t)$ of the regulating valve.

Based on the above-considered features of functioning ACS of ED of PC with cavitation processes in the pipeline, we worked out the search algorithms for the determination of the values of supply voltage frequency $f(t)$ (Figure D.2) and the degree of opening $\beta(t)$ of the regulating valve (Figure D.3), using the iteration method of the search of the minimum of the quality criterion with an alternating step [139, 141, 142].

A characteristic feature of the method consists in the determination of the quality criterion at every iteration step Δf_i (Figure D.2) and $\Delta\beta_i$ (Figure D.3). The values of the control actions f_i and β_i change according to them. It should be noted that the search for the value of the supply voltage frequency f_i and the degree of opening β_i of the regulating valve continues till the difference between the current I_i and the previous I_{i-1} values of the quality criterion obtained at the time interval $t_i - t_{i-1}$, is lower or equal to the assigned accuracy $\varepsilon \geq I_i - I_{i-1}/(t_i - t_{i-1})$. The value of the latter is determined from the planes (Figure 5.9), representing the alteration of the quality criterion due to the control actions for different values of water consumption. In the analyzed case the difference between the quality criterion at the point of the minimum and its previous value was 0.01, which enabled the assignment of the accuracy equal to $\varepsilon = 0.01$.

5.2.4. The Research of the Dynamic Processes in the Closed Loop System for the Automatic Control of the Electric drive of the Pumping Complex with a Cavitation Channel when the Pressure in the Pipeline is Stabilized

A block diagram of PC ED ACS closed by the head at the second section of the hydronetwork, taking into account the cavitation processes, is shown in Figure D.4. It includes: a variable-frequency electric drive (FC – IM system), a pump operating on the hydrodynamic network represented by three sections of the pipeline with a cavitation channel, an electromechanical device for the control of the cavitation processes, the blocks of the determination of the quality criterion and the formation of the control signals. The latter block includes:

- two calculation modules each of which is designed for the formation of the relevant control action and is equipped with a function of limitation;
- a block of synchronization whose basic function is the provision of the timely and even formation of the control signals by each of the calculation modules operating in parallel.

Figures 5.11–5.16 contain the curves representing the time variation of the head ((curves $H_1(t)$ – $H_3(t)$), the productivity (curves $Q_1(t)$ – $Q_3(t)$), the head losses (curves $\Delta H_1(t)$ – $\Delta H_3(t)$, $\Delta H_{kav}(t)$), the power (curves $N_N(t)$, $N_d(t)$, $N_{kav}(t)$, $N_{s1}(t)$– $N_{s3}(t)$, $N_g(t)$, $N_{g\,max}(t)$), the quality criterion (curve $I(t)$), the control actions (curves $f(t)$, $\beta(t)$), the moment on the motor shaft (curve $M_d(t)$), the angular rotation frequency (curve $\omega(t)$) of the pump ED and the moment (curve $M_s(t)$) of resistance developed by the turbomechanism.

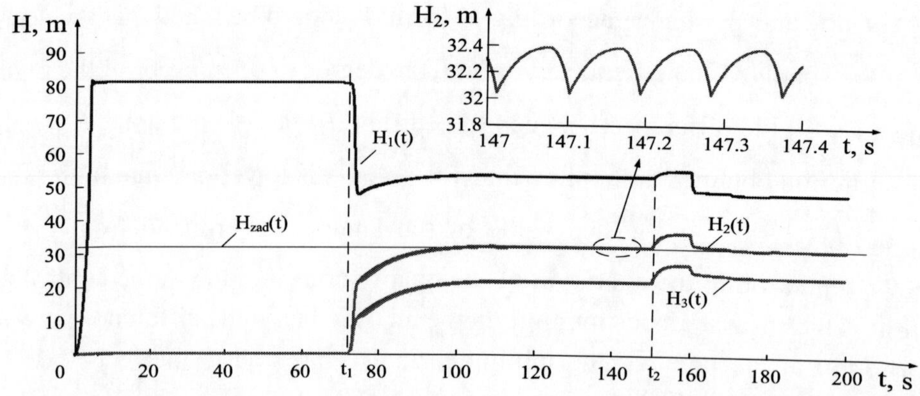

Figure 5.11. PC head variation curves: $H_1(t)$ – $H_3(t)$ – the head at the corresponding sections of the pipeline; $H_{zad}(t)$ – the required value of the head.

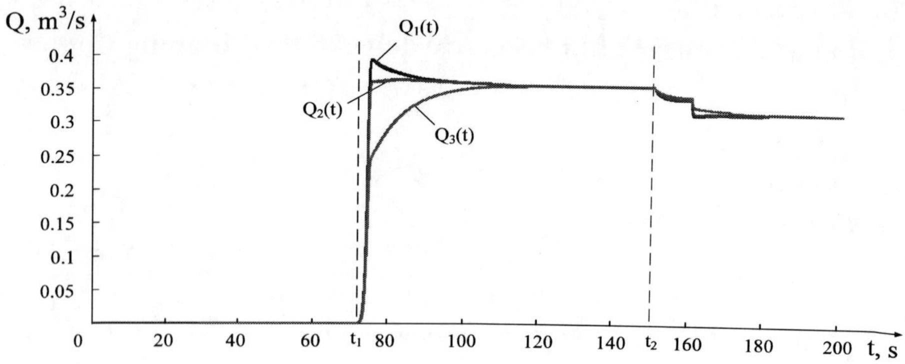

Figure 5.12. The curves representing PC discharge time variation: $Q_1(t)$ – $Q_3(t)$ – the discharge of the liquid at the corresponding sections of the pipeline.

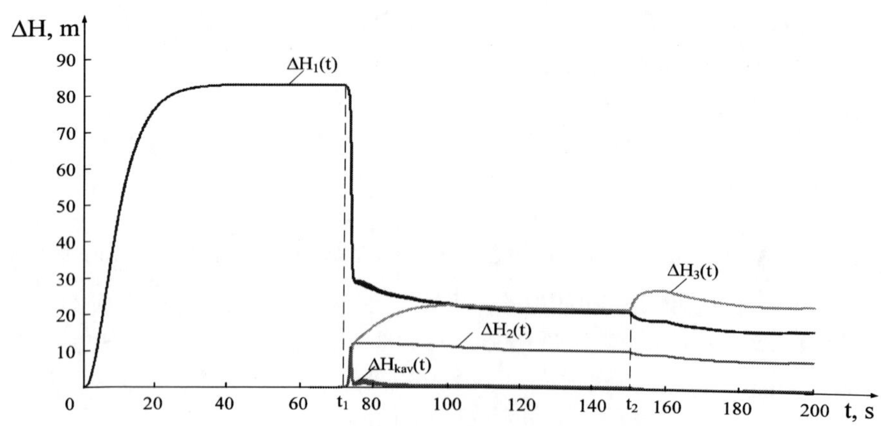

Figure 5.13. The curves representing the distribution of the head losses in PC: $H_1(t)$ – $H_3(t)$ – the head losses at the corresponding sections of the pipeline; $\Delta H_{kav}(t)$ – the head losses caused by the presence of the cavitation processes in the pipeline.

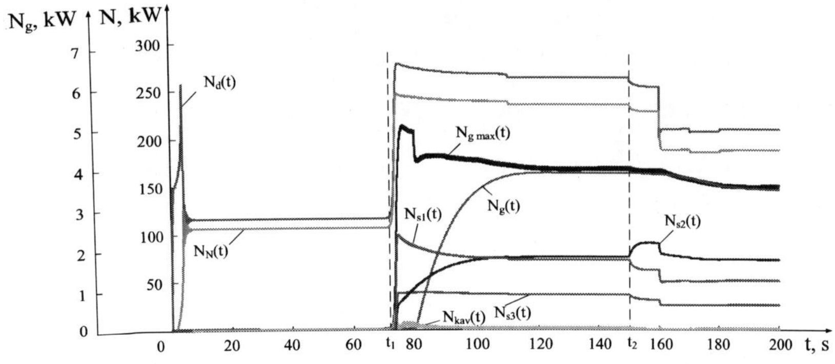

Figure 5.14. PC elements power variation curves: $N_d(t)$ – the power entering the induction motor on the side of the stator; $N_N(t)$ – the power of the pump; $N_{s1}(t)$ – $N_{s3}(t)$ – the power of the liquid flow at the corresponding sections of the pipeline; $N_{kav}(t)$ – the power losses caused by the presence of the cavitation processes in the pipeline; $N_{g\,max}(t)$, $N_g(t)$ – the maximal possible and the current power of the recuperation of the steam-gas mixture energy.

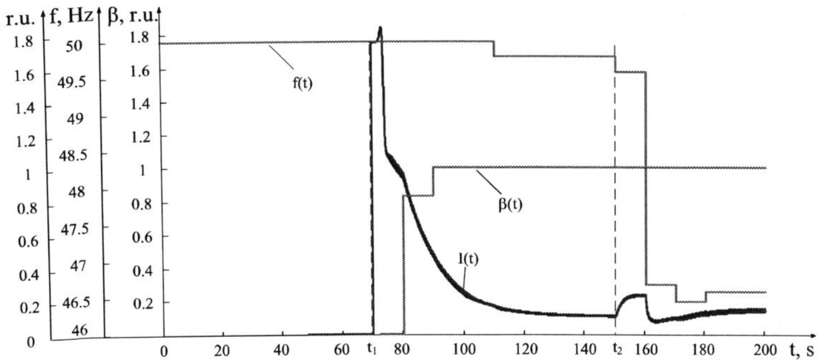

Figure 5.15. The curves representing the time variation of the IM supply voltage frequency $f(t)$, the degree $\beta(t)$ of opening the regulating valve and the quality criterion $I(t)$ in PC.

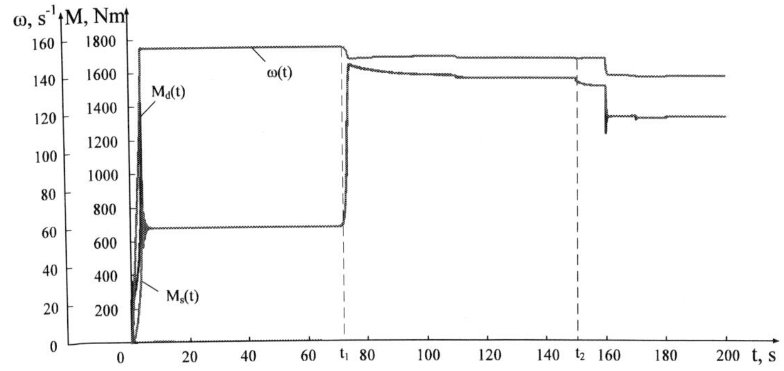

Figure 5.16. The curves representing the variation of the moment and the angular frequency of ED: $M_d(t)$ – the moment developed by IM; $M_s(t)$ – the moment of resistance created by the turbomechanism; $\omega(t)$ – the angular frequency of the pump ED rotation.

The increase of the pump load (opening of the stopcock) results in the growth of the resistance moment M_s, created by the turbomechanism and, accordingly, in the increase of power N_d consumed by ED and power N_N consumed by the pump, moment M_d, developed on PU shaft and the decrease of angular frequency ω of the pump ED rotation. During this variation of the mode PC operation is accompanied by the development of the cavitation processes at the first section of the pipeline as $\nu_i > \nu_{kr\_1}$, by meeting the $(H_{zad} - H_{tek\,i}) > H_{por}$, $(N_{g\,max} - H_{g\,i}) > N_{por}$ and by the violation of the $I_i - I_{i-1}/(t_i - t_{i-1}) \leq \varepsilon$ conditions. Based on the search algorithms (Figure D.2, Figure D.3), ACS of PC ED forms the required control actions (the supply voltage frequency $f(t)$ and the degree $\beta(t)$ of opening the regulating valve) with the time lag of $\Delta t = 10$ s. This process continues till the difference between the current I_i and the previous I_{i-1} values of the quality criterion (Figure 5.15), obtained at the time interval $t_i - t_{i-1}$, meets the condition $\varepsilon \geq I_i - I_{i-1}/(t_i - t_{i-1})$. In this case, the head and power losses caused by the presence of cavitation processes make less than 1% of the parameters acting at the cavitating section of the hydronetwork, respectively.

So, time moment $t_1 = 70$ s corresponds to opening of the stopcock in the PP pipeline and switching-on of the PC ED ACS, which is accompanied by the decrease of the head at the first section H_1 and the growth of the head at the second H_2 and the third H_3 sections of the hydronetwork located after the stopcock.

Time moment $t_2 = 150$ s is characterized by the variation of the consumer operation mode corresponding to the decrease of the value of water consumption. It is accompanied by the growth of the value of the head at the check point H_2, which results in meeting the $(H_{zad} - H_{tek\,i}) > H_{por}$, $(N_{g\,max} - H_{g\,i}) > N_{por}$ and the violation of the $I_i - I_{i-1}/(t_i - t_{i-1}) \leq \varepsilon$ conditions. In this case, the developed ACS of PC ED works for 40 s according to the exciting action of the previously proposed algorithms developed for the formation of the control actions. This process is accompanied by the reduction of the pump ED angular rotation frequency ω, consumed power N_N and moment M_d developed on the PU (Figure 5.14, 5.16, respectively), the reduction of PC head losses and the power losses caused by the presence of cavitation processes. In this case, the accuracy of the stabilization of the head in the pipeline, representing its deviation from the assigned parameter, is determined by the expression of the form:

$$\varepsilon_h = \frac{H_{zad} - H_{tek}}{H_{zad}},$$

where H_{zad}, H_{tek} – the required and the current values of the head at the check point of the pipeline.

It is obtained that in the considered case the accuracy of the stabilization of the head at the consumer is within the range of $(0.01-0.03)H_{zad}$.

Thus, the proposed structure of PC ED ACS enables the stabilization of the pressure at the consumer, the minimum of the power losses caused by the cavitation processes in the pipeline and the maximal value of the recuperated power of the steam-gas mixture.

CONCLUSION

1. The analysis revealed that in the operation and emergency modes of PC with ESDDL and ED power equal to 5.2 kW, the time of the pipeline valves control varies within the range of 80-137 s. It allows the determination of the uninterrupted power supply technical indices required for the choice: the time of the continuous operation and the output power whose values should not be less than 137 s and 5.2 kW, respectively. The above-mentioned values correspond to the parameters of the nomenclatural series of the commercial uninterrupted power supplies, which makes it possible to consider them as reserve sources of power supply in ESDDL. The technical indices and the time of the autonomous operation of the uninterrupted power supplies are given in Tables I1, I2 in Appendix I, respectively.
2. We have substantiated the structure of the criterion of the quality of the closed loop electromechanical system of the reduction of the dynamic loads in the pumping complex including a component increasing the head in the hydronetwork and the time of the control of the electric drive of the pipeline valves. We have determined that the value of the weighting coefficient at the time component has a constant value chosen depending on the operation mode of the pumping complex and the priority of the problems solved at the valves control. Therefore, in the pumping complex operation modes only requiring the minimization of the pressure increase in the pipeline network the weighting coefficient at the time component is insignificant and equal to $k_t = 0.0025$. In the emergency modes, where it is necessary to reduce the time of shutting the pipeline valves to prevent a long rotation of PU in the opposite direction at the counter flow the weighting coefficient is assumed equal to the value one order higher $k_t = 0.02$.
3. We have developed the procedures of the search for the optimal law of control of the electric drive of the pipeline valves. They allow, using the method of dynamic programming, the optimization of the control at every step of the variation of the relative degree of opening the valves, thereby eliminating the complete enumeration of all the possible variants of control.

4. We have determined the optimal laws of control of the variable-frequency electric drive of the stopcock in the operation and emergency modes providing the minimal dynamic loads in the hydronetwork and the most rapid control of the pipeline valves in the emergency (non-steady) situations related to the abrupt disconnection of the power supply of the pumping complex.
5. We have demonstrated that the use of a variable-frequency electric drive of the power of 5.2 kW, in comparison with the unregulated electric drive, at the optimal law of control, makes it possible to reduce the increase of the head in the pipeline by 2.8 times. It also allows decreasing the time of the valves control by 2 times in the pumping complex operation modes. In the emergency modes, at the abrupt disconnection of the power supply and shutting of the stopcock in the pressure header of the pumping complex, the optimal control of the electric drive allows the elimination of considerable dynamic loads caused by sharp actuation of the check valve and the provision of the head increase not exceeding 22 m (22% of the rated head in the hydrosystem). In this case, the time of shutting reduces by 3.4 times.
6. The obtained results made it possible to determine the uninterrupted power supply technical indices required for the choice: the time of the continuous operation and the output power whose values should not be less than 137 s and 5.2 kW, respectively, for the analyzed pumping complex.
7. We have substantiated the structure of the criterion of the quality of the closed loop automatic control system for the electric drive of the pumping complex taking into account the cavitation processes, including three components: the deviation by the head at the check point of the hydronetwork; the power losses caused by the presence of cavitation processes; the power of the recuperation of the steam-gas mixture energy.
8. We have developed a search algorithm and, based on it, for different modes of the considered pumping complex water consumption, we obtained the surfaces representing the quality criterion dependence on the control action. The presence of the surfaces allows the determination of the character of the variation of the quality criterion and the development of the algorithm of the search for its minimum based on the use of the iteration method.
9. We have proposed the pumping complex electric drives automatic control system enabling the improvement of the pumping plant power efficiency. It is improved by means of the decrease of the power losses caused by the cavitation processes in the pipeline at the stabilization of the head at the consumer with the accuracy of $(0.01-0.03)H_{zad}$ and simultaneous recuperation of the steam-gas mixture energy to 1.6% of the pump rated power.

Chapter 6

EXPERIMENTAL RESEARCH

6.1. THE STRUCTURE OF THE PUMPING UNIT LABORATORY COMPLEX

The obtained theoretical propositions were verified based on a small-scale experimental unit including [27, 143–149]:

- two WILLO centrifugal pump units PU1, PU2 that can be connected both in series and in parallel;
- the devices for the regulation of the rotation frequency of the induction electric motors IM1, IM2 of the pumps based on LENZE and MITSUBISHI frequency convertors FC1, FC2, respectively;
- a branched pipeline network containing horizontal and vertical sections of the pipeline;
- shutoff-control and hydro-protective valves in the form of Z1–Z5, Z7–Z32 globe valves with manual control and a check valve CV, respectively;
- discharge tanks T1, T2 of the volume of 1.08 m^3 and 0.2 m^3, respectively;
- Z6 regulating stopcock manufactured by ZETKAMA, with a rubberized wedge, installed in the head line of PC and meant for the variation of the technological parameters (the head and the discharge) by the alteration of the degree of its opening;
- control-measuring equipment: pressure sensors H1–H4 manufactured by OWEN and discharge sensors Q1, Q2 manufactured by ERGOMERA; current sensors CS1–CS4 manufactured by ALLEGRO MICROSYSTEMS; voltage sensors VS1–VS6, built on galvanically isolated amplifiers manufactured by AVAGO TECHNOLOGIES; wattmeters W1, W2; rotation frequency sensors TG1–TG3;

- a module of the control and data collection in the form of a board of an analog-to-digital and a digital-to-analog signal converter with a USB interface manufactured by L-CARD.

The pumping unit operates by the principle of a circulation system with static head, wherein the same tank T1 with water serves as the input and the output reservoir. Tank T2 installed at the height of 8.8 m in relation to the axis of the pump units location creates the static head (counter pressure) in the system. The unit also contains a centrifugal pump with an induction motor on one shaft; a branched pipeline with installed shutoff-control valves and discharge tanks; a regulated disk shutter with an electric drive; a frequency converter for measuring the pump electric drive rotation frequency; control-measuring devices (the sensors of current, voltage, rotation frequency, pressure and discharge).

The general view and the functional diagram of an electrohydraulic complex (EHC) physical model are given in Figures 6.1, 6.2, respectively.

The engineering performance of the electrohydraulic equipment is shown in Table 6.1.

Table 6.1. The engineering performance of the laboratory stand equipment

Parameter name	Parameter value
Rated power, W	830
Rated voltage, V	380
Rated current, A	1.7
Mains frequency, Hz	50
Rotation frequency, rpm	2900
Maximum discharge, m^3/hour	8
Maximum head, m	22
Rated power, W	550

Figure 6.3 contains a functional diagram of a control-measuring block (CMB) of the experimental EHC including: four current sensors CS_1 – CS_4; six voltage sensors VS_1 – VS_6 based on the Hall effect; pressure sensors PS_1 – PS_3 manufactured by OWEN and JUMO; discharge sensors DS_1, DS_2 (a two-channel ultrasonic Ergomer counter 125 is used in the laboratory complex); velocity sensor VS (tachogenerator mounted on one shaft with the motor). The sensors measuring electrical, technological and mechanical parameters of EHC are connected with a personal computer (PC) by means of an LCard E440-14 analog-to-digital converter (ADC).

This experimental unit enables the solution to a complex of research and applied problems: the research of different laws of the pipeline valves shutting influence on the values of the dynamic loads in PC, the cavitation processes influence on the energy characteristics of the complex, the assessment of the energy conversion processes in PC at the occurrence of a non-steady mode of the operation of the equipment.

Figure 6.1. The general view of the experimental PC small-scale model.

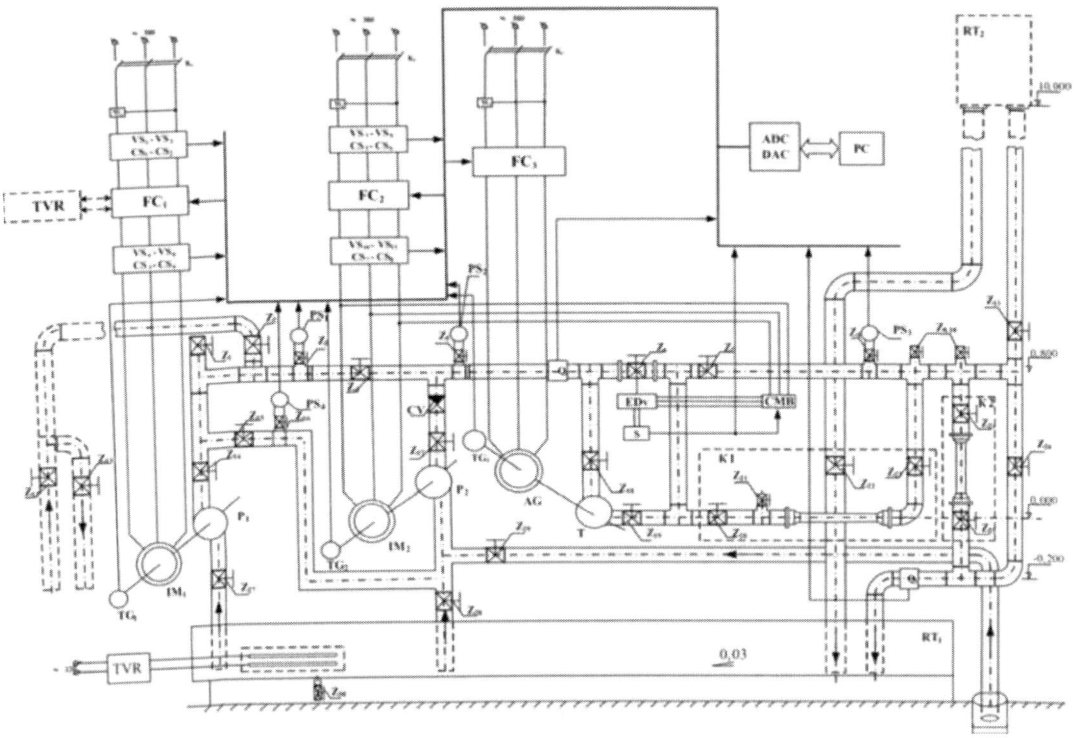

Figure 6.2. The functional diagram of a PC experimental model.

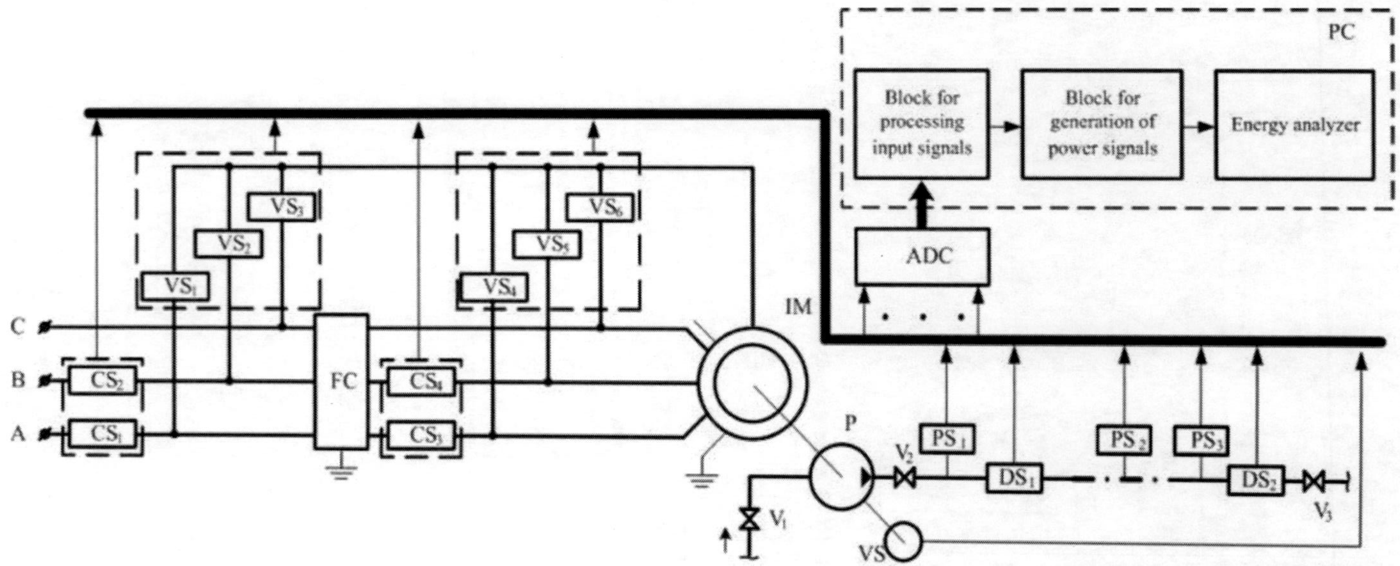

Figure 6.3. The functional diagram of the control-measuring block of the experimental plant.

6.2. The Dynamic Processes in the Pumping Complex at the Regulating Stopcock Control

To research the different laws of the pipeline valves shutting influence on the values of the dynamic loads in PC, based on the experimental unit, regulating stopcock Z_6 is equipped with an electromechanical system of the decrease of dynamic loads (ESDDL) using a variable-frequency electric drive whose appearance is shown in Figure 6.4.

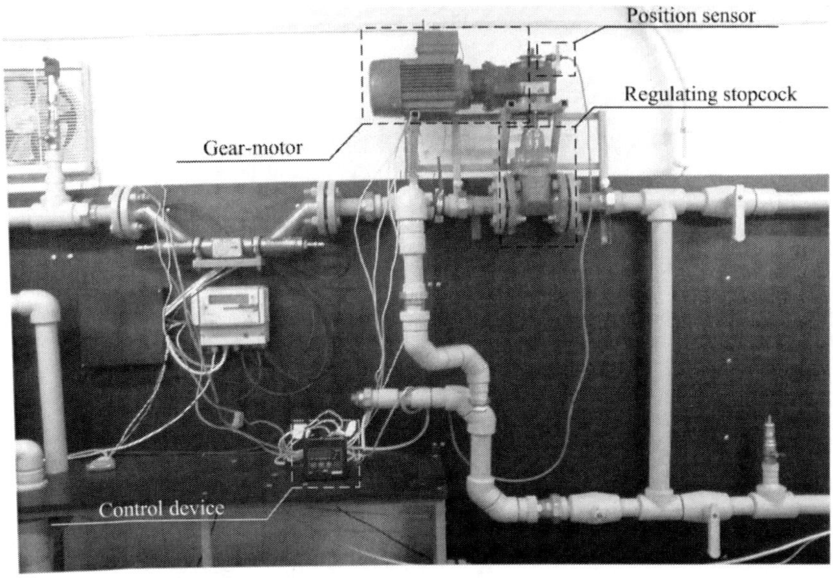

Figure 6.4. ESDDL as a part of the laboratory PC.

The functional diagram of ESDDL in PC is shown in Figure 6.5. The technical characteristics of the equipment used in the system are given in Tables E3 – E7 in Appendix E.

The functional diagram (Figure 6.5) of the system of the decrease of the dynamic loads in PC includes: a regulating stopcock proper (Figure 6.6, a); a worm gear motor G (Figure 6.6, b) with a cylindrical step manufactured by DAVID BROWN, equipped with a three-phase induction motor IM3; a three-phase frequency converter FC2 manufactured by MITSUBISHI (Figure 6.6, c), enabling the control of the time and the trajectory of shutting the stopcock Z_6 by variation of IM3 supply voltage frequency; a control device CoD manufactured by OWEN (Figure 6.6, d), intended for the indication and the determination of the current degree of the stopcock opening according to the number and the period of passing impulses coming from the position sensor PS on the shaft of the stopcock; an active-inductive position sensor PS of the stopcock working element manufactured by CARLO GAVAZZI (Figure 6.6, e), allowing the formation of rectangular impulses whose number is proportional to the degree of its opening; an automatic interface converter IC manufactured by OWEN (Figure 6.6, f), intended for reading the values of the current degree of opening of the stopcock from CoD by means of a computer.

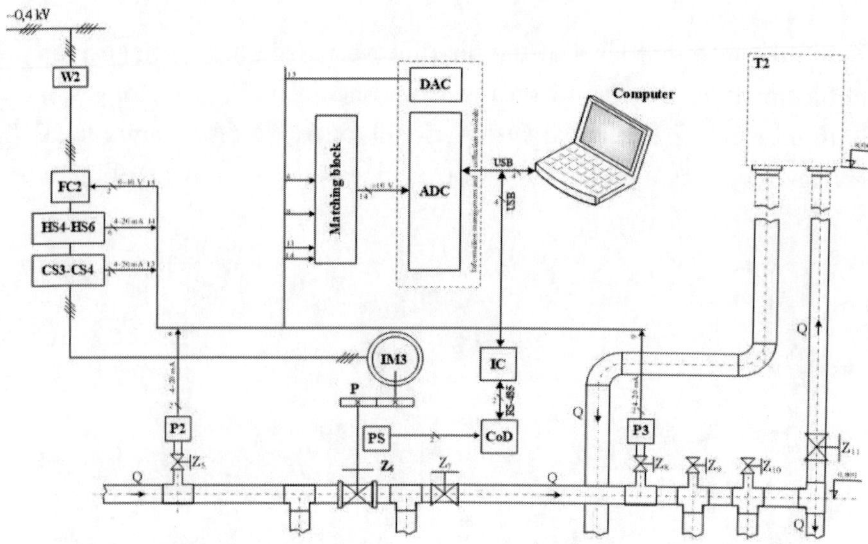

Figure 6.5. A fragment of the functional diagram of the physical model of PC with ESDDL.

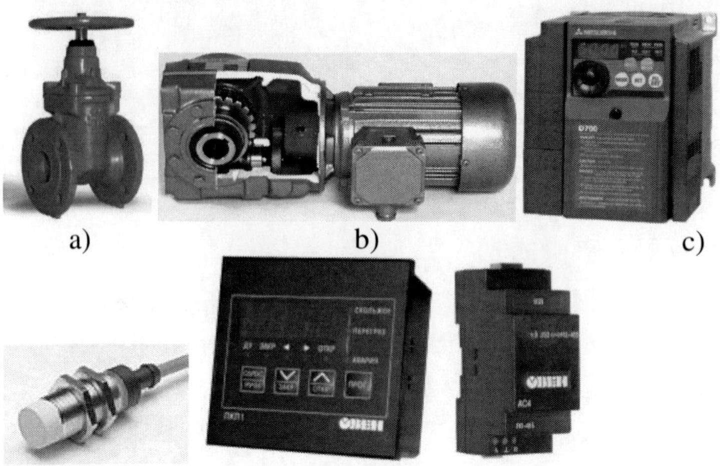

a) b) c)

d) e) f)

Figure 6.6. The appearance of ESDDL equipment in PC: a) stopcock; b) gear-motor; c) frequency converter; d) position sensor; e) control device; f) interface converter.

Figure 6.7 contains a schematic circuit of the connections of ESDDL elements in the laboratory PC.

A PC physical model makes it possible to carry out a lot of research directed to the determination of the dynamic loads in the pipeline network at the stopcock control in the pumping equipment emergency modes caused by an abrupt disconnection of the power supply and occurrence of liquid counter flow. The results of the research suggest the conclusion about the formation of the stopcock shutting laws at which the surge occurring in the pipeline network is accompanied by the minimal dynamic loads (pressure hits). The algorithm of the performance of the experimental research of the dynamic loads in PC at an abrupt disconnection of PU1 power supply and occurrence of the liquid counter flow is shown in Figure 6.8 [145].

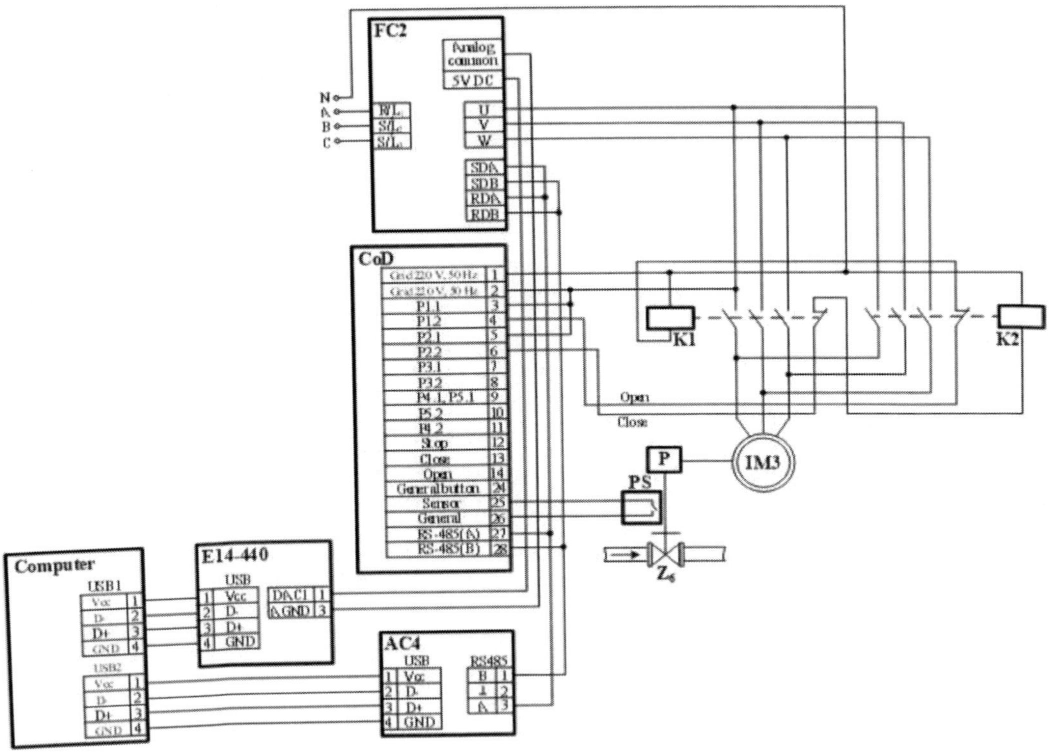

Figure 6.7. The schematic circuit of connecting the ESDDL elements.

To provide the liquid counter flow in PC at an abrupt disconnection of PU1 power supply tank T2 is filled with water. In this case, globe valves Z_4, Z_5, Z_7, Z_8, Z_{11}, Z_{14}, Z_{28} open, the other globe valves Z_1–Z_3, Z_9, Z_{10}, Z_{12}, Z_{13}, Z_{15}–Z_{27}, Z_{29}–Z_{32} and regulating stopcock Z_6 are shut. Pumping unit PU1 disconnects from frequency converter FC1 and switches to the operation from the alternating current mains. To form different laws of the stopcock shutting induction electric motor IM3 is connected to frequency converter FC2. In this case, frequency converter FC2 is adjusted to the mode of the alteration of the supply voltage frequency of the supply from the external analog signal 0-5 V.

To register the signals of the pressure of P2 before and after P3 of the stopcock, discharge Q1 in the pipeline network, voltage VS4-VS6 and current CS3, CS4 in the stator circuit of the induction electric motor IM3 we used LGRAPH2 software created by L-CARD. It is intended for the registration, visualization and processing the analog signals recorded by means of measuring module E14-440.

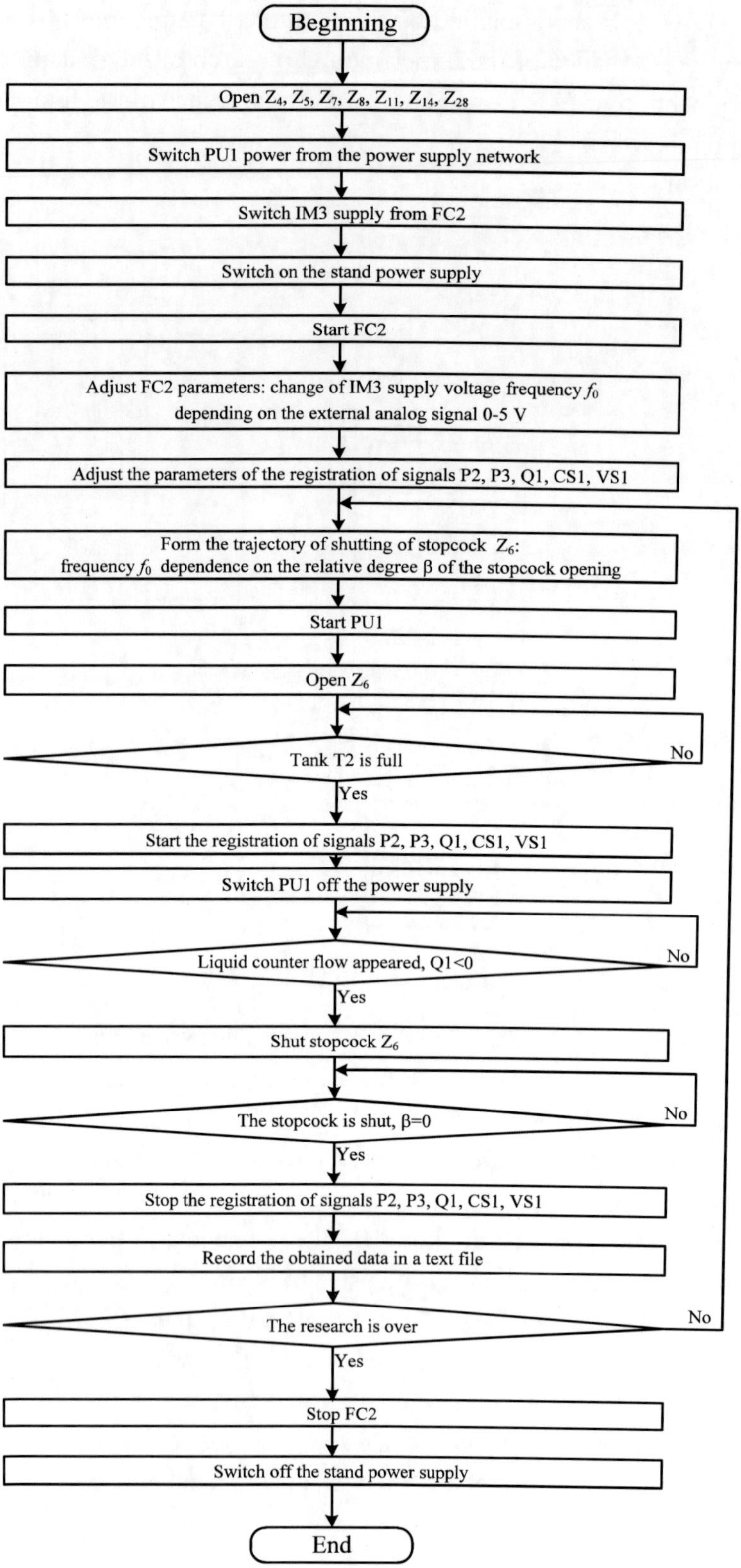

Figure 6.8. The algorithm of the research at the laboratory stand.

Figure 6.9 shows the interface of the main window of LGRAPH2 program, including five coordinate axes for mapping the graphs of the transient processes and the elements for the control of the record process, scaling and adjusting the signals mapping. Figure 6.10 shows a subsidiary window of LGRAPH2 program, allowing the assignment of the registration parameters: the frequency of ADC conversion for the measuring channel, the number and the range of the input signals for every measuring channel, the type of connection of the signals, the time of registration, etc.

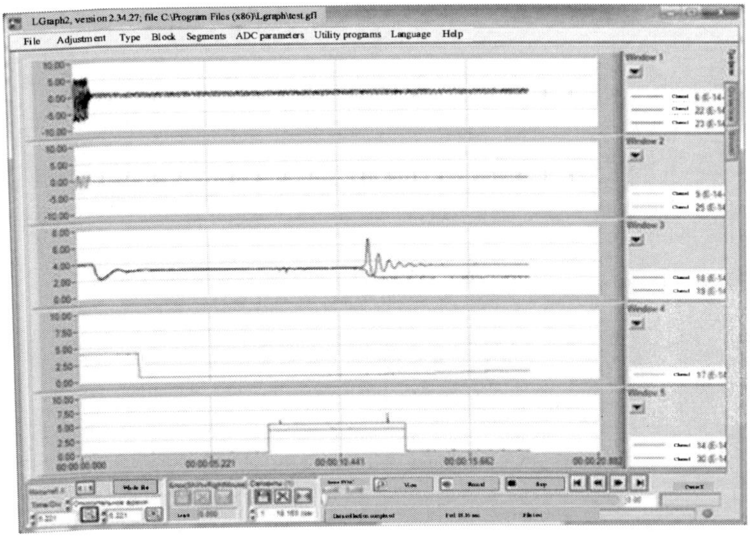

Figure 6.9. The interface of the main window of LGRAPH2 program for the registration and mapping of the transient processes in PC.

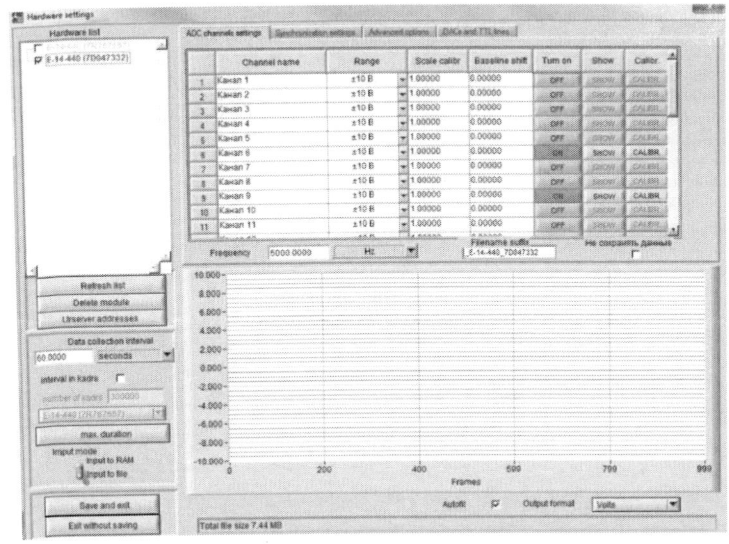

Figure 6.10. A subsidiary window of LGRAPH2 program for the assignment of the parameters of the registration of the transient processes in PC.

Different laws of the stopcock shutting are assigned by means of special software developed in the medium of graphic programming LabVIEW. The interface of the developed software (Figure 6.11) makes it possible to:

- remotely control ESDDL in PC by RS-485 interface via a computer;
- form the trajectory of shutting/opening of the stopcock – assign the value of supply voltage frequency f_0 of IM3 of the regulating stopcock at different relative degrees β of its opening in the form of a table and a graph;
- assign and map the current relative degree β of the stopcock opening in percent;
- open, shut and stop the stopcock at any time moment by means of buttons;
- determine the direction of the stopcock opening and its final positions;
- give signals of the occurrence of overload or an emergency situation (slipping or jamming of the shaft, etc.).

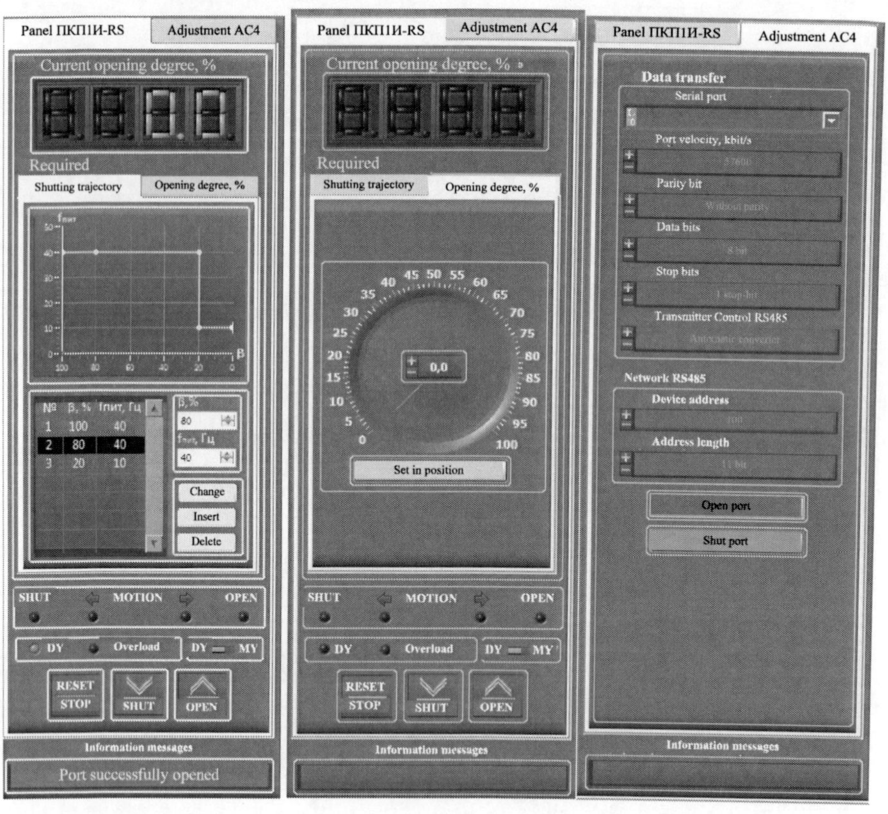

a) b) c)

Figure 6.11. The interface of the software used in the experimental research: a), b) main marker; c) subsidiary marker.

To fill tank T2 pumping unit PU1 is started and regulating stopcock Z_6 is opened. After filling tank T2 pumping unit PU1 is switched off the power supply. The appearance of the water counter flow in the pipeline network is determined by the condition of negative discharge Q1. Stopcock Z_6 is shut at Q1 < 0. To record the obtained transient processes in the text file a subsidiary window of LGRAPH2 program, shown in Figure 6.12 is used.

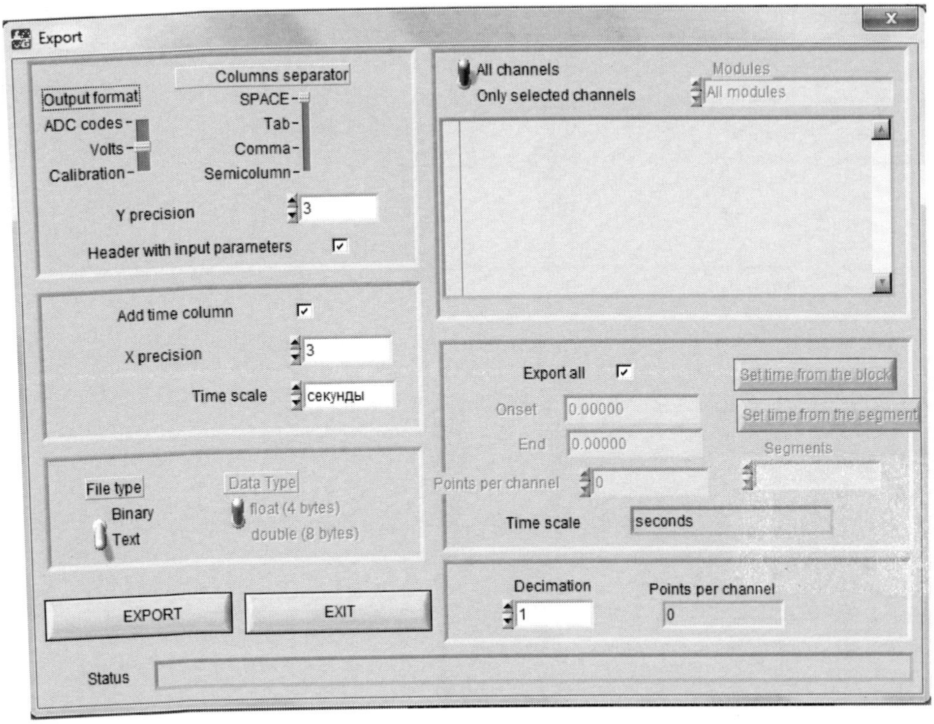

Figure 6.12. A subsidiary window of LGRAPH2 program for the record of the transient processes in PC in the text file.

To analyze the dynamic loads in the pump complex at a sudden outage of the pumping unit power supply and shutting of the regulating stopcock the following cases are considered: shutting of the stopcock at the uniform control with invariable frequency f_0 of the supply voltage ($f_0 = 40; 20; 10; 5$ Hz) along the whole trajectory of shutting; shutting of the stopcock at the irregular control with the supply voltage frequency f_0 equal to 40 Hz at section $0.2 \le \beta \le 1$ and variable frequency (5, 10, 20 Hz) at the section $\beta \le 0.2$.

Figures 6.13, 6.14 show pressure $H_v(t)$ and current $I_{ph}(t)$ curves of IM stator phase for the case of a sudden pumping unit hitch and stopcock shutting in the head pipeline at a uniform (Figure 6.4) and irregular (Figure 6.5) control. In the former case the supply voltage frequency $f_0(t)$ is 40 Hz at the relative rate of the stopcock opening $0 \le \beta \le 1$; when an irregular pace is used – at the section $0.2 \le \beta \le 1$ the frequency is 40 Hz and 5 Hz – at the section $\beta \le 0.2$.

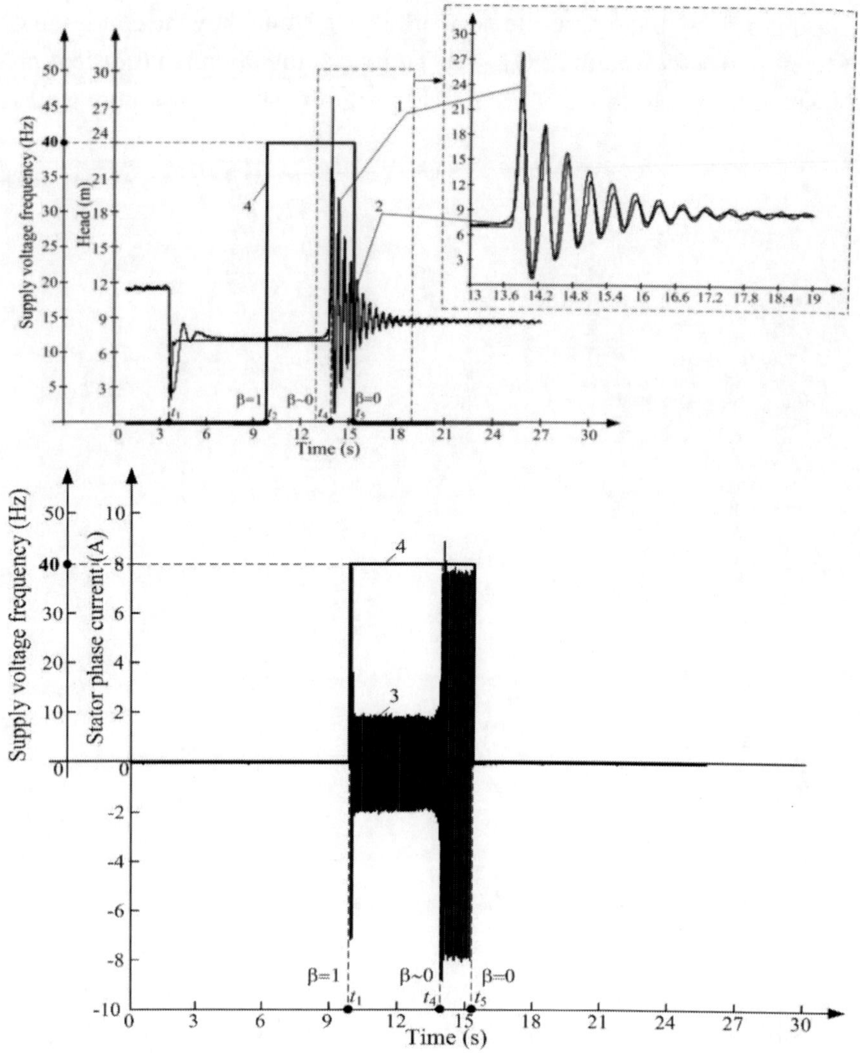

Figure 6.13. The plots of the transient processes in the pump complex at uniform shutting of the stopcock with the supply voltage frequency f_0 of 40 Hz: 1, 2 – the curves of head $H_v(t)$ change at the stopcock, obtained by the experiment and by modeling, respectively; 3 – the curve of the change of the stopcock IM stator phase current $I_{ph}(t)$; 4 – the curve of the change of the supply voltage frequency $f_0(t)$.

In Figures 6.13, 6.14 the following designations are used: t_1 – the moment of sudden disconnection of the pumping unit from the electric mains; t_2 – the moment of connecting IM to the frequency converter and the start of regulating stopcock shutting corresponding to relative rate of its opening $\beta = 1$; t_3 – the moment of the variation of IM supply voltage frequency f_0 corresponding to the relative rate of the stopcock opening $\beta = 0.2$; t_4 – the moment of the highest increase of pressure ΔH in the pipeline network corresponding to practically complete shutting ($\beta \sim 0$) of the stopcock; t_5 – the moment of IM disconnection from the frequency converter and the finish of the regulating stopcock shutting corresponding to the relative rate of its opening $\beta = 0$.

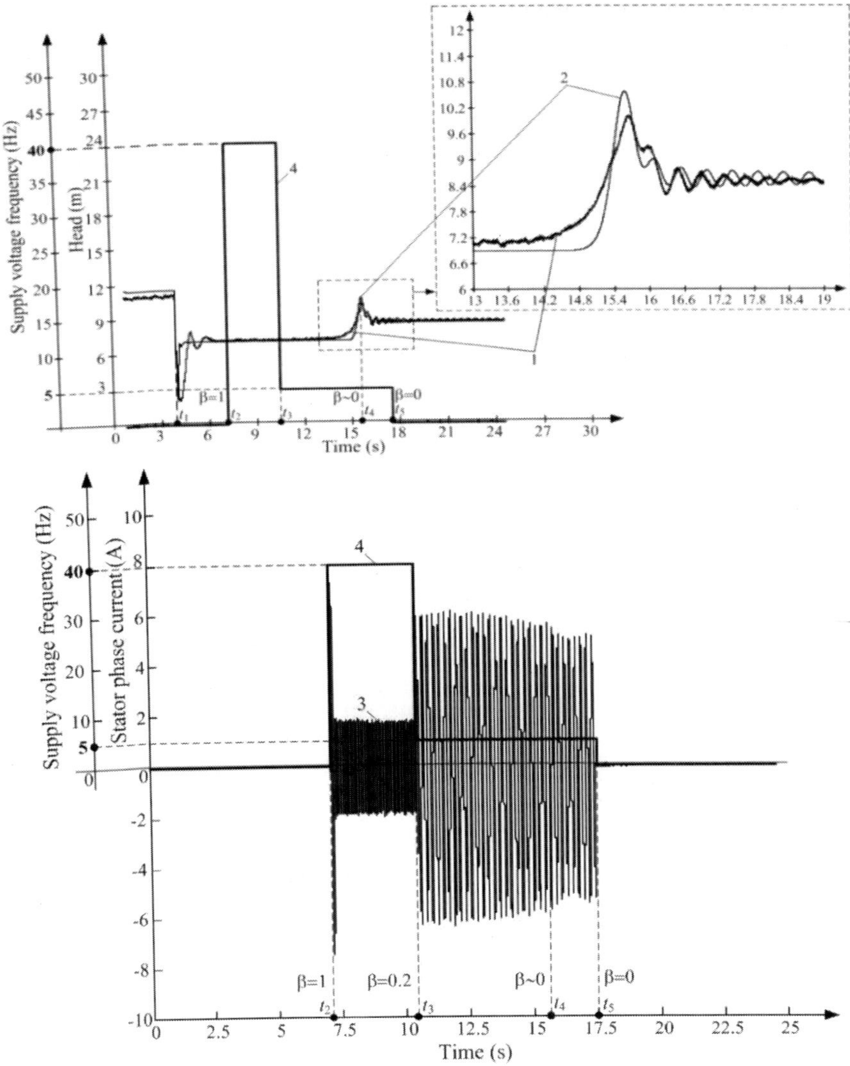

Figure 6.14. Plots of the transient processes in the pump complex at irregular shutting of the stopcock with the supply voltage frequency $f_0(t)$ of 40 Hz at the section $0.2 \le \beta \le 1$ and 5 Hz at the section $\beta \le 0.2$: 1, 2 – the curves of head $H_v(t)$ change at the stopcock obtained by the experiment and by modeling, respectively; 3 – the curve of the change of the stopcock IM stator phase current $I_{ph}(t)$; 4 – the curve of the change of the supply voltage frequency $f_0(t)$.

Figures 6.13, 6.14 show pressure curves $H_v(t)$ obtained by the experiment using a pumping unit physical model (curve 1) and by modeling (curve 2).

The indices of the transient processes at different stopcock control are given in Table 6.2.

Table 6.2. The indices of the transient processes in the pump complex at different stopcock control

Control method	Supply voltage frequency f_0, Hz	Shutting time t_{sh}, s	Increase ΔH of the head, m
uniform	40	5.6	19.62
	20	10	8.3
	10	17.4	3.9
	5	34	1.6
irregular	40; 20	6.2	8.3
	40; 10	8	3.9
	40; 5	9.7	1.6

The reduction of the supply voltage frequency at time moment t_4 at $\beta = 0.2$ and a considerable increase of the pressure in the pipeline at time moment t_5 are caused by the fact that at the section $\beta \le 0.2$ for the stopcocks there is a significant increase of their hydraulic resistance coefficient ξ (at $0.2 \le \beta \le 1$ value ξ practically does not change).

At the time interval from t_5 to t_6 the full compression of the operating element of the stopcock till its complete shutting takes place. To provide this the frequency converter FC2 is adjusted in such a way that at low supply voltage frequencies f_0 at the section $\beta \le 0.2$ it provides the value of the induction motor IM3 torque that exceeds the rated one by 17.5%.

The analysis of the obtained results demonstrated that at the uniform shutting of the stopcock, when the supply voltage frequency is 40 Hz, the pipeline network pressure is 3.2 times as high as the pump rated head. When the pipeline valves are controlled at the supply voltage frequency of 5 Hz, the pressure is 12.2 times as high, but the shutting time is six times as large. The irregular control is optimum. In this case, the supply voltage frequency is 40 Hz at section $0.2 \le \beta \le 1$ and 5 Hz at section $\beta \le 0.2$. The pressure increase is inconsiderable $\Delta H = 1.6$ m, and the valve shutting time decreases by 3.5 times compared with the last analyzed case.

When the stopcock head curves (Figures 6.13, 6.14) obtained on PC physical and mathematical models were compared, the determination coefficient was higher than 0.86, which confirms the adequacy of the theoretical results of the paper.

6.3. The Pumping Complex Characteristics at the Cavitation Control by Means of the Variable-Frequency Electric Drive

Let us consider in more detail the circuit designs intended for the research of the cavitation processes in the experimental hydrotransport complex, their influence on PC characteristics.

It is known that the variation of the temperature of the pumped liquid, PC operation mode, the pressure difference in the hydrosystem result in the occurrence of cavitation in the intake pipe in the pump impeller, at the local resistances of the hydraulic network (at narrowing, turns, at the stopcock output). With this purpose, the experimental complex is equipped with [134, 143–149]:

- a regulated electric heating element (Figure 6.15);
- stopcocks at the pumps intake pipes (Figure 6.15);
- transparent Venturi tubes (Figure 6.16) of different cross sections in the narrow part.

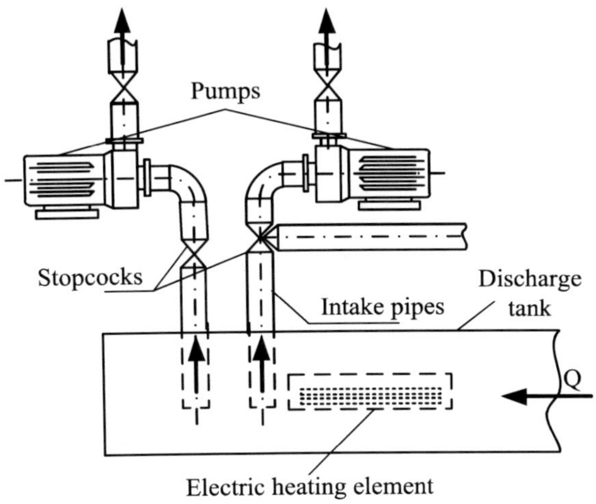

Figure 6.15. A fragment of the physical model functional diagram.

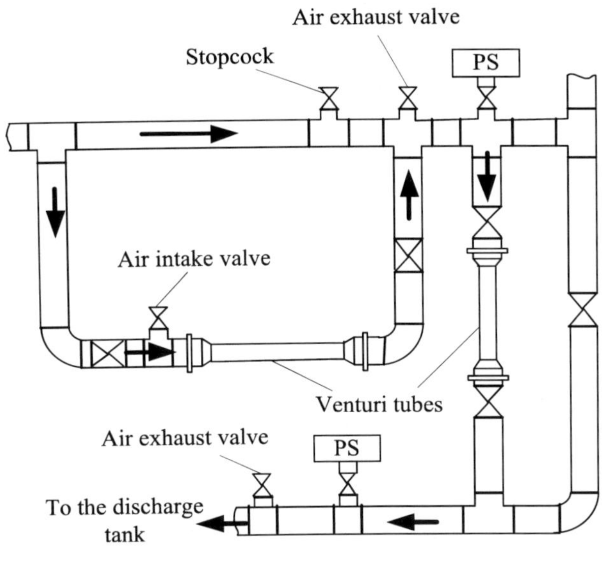

Figure 6.16. The points of location of the Venturi tubes.

The degree of the development of cavitation in the pump is changed by regulating stopcocks mounted on the intake pipes of the corresponding turbomechanisms.

A tube electric heater of the power of 2 kW changes the temperature modes of the physical model liquid motion.

PC physical model allows research directed to the study of the cavitation parameters influence on the characteristics of the electromechanical, pumping and piping equipment; the boundaries of the occurrence of cavitation modes at the variation of the consumer characteristics, temperature modes and the properties of the transported liquid.

The research of the dynamic processes in PC physical model (Figure 6.16) was performed for the similar load modes of the pump ED operation in the absence and presence of cavitation processes in the pipeline (liquid passing along the Venturi tubes).

The performed research resulted in obtaining experimental and calculation curves representing the time variation of:

- the pump output pressure in the absence $p_1(t)$ (Figure 6.17, a) and presence $p_{1\_k}(t)$ (Figure 6.17, b) of the cavitation processes in the pipeline;
- the discharge at the pump output $Q_1(t)$, $Q_{1\_k}(t)$ and in the pipeline $Q_2(t)$ in the absence (Figure 6.18, a) and presence (Figure 6.18, b) of the cavitation processes in the pipeline;
- the pump ED angular rotation frequency $\omega(t)$, $\omega_k(t)$ (Figure 6.19, Figure 6.20);
- hydraulic power $N_n(t)$, $N_{n\_k}(t)$, at the pump output (Figure 6.21);
- torque $M_d(t)$, $M_{d\_k}(t)$ on the pump ED shaft (Figure 6.22);
- phase voltage $u_a(t)$, $u_b(t)$, $u_c(t)$, $u_{a\_k}(t)$, $u_{b\_k}(t)$, $u_{c\_k}(t)$, applied to the stator windings of the pump induction electric drive (Figure 6.23);
- currents $i_a(t)$, $i_b(t)$, $i_c(t)$, $i_{a\_k}(t)$, $i_{b\_k}(t)$, $i_{c\_k}(t)$, flowing in the stator windings of the pump ED (Figure 6.24, Figure 6.25);
- power $N_d(t)$, $N_{d\_k}(t)$, supplied to the stator windings of the induction motor (Figure 6.26).

The comparative analysis of curves $p_1(t)$ and $p_{1\_k}(t)$ (Figure 6.17), representing the variation of the pressure at the pump output allowed singling out the sections of periodic growth and decrease of the technological parameters in the presence of the cavitation processes in the pipeline. We obtained that the amplitude and frequency of the pressure variations in the cavitation mode made $\Delta p_{1\_k} = 4$ kPa and $f_{kav} = 1.25$ Hz, respectively. Besides, it should be noted that the cavitation mode of PC operation is characterized only by curve $Q_{1\_k}(t)$ (Figure 6.18), representing the variation of the liquid discharge at the pump output. It results from the fact that the presence of more than 30% of air (cavitation caverns) in the liquid flow passing through the measuring tube of the ultrasonic flow meter Q_2 (Figure E.1) is characterized by temporary unavailability of the sensor.

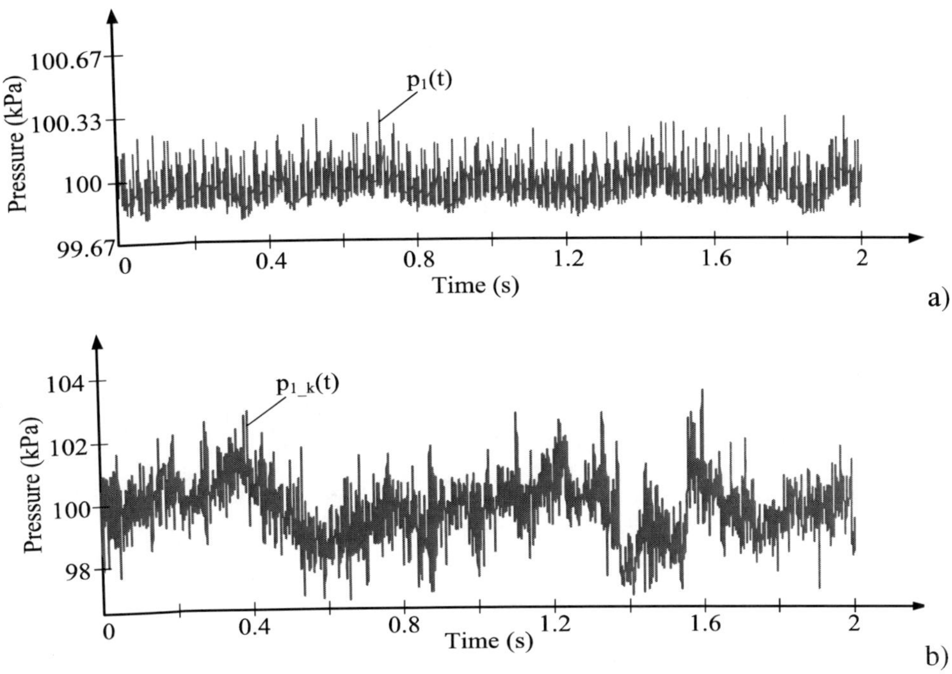

Figure 6.17. Curves representing the variations of pressure $p_1(t)$, $p_{1\_k}(t)$ at the pump output in the absence (a) and presence (b) of cavitation processes in the pipeline, respectively.

Curves $\omega(t)$, $\omega_k(t)$ (Figure 6.19), representing the variation of the angular rotation frequency of the pump ED, are characterized by noisiness caused by the presence of a brush-collector device in the speed sensor design design. Also by high-frequency pulsations of the variations of the values of the technological parameters caused by the special features of the centrifugal pumps design [64, 65, 78–80]. That is why, to analyze the cavitation processes influence on the pump ED mechanical characteristics, we carried out the filtration of the components forming the curves representing the PC impeller angular rotation frequency exceeding the frequency of 10 Hz (Figure 6.20). The obtained results made it possible to determine the presence of the variations of the rotation frequency of the pump ED in the cavitation mode of PC operation: $\Delta\omega_k = 8-10\ \text{s}^{-1}$.

The analysis of curves $N_n(t)$, $N_{n\_k}(t)$, $M_d(t)$, $M_{d\_k}(t)$ (Figures 6.21, 6.22) revealed that the oscillating character of the technological parameters variation caused by the presence of the cavitation processes in the pipeline, results in explicit periodic oscillations of hydraulic power N_{n_k} at the turbomechanism output and torque M_{d_k} on the pump ED shaft.

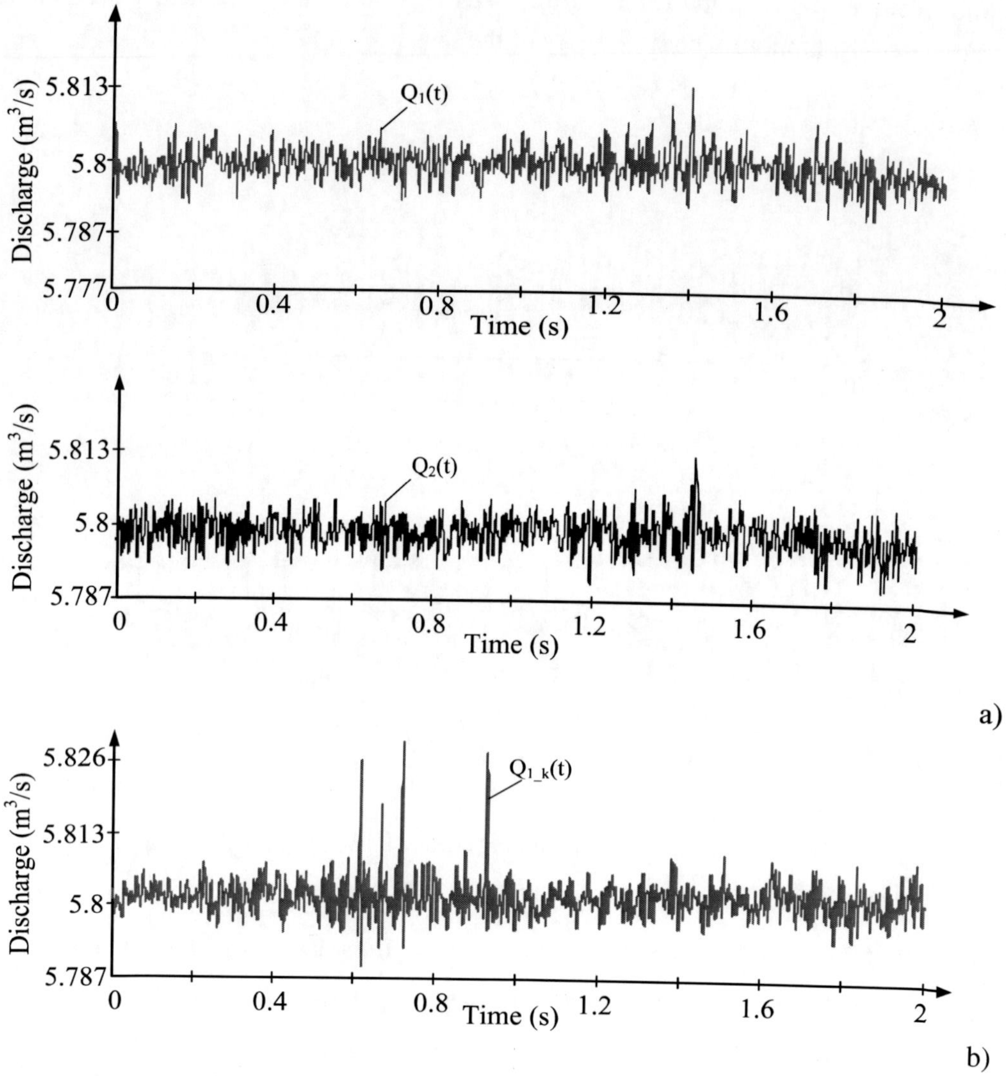

Figure 6.18. Curves representing the variations of the discharge at the pump output $Q_1(t)$, $Q_{1\_k}(t)$ and in the pipeline $Q_2(t)$ in the absence (a) and presence (b) of cavitation processes in the pipeline, respectively.

So, the hydraulic power losses (Figure 6.20, b), caused by the presence of the cavitation processes under the conditions of PC small-scale physical model at the frequency $f_{s\_1} = 50$ Hz of the voltage supplying PC made 1.1% of the pump rated power. Obviously, in real pumping plants, characterized by a complex configuration of the pipeline, a big volume of the transported liquid, caused by the presence of cavitation processes, may reach high values.

Due to the high inertia of the cavitation processes in comparison with the electromagnetic processes taking place in the pump ED, the further analysis of the characteristics of the electrohydraulic complex (Figures 6.23–6.26) was performed for the moment of the cavitation cavern collapse.

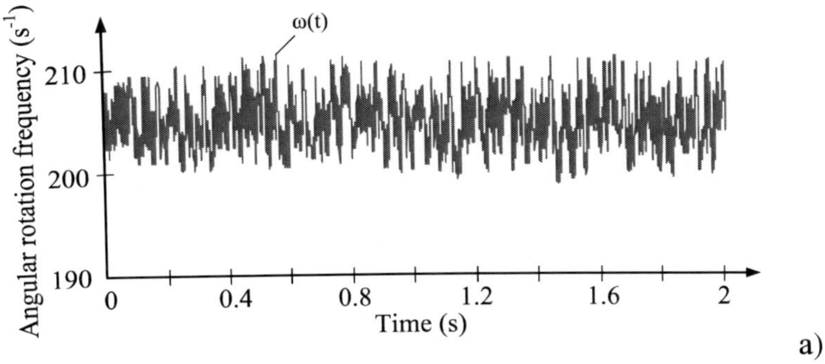

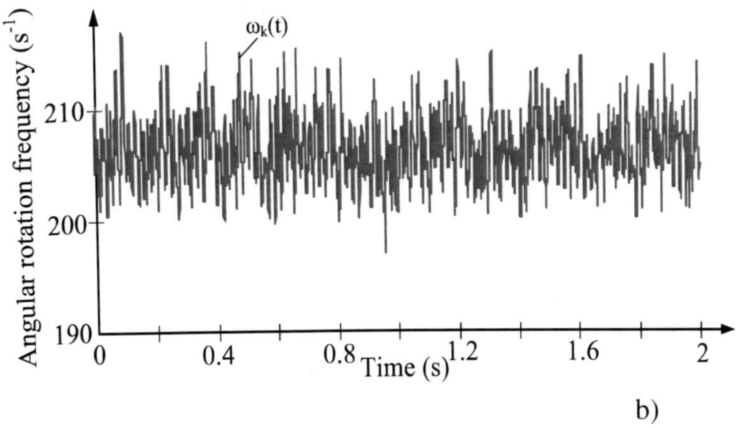

Figure 6.19. Curves representing the variation of angular rotation frequency $\omega(t)$, $\omega_k(t)$ of the pump ED in the absence (a) and presence (b) of cavitation processes in the pipeline, respectively.

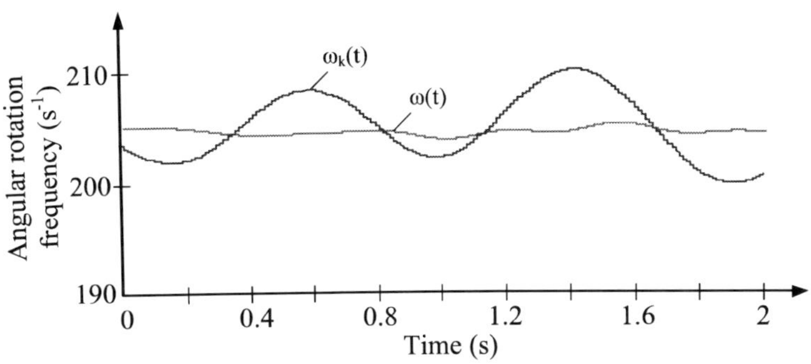

Figure 6.20. The filtered curves representing the variation of angular rotation frequency $\omega(t)$, $\omega_k(t)$ of the pump ED in the absence and presence of cavitation processes in the pipeline, respectively.

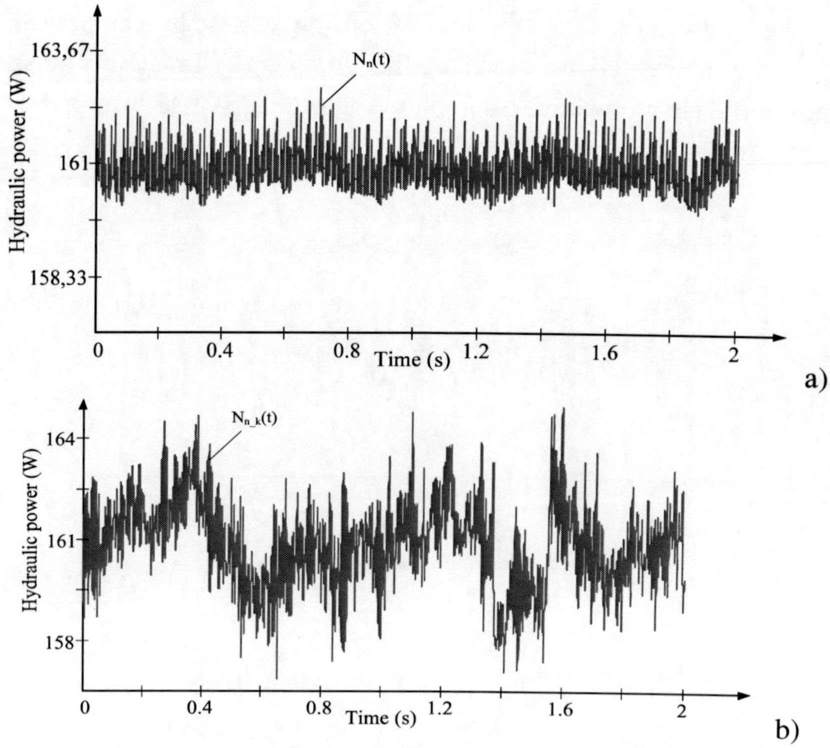

Figure 6.21. Curves representing the variation of hydraulic power $N_n(t)$, $N_{n\_k}(t)$ at the pump output in the absence (a) and presence (b) of cavitation processes in the pipeline, respectively.

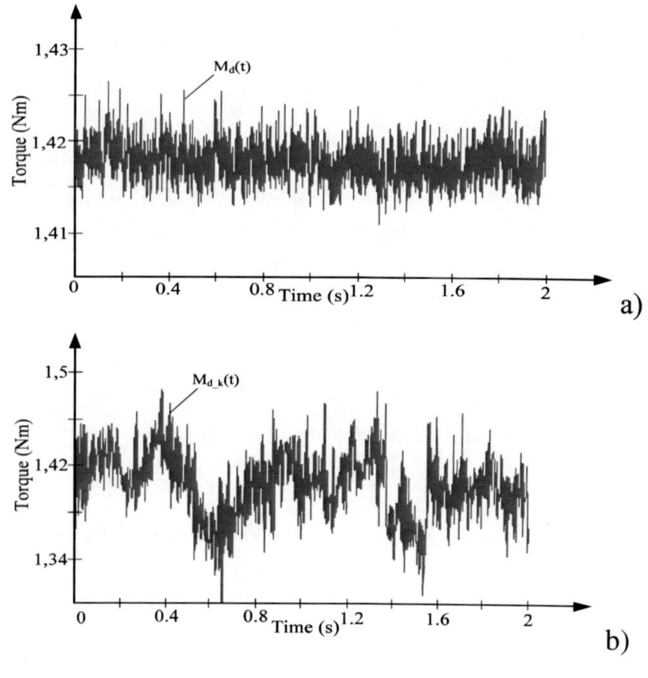

Figure 6.22. Curves representing the variation of torque $M_d(t)$, $M_{d\_k}(t)$ on the pump electric motor shaft in the absence (a) and presence (b) of cavitation processes in the pipeline, respectively.

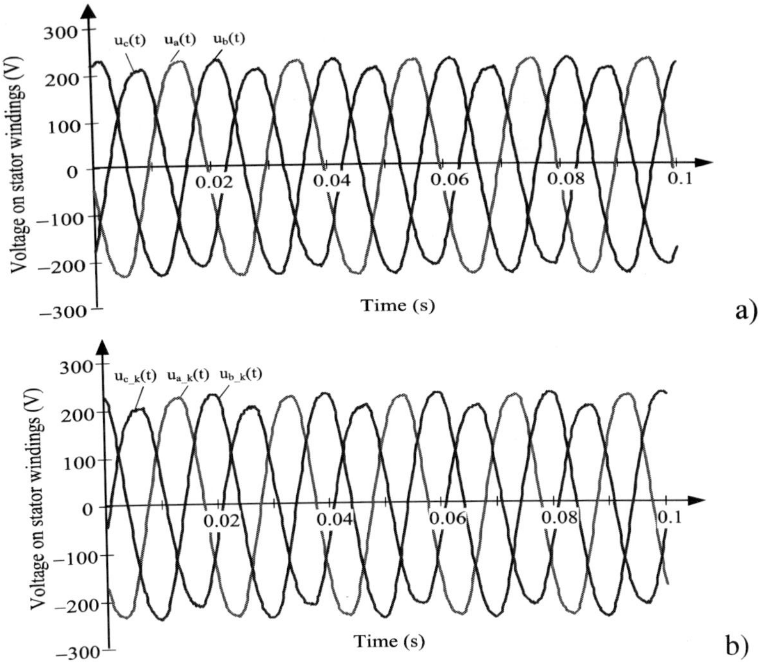

Figure 6.23. Curves representing the variation of voltage $u_a(t)$, $u_b(t)$, $u_c(t)$, $u_{a\_k}(t)$, $u_{b\_k}(t)$, $u_{c\_k}(t)$, applied to the stator windings of the pump induction electric motor in the absence (a) and presence (b) of cavitation processes in the pipeline, respectively.

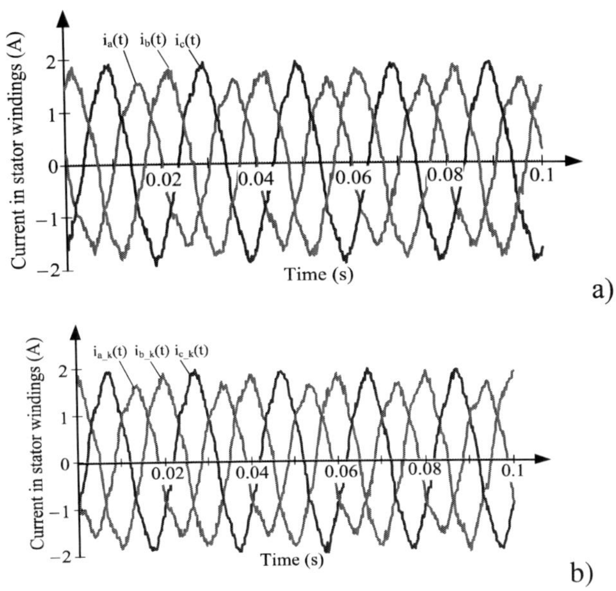

Figure 6.24. Curves representing the variation of current $i_a(t)$, $i_b(t)$, $i_c(t)$, $i_{a\_k}(t)$, $i_{b\_k}(t)$, $i_{c\_k}(t)$ in the pump electric motor stator windings in the absence (a) and presence (b) of cavitation processes in the pipeline, respectively.

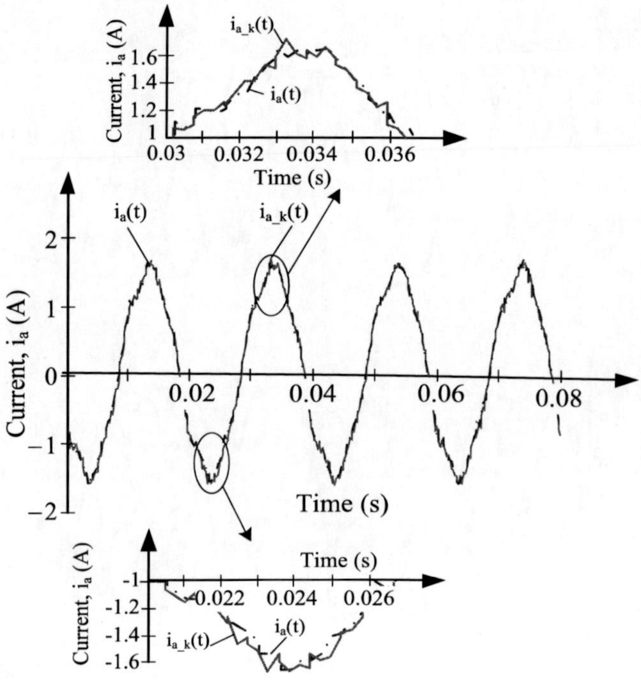

Figure 6.25. Curve representing the variation of current in the pump electric motor stator winding in the absence $i_a(t)$ and presence $i_{a\_k}(t)$ of cavitation processes in the pipeline, respectively.

So, the comparative analysis of the curves (Figure 6.25), representing the variation of the current in the pump ED stator winding in the absence $i_a(t)$ and the development $i_{a\_k}(t)$ of the cavitation processes in the pipeline, made it possible to indicate insignificant sharp bends in the current curve $i_{a\_k}(t)$.

For the known signals of the current and voltage in ED phases we obtained curves $N_d(t)$, $N_{d\_k}(t)$ (Figure 6.26), representing the variation of the instantaneous power supplied to the pump electric motor stator, W:

$$\begin{aligned} N_d(t) &= N_{d\_a}(t) + N_{d\_b}(t) + N_{d\_c}(t) = \\ &= i_a(t)u_a(t) + i_b(t)u_b(t) + i_c(t)u_c(t), \end{aligned} \tag{6.1}$$

where $N_{d\_a}(t)$, $N_{d\_b}(t)$, $N_{d\_c}(t)$ – the instantaneous power in phases a, b, c, respectively, W; $i_a(t)$, $i_b(t)$, $i_c(t)$ – the instantaneous current flowing in phases a, b, c, respectively, A; $u_a(t)$, $u_b(t)$, $u_c(t)$ – the instantaneous voltage in the corresponding phases of the electric motor, V.

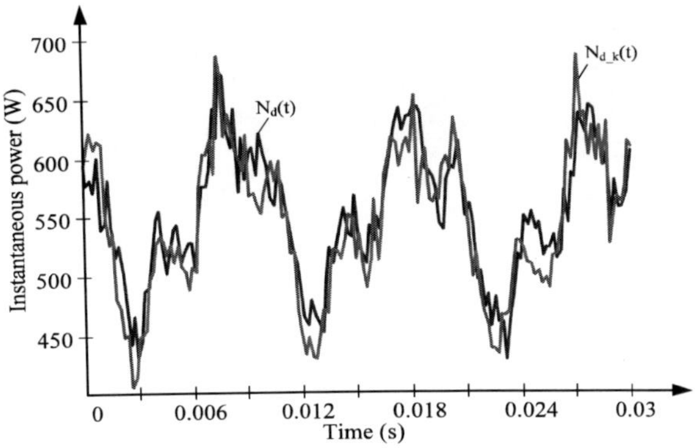

Figure 6.26. Curves representing instantaneous power $N_d(t)$, $N_{d\_k}(t)$, supplied to the pump ED stator in the absence and presence of cavitation processes in the pipeline, respectively.

It is known that the cavitation phenomenon develops when cavitation current number χ is equal or less than its critical value χ_{kr}, determining the boundary of the development of the cavitation processes [148]. For the analysis of the boundaries of the non-cavitation mode of PC physical model, when the pump with a controlled rotation frequency operates on the pipeline network with a vertical Venturi pipe, we indicated three hydronetwork sections whose parameters are given in Table. 6.3.

Table 6.3. The parameters of PC physical model

The parameters of Willo MHI 402 pump				
Head H_N, m	Discharge Q_N, m³/h	Power N_N, W	Hydraulic resistance R_N, s²/m⁵	Head, developed by the pump at zero supply H_0, m
18	5	550	4087562.43	20.8
The parameters of the hydronetwork sections				
Hydrodynamic resistance, s²/m⁵	Counter pressure H_{st}, m	Cavitation critical number χ_{kr}	Value of the pump ED critical relative rotation frequency ν_{kr}	
$R_{s1} = 3.93 \cdot 10^5$	0	50	–	
$R_{s2} = 3.01 \cdot 10^6$	1	135	0.671	
$R_{s3} = 1.41 \cdot 10^5$	0	5	–	

The head-discharge characteristics of the pump and the considered sections of the pipeline are shown in Figure 6.27. Their analysis revealed that in the considered case PC operation mode is determined by point A with the head $H_A = 10.19$ m and the supply $Q_A = 5.8$ m³/h.

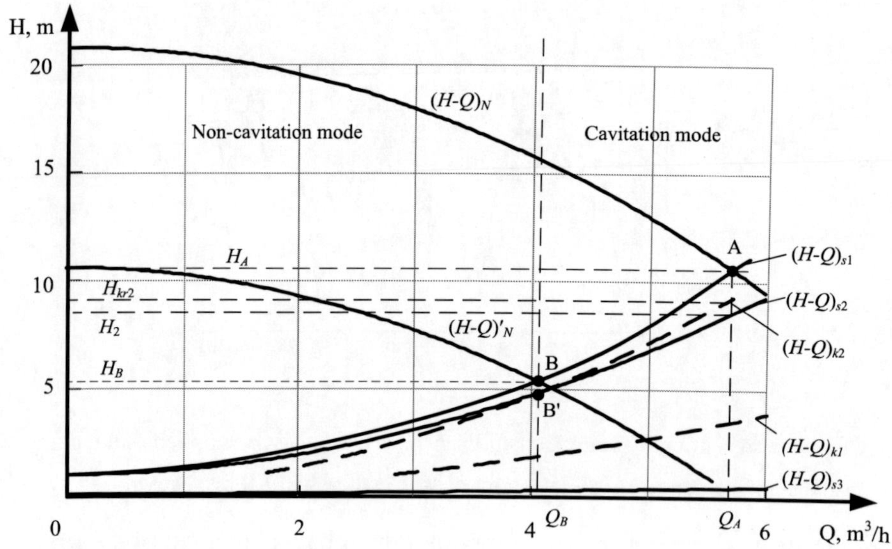

Figure 6.27. PC head-discharge characteristics.

Using expressions (3.4) and (3.5) for every section of the pipeline network, we obtained curves $(H-Q)_{kl}$, $(H-Q)_{k2}$, representing the boundary of the non-cavitation mode of operation of each of the considered sections of the hydrodynamic network. It is obvious from their analysis that the current mode of the pumping station operation corresponding to discharge $Q_A = 5.8$ m³/h, results in meeting the condition of $H_2 \le H_{kr2}$ and the development of the cavitation processes at the second section of the hydrodynamic network.

Using expression (3.7) for the considered PC, we obtained the value of the pump relative critical rotation frequency providing the non-cavitation mode of the liquid motion at the second section of the hydraulic network: $\nu_{kr} = 0.671$. Then point B at the intersection of the pump head-discharge characteristic $(H-Q)'_n$ at the operation with relative rotation frequency equal to ν_{kr}, and network characteristic $(H-Q)_{s1}$ corresponds to the discharge value of $Q_B = 4.2$ m³/h. At point B' (Figure 6.20) condition $H_2 = H_{kr2}$ is met, which determines the boundary of the non-cavitation mode of the liquid motion long the pipeline with a vertical Venturi pipe.

To confirm the above stated calculation parameters of PC non-cavitation operation at the liquid passing through the Venturi pipe we obtained the experimental curves representing the variation of:

- the discharge at the pump output $Q_{1\_36}(t)$ (Figure 6.28, a) and in the pipeline $Q_{2\_36}(t)$ (Figure 6.28, b);

- the pressure at the pump output $p_{1\_36}(t)$ (Figure 6.29);
- the angular rotation frequency $\omega_{36}(t)$ of the pump ED (Figure 6.30);
- the hydraulic power $N_{n\_36}(t)$, at the pump output (Figure 6.31);
- moment $M_{d\_36}(t)$ on the pump ED shaft (Figure 6.32);
- phase voltage $u_{a\_36}(t)$, $u_{b\_36}(t)$, $u_{c\_36}(t)$, applied to the stator windings of the pump induction electric motor (Figure 6.33);
- currents $i_{a\_36}(t)$, $i_{b\_36}(t)$, $i_{c\_36}(t)$, flowing in the stator windings of the pump electric motor (Figure 6.34);
- power $N_{d\_36}(t)$, supplied to the stator windings of the induction motor (Figure 6.35).

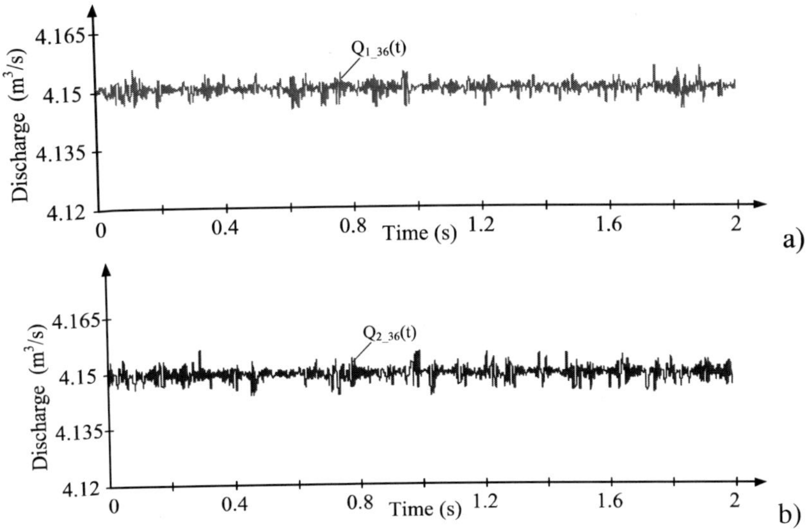

Figure 6.28. Curves representing the variations of the discharge at the pump output $Q_{1\_36}(t)$ (a) and in the pipeline $Q_{2\_36}(t)$ (b) at PU supply voltage frequency 36.4 Hz.

PC experimental and calculation parameters at the analysis of the pump ED rotation frequency influence on the cavitation processes in the pipeline are given in Table 6.4.

Table 6.4. The experimental values of PC basic parameters

Parameter name	Experimental values			
	1	2	3	4
Supply voltage frequency f_s, Hz	50	45	40	36.4
Pressure at the pump output p_1, kPa / harmonic coefficient k_{gar}	100/8.4	76	55.7	43/5.2
PU discharge Q, m³/h	5.8	5.22	4.55	4.15

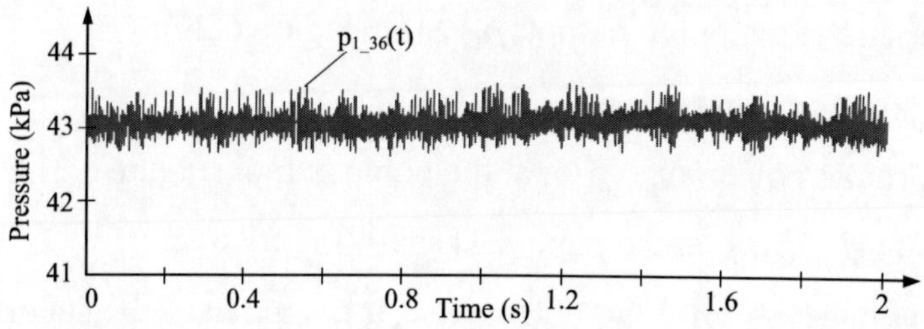

Figure 6.29. A curve representing the variation of pressure $p_{1\_36}(t)$ at the pump output at PU supply voltage frequency 36.4 Hz.

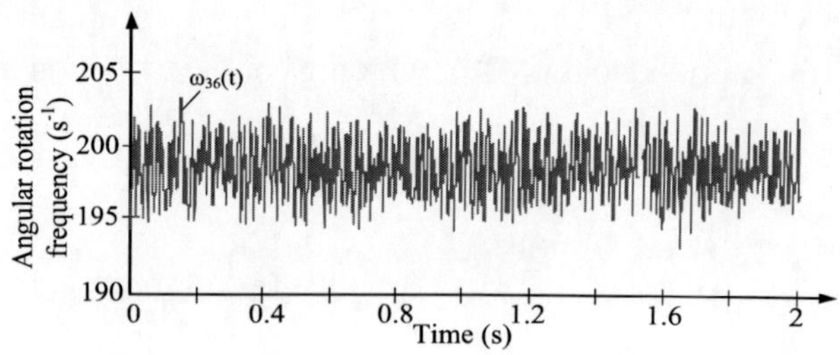

Figure 6.30. A curve representing the variation of the angular rotation frequency $\omega_{36}(t)$ of the pump ED at PU supply voltage frequency 36.4 Hz.

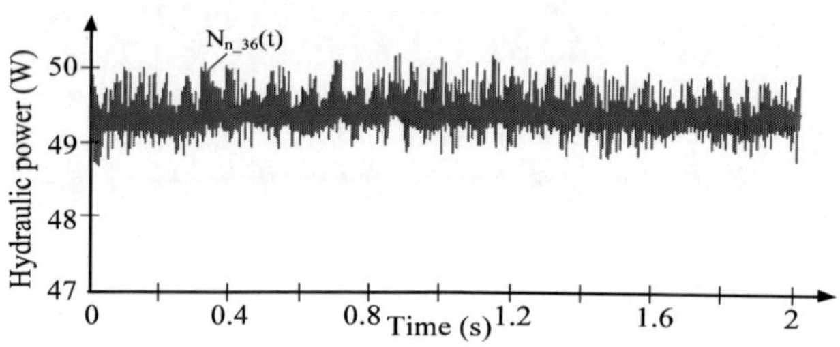

Figure 6.31. A curve representing the variation of hydraulic power $N_{n\_36}(t)$ at the pump output at PU supply voltage frequency 36.4 Hz.

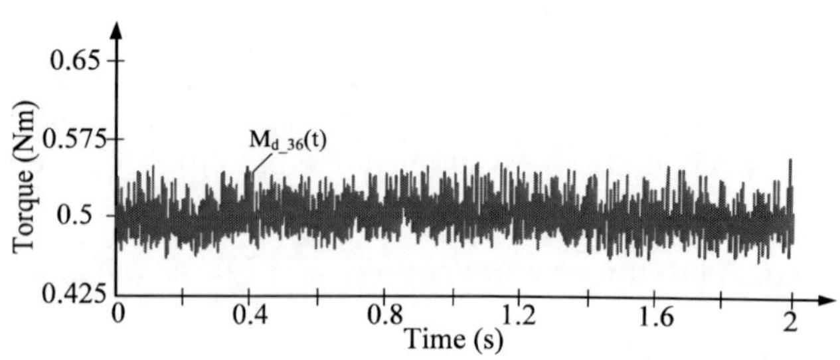

Figure 6.32. A curve representing moment $M_{d\_36}(t)$ on the pump ED shaft at PU supply voltage frequency 36.4 Hz.

The comparative analysis of the curves (Figure 6.17–6.26, Figure 6.28–6.35), obtained at the liquid flow passing through the Venturi tube, proved the absence of the cavitation oscillations in PC technological, mechanical and energy parameters at the pump ED angular rotation frequency ω_{36} corresponding to PU supply voltage frequency 36.4 Hz.

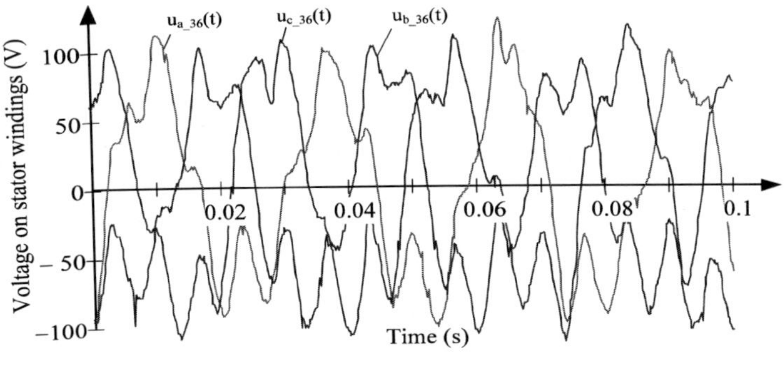

Figure 6.33. Curves representing the variation of voltage $u_{a\_36}(t)$, $u_{b\_36}(t)$, $u_{c\_36}(t)$ applied to the stator windings of the pump induction electric motor at PU supply voltage frequency 36.4 Hz.

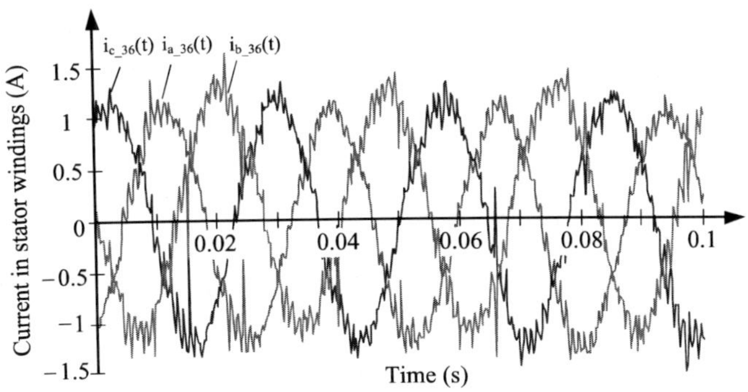

Figure 6.34. Curves representing the variation of current strength $i_{a\_36}(t)$, $i_{b\_36}(t)$, $i_{c\_36}(t)$ in the pump ED stator windings at PU supply voltage frequency 36.4 Hz.

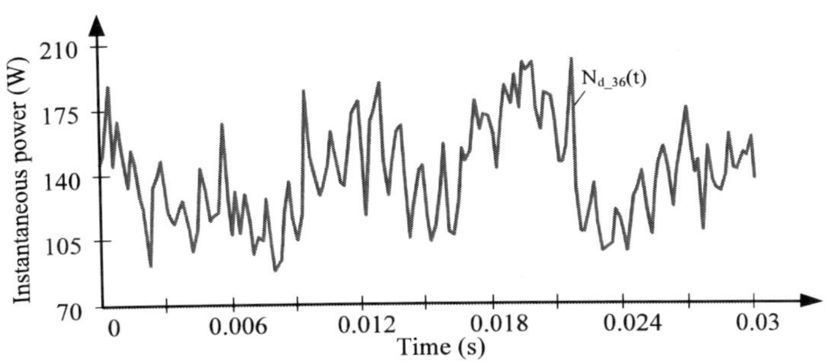

Figure 6.35. Curves representing instantaneous power $N_{d\_36}(t)$, supplied to the pump ED stator at PU supply voltage frequency 36.4 Hz.

The above said is confirmed by the decrease of the value of the coefficient of higher harmonics in the pressure signal (Table 6.4):

$$k_{gar} = \sqrt{\sum_{m=1}^{M} p_{j\_m}^2} \Big/ p_{j\_0}, \tag{6.2}$$

where p_{j_0}, p_{j_m} – the amplitude value of the zero and *m*-th harmonics of the pressure signal, respectively; m – the harmonic number; M – the harmonics quantity; j – the index characterizing the point of the variation of pressure in PC.

The obtained experimental value of the pump relative critical rotation frequency providing the non-cavitation operation of the hydraulic system:

$$\nu_{kr\_eksp} = \frac{\omega_{36}}{\omega_n} = \frac{199.1}{303.5} = 0.656, \tag{6.3}$$

where ω_{36} – the value of the pump ED angular rotation frequency at $f_{s\_4} = 36.4$ Hz; ω_n – the rated value of the pump ED angular rotation frequency.

The error of the results obtained by calculation ν_{kr} and experiment ν_{kr_eksp} made 2.23%.

To confirm the adequacy of the proposed method of the analysis of the operation modes of PC with a variable-frequency ED based on the analytical representation of the cavitation channel in electrohydraulic complex models we obtained pressure variation curves shown in Figure 6.36.

The analysis (Figure 6.36) revealed that experimental $p_{1\_e}(t)$ and model $p_{1\_m}(t)$ curves contain sections of sharp growth and decrease of the pressure caused by the formation and subsequent collapse of cavitation caverns. The frequency and the amplitude of the pressure oscillations at the turbomechanism output are $f_{kav} = 1.25$ Hz and $\Delta p_1 = 4$ kPa, respectively. In this case, the coefficient of the determination of the obtained results is $R^2 = 0.82$, which proves the adequacy of the proposed method.

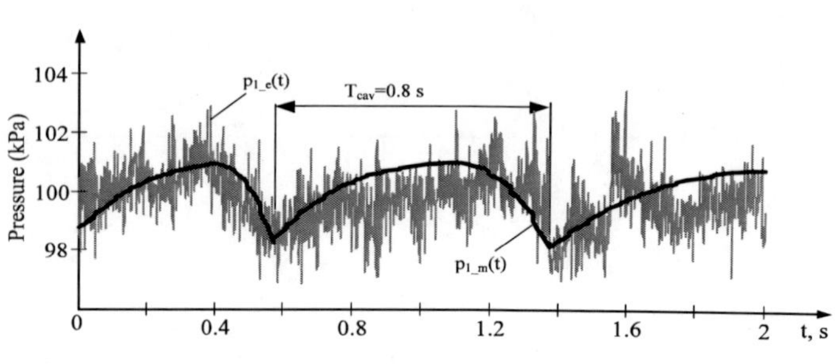

Figure 6.36. Curves representing the variation of the pressure at the pump output in the presence of the cavitation processes in the Venturi tube.

6.4. The Harmonic Analysis of the Power in a Pumping Complex with Periodic Nonlinear Processes

For the generation and the frequency analysis of the signals of power (electric, hydraulic) CMB structure provides a software module including: a block for processing the input signals directly coming from the sensors; a block for the generation of the power signals obtained as a product of the corresponding input signals; an energy analyzer allowing the presentation of the currents and voltages curves (heads and discharges) by a generalized Fourier series, and also the approximation of the signals of power (electric, hydraulic) in the form of a sum of the orthogonal harmonic components.

Closing of the valve in the pump suction pipe provides the growth of the pumpdown area located at the input of the impeller and causing the occurrence of the cavitation pulsations in the hydrosystem. To analyze the cavitation processes, influence on PC energy characteristics a constant value of hydraulic power $p_{con}(t)$ is maintained in the consumer network by means of a controlled electric drive.

Figures 6.37–6.39 contain curves of the variation of the head, discharge and the hydraulic power at the consumer (at the check point of the pipeline network) at the different degree n_{kavcur} of the development of the cavitation processes in the hydrosystem where $n_{kavcur} = V_{cur}/V_0$, $V_0 = 0.0008$ m^3 – the initial value of the volume of the cavitation cavity [126–128].

The analysis revealed that at a higher degree of the valve closing at PU input the cavitation fluctuation period T_{kav} grows, which is caused by the increase of the volume of the cavitation caverns moving along the pipeline. So, in PC with $n_{kav5} = 31.3$ the head fluctuations are more explicit than in the system with $n_{kav1} = 3.7$, in this case, the frequency of cavitation self-fluctuations is $f_{kav5} = 0.23$ Hz and $f_{kav5} = 0.5$ Hz, respectively.

Figures 6.37, 6.39 contain the frequency spectra of the head and the hydraulic power signals confirming the occurrence of the low-frequency components typical of the considered cavitation processes in the pipeline.

The variation of the hydrosystem load condition resulting from the development of the cavitation self-fluctuations influences not only PC hydraulic characteristics but also causes the variation of current $i(t)$ and electric power $p_{el}(t)$ supplied to the motor stator windings. To compensate for the cavitation power losses and maintain the required value of hydraulic power $p_{con}(t)$ at the consumer the supply voltage frequency f_{zad} at FC output is changed by means of a variable-frequency ED. So, at $n_{kav1} = 3.7$, $f_{zad} = 44.4$ Hz; at $n_{kav5} = 31.3$, $f_{zad} = 50$ Hz.

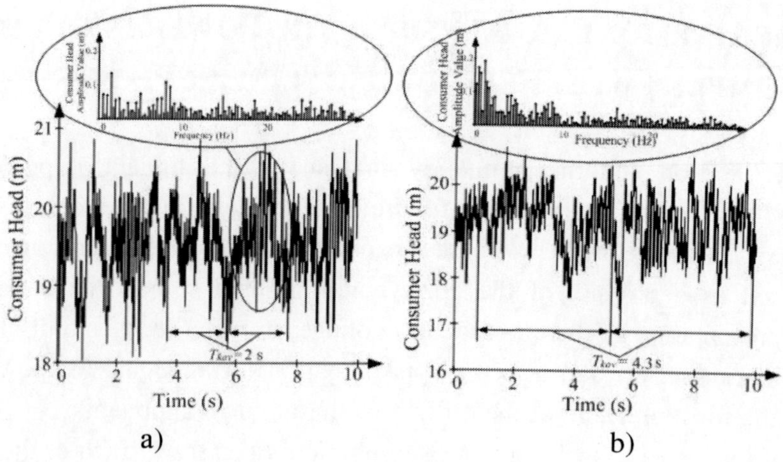

Figure 6.37. The curves of the head variation in the pipeline network in the system with n_{kav1} (a) and n_{kav5} (b).

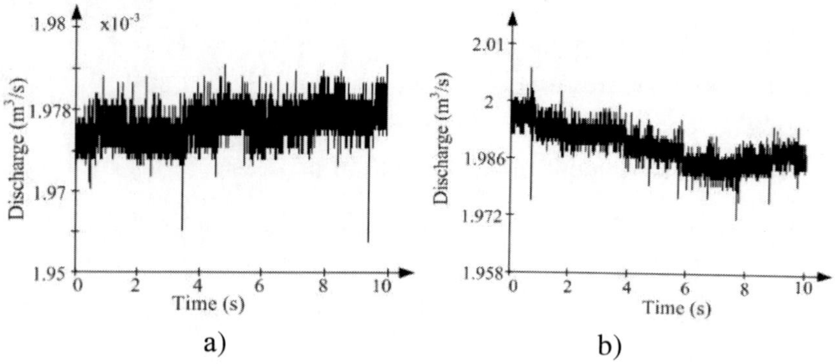

Figure 6.38. The curves of the discharge variation in the pipeline network in the system with n_{kav1} (a) and n_{kav5} (b).

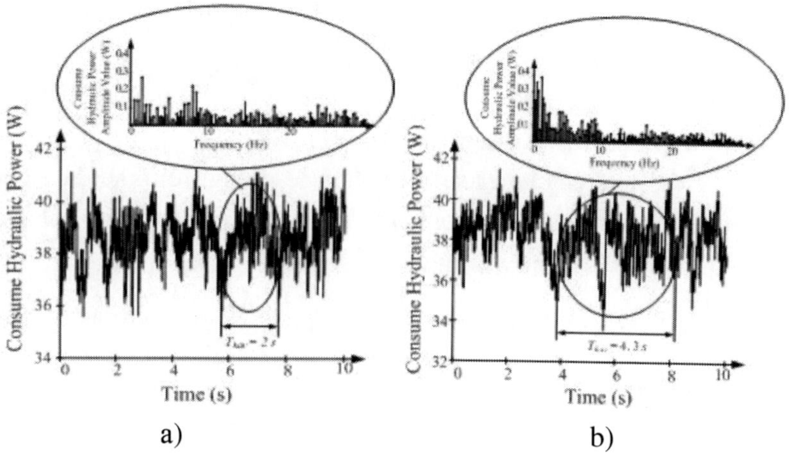

Figure 6.39. The curves of the hydraulic power variation in the pipeline network in the system with n_{kav1} (a) and n_{kav5} (b).

Figures 6.40–6.42 contain curves of phase voltages $u_A(t)$, $u_B(t)$, $u_C(t)$, currents $i_A(t)$, $i_B(t)$, $i_C(t)$ and powers $p_A(t)$, $p_B(t)$, $p_C(t)$ for different degrees of the cavitation development.

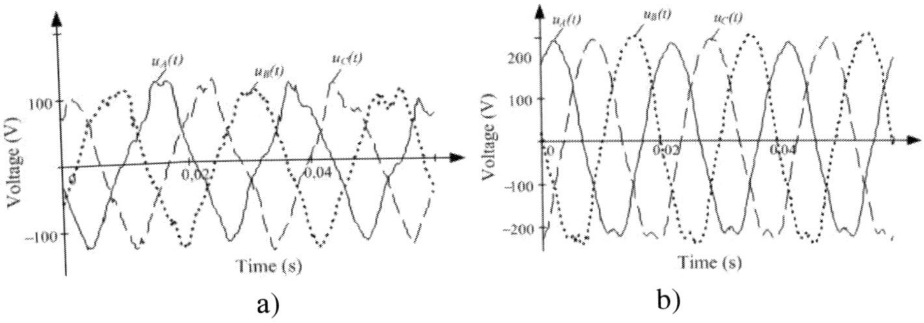

Figure 6.40. The curves of the variation of the phase voltage signals in the system with n_{kav1} (a) and n_{kav5} (b).

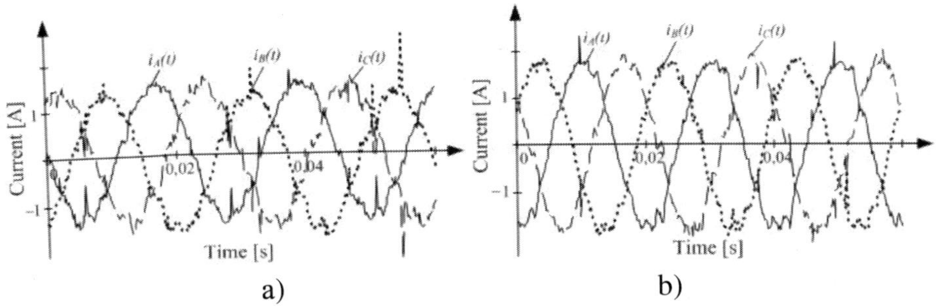

Figure 6.41. The curves of the variation of the phase current signals in the system with n_{kav1} (a) and n_{kav5} (b).

Using expression:

$$p_{el}(t) = p_{el\,A}(t) + p_{el\,B}(t) + p_{el\,C}(t) = \\ = u_A(t)i_A(t) + u_B(t)i_B(t) + u_C(t)i_C(t) \quad (6.4)$$

the curves of the variation of total electric power $p_{el}(t)$ supplied to IM stator are obtained (Figures 6.43, 6.44).

The analysis revealed that the cavitation processes in PC technological contour are characterized by the increase of the variable component in the signals of the electric and hydraulic powers (Figures 6.39, 6.44), availability of the low-frequency harmonic components in the spectra [150, 151].

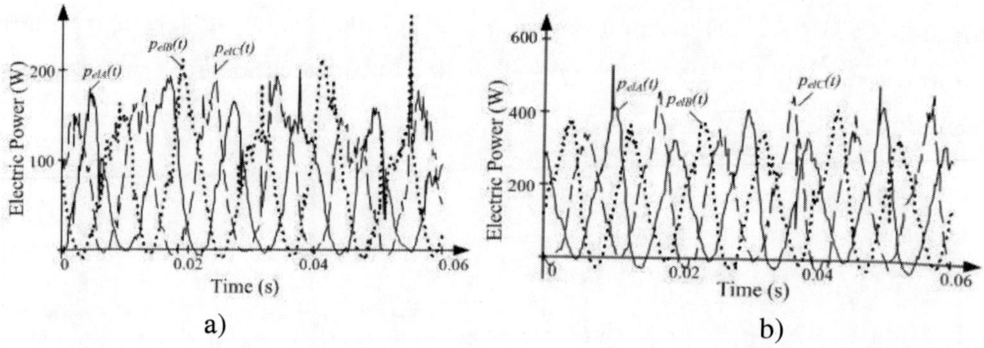

Figure 6.42. The curves of the variation of the phase electric power signals in the system with n_{kav1} (a) and n_{kav5} (b).

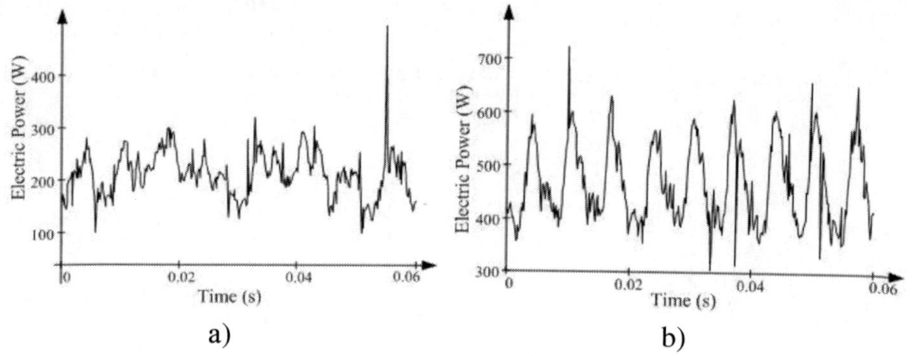

Figure 6.43. The curves of the variation of the total electric power signal in the system with n_{kav1} (a) and n_{kav5} (b) at the interval of time multiple of triple period of supply frequency.

With the aim of the automation of the frequency analysis of the electrical (current, voltage), technological (head, discharge) as well as the power (electrical and hydraulic power) signals an energy analyzer is provided in CMB structure [152]. The analyzer operation is demonstrated by the windows of the faceplates of the automated module (Figures 6.45–6.48).

By way of example, Figures 6.47, 6.48 demonstrate the curves of the variation of the phase voltages, currents and phase powers as well as the curves of the head, discharge and hydraulic power at the output of the pump at the development of the cavitation processes in the system with frequency $f_{zad} = 0.3$ Hz. The spectral analysis of signals $H_p(t)$ and $p_{hp}(t)$ revealed that the cavitation fluctuations are accompanied by the occurrence of the variable components of the head and power, namely their low-frequency component (Figure 6.48). It results in the growth of the mean-root-square value of the power and, correspondingly, in the decrease of PC energy effectiveness.

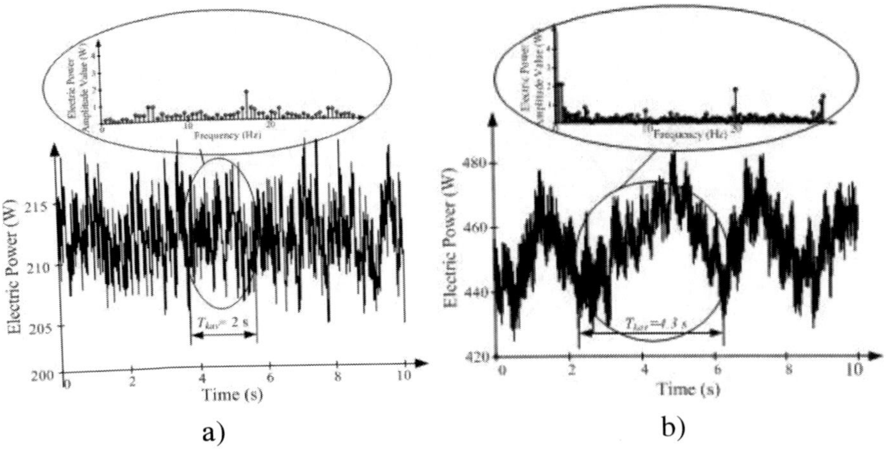

Figure 6.44. The curves of the variation of the total electric power in the system with n_{kav1} (a) and n_{kav5} (b) at the whole interval of measurement.

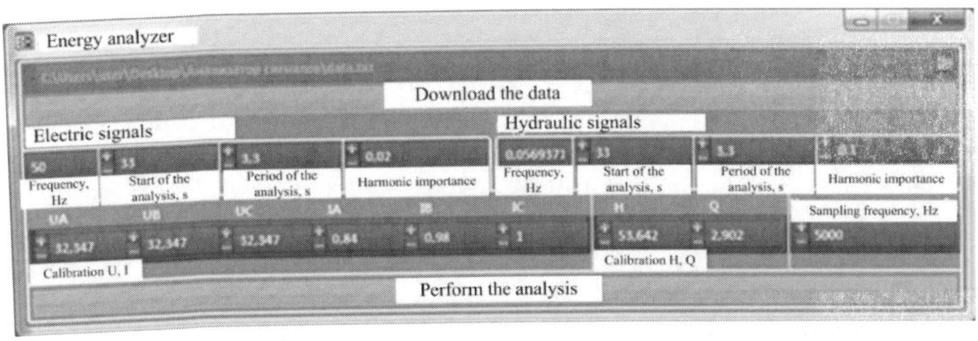

Figure 6.45. The window for the assignment of the energy analyzer initial parameters.

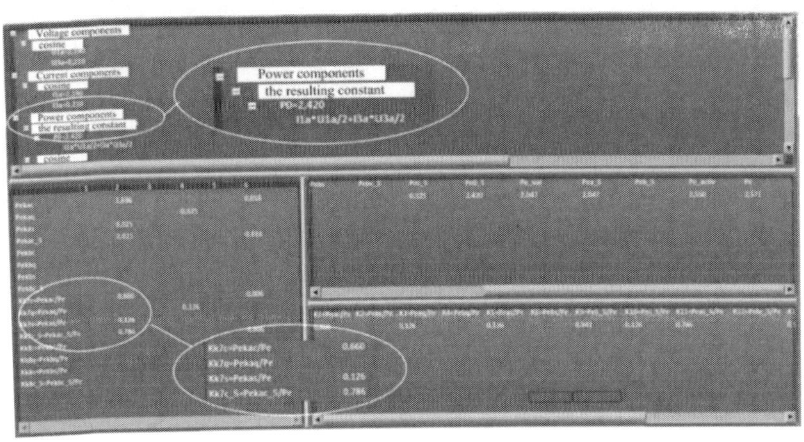

Figure 6.46. The window for the reflection of the orthogonal components of the voltage, current and power as well as their mean-root-square assessments.

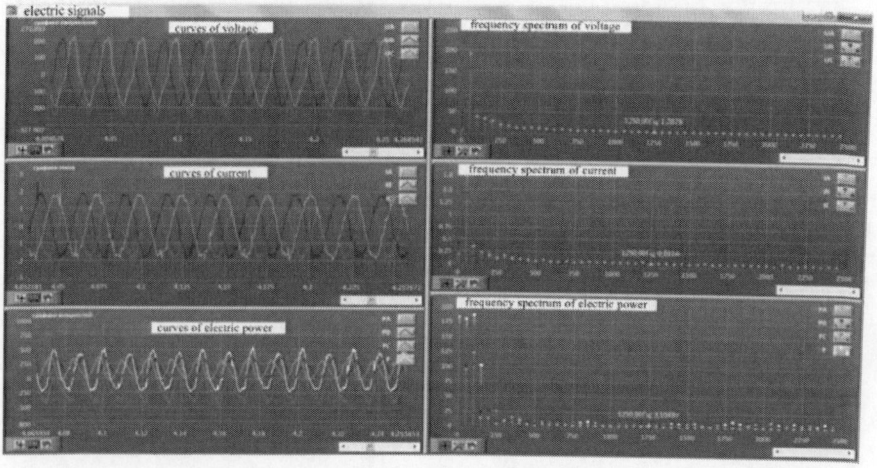

Figure 6.47. The window of the analyzer of the electric signals spectrum.

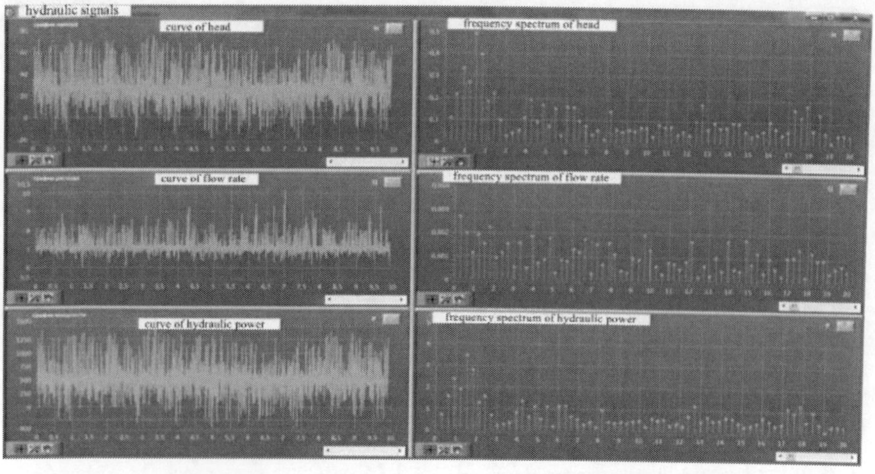

Figure 6.48. The window of the analyzer of the hydraulic signals spectrum.

The efficient power obtained as a mean-root-square value of the power time function (electrical, mechanical, hydraulic) at the i-th element of the energy conversion serves as a measure of the assessment of the quality of the energy processes in PC [153, 154]:

$$P_{ei} = \sqrt{1/T \int_0^T p_i^2(t)dt} = \\ = \sqrt{\left(\sum_{k=1}^{K} P_{k0i}\right)^2 + \left(\sum_{k=1}^{K} P_{kai}\right)^2 \Big/ 2 + \left(\sum_{k=1}^{K} P_{kbi}\right)^2 \Big/ 2}. \tag{6.5}$$

One of the indices characterizing a continuous process of the energy conversion between the power supply and the consumer, as proposed in [151–154], is represented by a coefficient of throughput of PC energy channel, determined as the relation of the effective powers at the output and the input of the system:

$$k_{cap} = P_{e\,con} / P_{e\,el} \, , \tag{6.6}$$

where P_{ehcon}, P_{eel} – the effective power of the consumer and PC power supply, respectively.

Essentially, the mentioned expression is analogous to the efficiency, characterizing the energy condition of the system in the steady operation mode. However, unlike the efficiency, k_{cap} makes it possible to take into account the availability of the power variable components in PC power channel, caused by the introduction of various energy storage elements, the occurrence of nonlinear processes in energy-conversion contour that reflect the special feature of the operation of technological mechanism electric drive.

Under the real conditions the throughput coefficient $k_{cap} < 1.0$. As the number of the elements in the energy channel increases, effective power P_{ehcon} decreases, which, consequently, results in the deterioration of the PC energy channel throughput and more significant reduction of k_{cap}.

A coefficient of PC energy controllability is proposed for the comparative assessment of the processes of energy conversion in various operating modes of PC [153, 154]:

$$k_{pc} = P_{e\,i\,id} / P_{e\,i\,r} \, , \tag{6.7}$$

where $P_{e\,i\,id}$, $P_{e\,i\,r}$ – the effective power in an ideal (in the absence of most typical nonlinearities in PC) and real (at occurrence of nonlinear hydrodynamic processes) systems at the i -th element of energy conversion, respectively.

It should be noted that it is possible to assess the energy controllability of both separate elements based on P_{ehp}, P_{ehnetj}, $P_{e\,con}$ under corresponding operating modes, and the whole system using for the analysis the values of effective input electric power in ideal $P_{eel\,id}$ and real $P_{eel\,r}$ PC.

The proposed approach to the assessment of PC energy efficiency enables the energy processes analysis when they change in real time with the retention of the complete information about the initial signals generation power. It also allows taking into account the manifestation of the specific properties of the electromechanical equipment, the occurrence of the power additional variable components loading the system power channel.

Figure 6.49 contains the curves of the variation of k_{pc} and k_{cap} from degree n_{kav} of the development of the cavitation processes in the hydrosystem.

Table 6.5 contains the values of the effective powers and the indices of the energy effectiveness of PC.

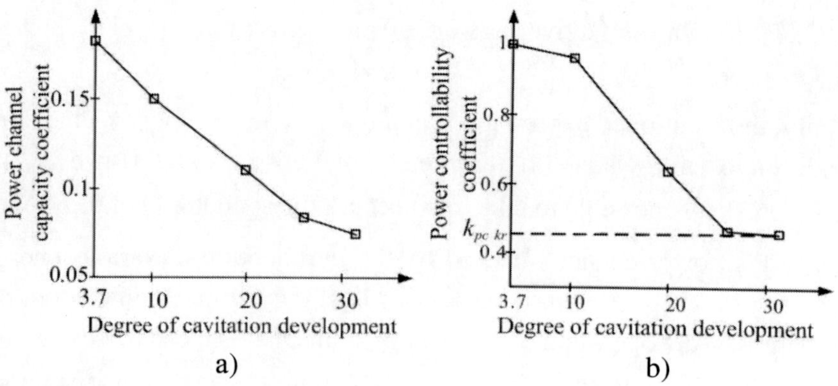

Figure 6.49. The dependences on the degree of the cavitation development of the coefficients of throughput (a) and energy controllability (b).

Table 6.5. The effective power and the indices of the energy effectiveness of PC

Parameter	$n_{kav1}=3.7$	$n_{kav2}=7$	$n_{kav3}=19.7$	$n_{kav4}=26$	$n_{kav5}=31.3$
P_{eel}	216.937	225.976	341.063	467.647	478.89
P_{ep}	51.82	51.433	51.468	51.893	51.457
P_{econ}	38.402	38.41	38.419	38.461	38.469
k_{cap}	0.182	0.17	0.113	0.084	0.08
k_{pc}	1.0	0.96	0.636	0.464	0.452

It follows from the analysis of the obtained curves that at a low value $n_{kav1}=3.7$, corresponding to the mode in which cavitation phenomena are practically absent, the coefficient of throughput of the PC power channel takes the maximum value $k_{cap1}=0.182$. The intensive development of the cavitation self-fluctuations at degree $n_{kav5}=31.3$ is accompanied by the reduction of coefficient k_{cap} by 2.2 times ($k_{cap5}=0.082$), which is caused by the growth of the variable components in the signals of the head and hydraulic power. Aiming at the compensation for the deviation of the consumer hydraulic power from the assigned value, the rotation frequency of the pump electric motor is increased, which results in the growth of the electric power supplied to ED of the pump unit (Table 6.5).

A somewhat different situation is observed during the analysis of the energy controllability of PC with the nonlinear processes in the pipeline. It is obtained that at small degrees of n_{kav1}, n_{kav2} the system is practically insensitive to the occurrence of the cavitation processes in the pipeline – the coefficient of energy controllability $k_{pc} = 1.0 - 0.96$; at high degrees of the cavitation development n_{kav3}, n_{kav4} coefficient k_{pc} begins to drop sharply and at n_{kav5} it reaches the critical value $k_{pc\,kr} = 0.45$. This is a mode at which a cavitation breakdown in the pump unit operation is possible, which requires taking relevant measures to eliminate probable emergency modes of the operation of the equipment.

Thus, the cavitation processes in the pipeline network result in the redistribution of the power losses in PC power channel, are accompanied by the increase of the power variable component, which reduces the effectiveness of the used equipment.

CONCLUSION

1. The analysis of the dynamic processes in the pumping complex revealed that at the stopcock shutting according to the linear law at the supply voltage frequency equal to f_0 = 40 Hz, one can observe the highest increase of the pressure in the pipeline network, that is 2.5 times as high as the pump rated head. In its turn, the stopcock shutting by the linear law at the supply voltage frequency equal to f_0 = 5 Hz results in 12-fold decrease of the pressure growth, but the shutting time is 6 times as big.
2. The research revealed that to reduce the dynamic loads in the hydronetwork at the pipeline valves shutting it is necessary to form a nonlinear law of its electric drive control, whereat, on the section with the relative degree of the regulating stopcock opening $0.2 \leq \beta \leq 1$, the supply voltage frequency is 40 Hz, and at $\beta \leq 0.2$ – 5 Hz. In this case, the pressure in the pipeline network is by 18% higher than the pressure in the steady mode of the pipeline complex operation before the regulating stopcock shutting.
3. The obtained experimental characteristics of the pumping complex at different laws of the shutoff-control pipeline valves shutting confirmed the feasibility of the proposed approach to the decrease of the dynamic loads in the pumping complex. The formation of the nonlinear law of the regulating stopcock shutting with the supply voltage frequency in the range of 5-20 Hz at $\beta \leq 0.2$ allows regulating the increased pressure in the pipeline network within the limits of (18-94)% of the steady value. In this case, the time of the pipeline valves control reduces by 3.5-5.5 times in comparison with the longest shutting time t_{sh} = 34 s, corresponding to the supply voltage frequency of 5 Hz.

4. The analysis of the dynamic processes in the pumping plant during the development of cavitation processes in the hydronetwork revealed that the pressure oscillations at the pump output are characterized by the frequency of 1.2 Hz and the amplitude of 4 kPa. The mentioned cavitation processes result in the variation of the technological and energy parameters of the pumping plant. For the analyzed pumping plant, we have determined the range of the admissible values of the pump rotation frequency (0-0.656) from the rated value, determining the non-cavitation mode of the pumping complex operation. When the pump rotation frequency reduces, the cavitation oscillations essentially decrease and at ω=199.1 s^{-1} there is no cavitation in the hydrosystem.
5. Using a physical model of the pumping complex, we confirmed the adequacy of the analytical representation of the cavitation channel in the models of an electrohydraulic complex with a regulated electric drive. We have obtained that the coefficient of the determination at the comparative analysis of the experimental and model curves representing the pressure variation at the pump output is $R^2 = 0.82$.
6. The use of the indices based on the determination of the power root-mean-square values both at the separate elements and at the system overall is justified for the assessment of the processes of the energy conversion in the pumping complex. Therefore, we have obtained that at the appearance of the nonlinear hydrodynamic processes in the technological outline of the pumping complex the power variable component grows. It results in the increase of the value of the effective electric power supplied to the electric motor stator and, consequently, in the decrease of the throughput coefficient of the power channel of the electrohydraulic system. At the intense development of the cavitation processes in the hydrosystem the energy controllability of the pumping complex significantly deteriorates. If this index approaches the critical value, it is necessary to form corresponding control signals on the electric drive of the pump unit or the pipeline valves to provide the cavitation protection of the hydraulic equipment.
7. To automate the procedure of the power frequency analysis we have developed an energy analyzer making it possible to research the energy processes in the pumping complex at their variation in the real time. With that full information of the initial signals forming the power are preservated.
8. The obtained result deserves especial attention in the problems of the creation of the systems of the parameter control and the protection of the pumping complexes in the emergency operation modes (surge, cavitation), enables monitoring of the state of the pumping and the pipeline equipment aiming at the timely replacement or repair of the elements of the technological complex.

CONCLUSION

1. It has been shown that in the process of pumping complexes operation, the alteration of the operation modes, there occur wave phenomena of different nature (surges, pressure pulsations, cavitation self-oscillations, etc.) in the hydrosystem, which is caused by a number of factors: the emergency switch-off of the electric power of the operating pump units; the actuation of the check valves, the quick shut or opening of the protection or shut-off and control valves; the decrease of the pressure in the pressure header, the change of the temperature of the pumped liquid in the hydrosystem etc. That is why the modern pumping complexes are characterized by the low indices of reliability and high breakdown rate. It has been determined that the conventional methods and equipment for the decrease of the dynamic loads do not enable the considerable reduction of the amplitude of the head fluctuations in the pipeline network and do not consider the problem of the control of the pipeline valves electric drive aiming at minimization of the dynamic loads in the hydrosystem. They do not take into account the real hydraulic characteristics of the valves. They do not provide the control at sudden disappearance of the electricity supply to the pumping complex and base on the increase of the shutting time at the linear law of shutting or the formation of the stepped trajectory of the shut-off and control valves shutting via the use of the uncontrolled electric drive. The conventional methods of the cavitation reduction based on the pipeline protection devices, putting additives into the liquid, injection of liquid or air into the cavitation area, are directed to the decrease of the destructive action of the cavitation processes and involve considerable material expenditure. The performed analysis revealed that the decrease of the rotation frequency of the pump impeller results in the reduction of the maximal value of the amplitude of the cavitation fluctuations of the pressure in the electrohydraulic system.

2. It has been proved that to decrease the dynamic loads in the pipeline network of the pumping complex it is necessary to form the nonlinear law of the stopcock shutting taking into account the hydraulic resistance coefficient factual dependence on the relative degree of opening along the whole interval of its operating device movement. We have developed the block diagram of the model of the pumping complex with the stopcock controlled electric drive taking into account the propagation of the surge waves occurring at the stopcock control in the hydronetwork.
3. We have proposed the cavitation channel mathematical model taking into account the presence of cavitation processes in the hydrosystem at the change of the temperature mode of the liquid motion and the technological parameters of the pumping plant. It has been demonstrated that the cavitation processes time constant depends on the degree of the development of cavitation, the velocity mode of the motion and the aeration of the liquid flow. We have worked out the block diagrams of the models of pumping complexes with different operation characteristics of the pump and pipeline equipment, with the possibility of the operation of the pump units equipped with a controlled electric drive for the pipeline network of a complex configuration with installed shut-off and control valves. We have proved that the cavitation processes in the pumping complex are accompanied by the considerable release of a volume of steam-gas mixture whose power is (23-25)% of the power of the liquid flow, which confirms the reasonability of the use of the energy of the cavitation caverns.
4. We have substantiated the structure of the electromechanical system of the reduction of the dynamic loads in the pipeline network based on the variable-frequency electric drive of the stopcock with the use of the reserve electricity supply from the uninterrupted power supply or an active dissipator of the hydraulic flow energy. This makes it possible to eliminate the inadmissible increases of pressure in the pipeline, actively use the energy of the liquid reverse flow, to increase of reliability of the operation of the completely pumping complex. We have developed the algorithm of the operation of the system at abrupt disappearance of the electricity supply to the pumping complex, which allows the search and the formation of such a law of control of the stopcock electric drive whereat the head in the pipeline network changes within the admissible limits. We have specified the choice of the reduction rate of the gear of the variable-frequency electric drive of the stopcock taking into account its nonlinear hydraulic and mechanical characteristics, the condition of overload by the current of the frequency converter invertor whereat the minimization of the increase of the head in the hydronetwork and the time of the valves control is provided.
5. We have worked out the method for the determination of the limits of the non-cavitation operation of the pumping complex with a variable-frequency electric

drive. This method based on the analysis of the joint modes of the pumping plant operation for the pipeline network and allowing the determination of the range of the admissible values of the variation of the pump electric drive rotation frequency at the regulation of the discharge within the required limits.

6. We have substantiated the structure of the electromechanical device of the control of the cavitation processes in the pumping complex based on the automatic air bleeder, a receiver and an air turbine with an electric generator on one shaft. We have provided the algorithm of the calculation of the parameters and the choice of the elements of the system taking into account the operation characteristics of the used equipment, the mode of the pumping complex operation and the degree of the development of the cavitation processes. It has been demonstrated that at the change of the discharge within the limits of (0.188-0.12) m^3/s, the power of the recuperation of the steam-gas mixture energy reaches 25% of the hydraulic power of the liquid flow.
7. We have demonstrated that the representation of the pumping complex by an equivalent circuit based on the method of the electrohydraulic analogy allows the determination of the hydraulic power and the power losses at any section of the hydrosystem at the development of nonlinear processes in the pipeline network.
8. We have determined that the nature of the energy processes taking place in the pumping complex consists in periodic steady fluctuations of the power at the analyzed time interval. Due to this, the synthesis of the time function of the power (electric, hydraulic) by the harmonic series formed based on the orthogonal components of the voltage and currents, heads and discharges is admissible.
9. We have proposed the energy model of the pumping complex in the form of the equation of balance of the harmonic components of the hydraulic power between the power supply and the elements of the hydrosystem at the periodic alteration of the process variables. Such an approach is the basis for the solution to the problems of the determination of the parameters of the electrohydraulic complexes. Their technological schemes take into account the group operation of the pump units with variable-frequency electric drives and uncontrolled electric drives of the pumps, the occurrence of periodic processes both in the turbomechanisms and at the sections of the pipeline network. The considered method makes it possible to single out the power variable component determined by the nonlinear periodic processes in the hydronetwork, to obtain the frequency characteristic of the energy processes in the hydrosystem.
10. We have substantiated the structure of the integral criterion of the quality of the closed-loop electromechanical system of the reduction of the dynamic loads in the pumping complex with a variable-frequency electric drive of the stopcock, which includes two components: the deviation by the head in the pipeline network and the time of the stopcock control. It has been determined that the weight coefficient at the

head component varies in time and the weight coefficient at the time component, which is of constant value, is chosen depending on the operation mode of the pumping complex and the priority of the problems solved during the valve control.

11. We have determined the optimal laws of control of the stopcock electric drive for the closed-loop electromechanical system of the reduction of the dynamic loads in the emergency and operation modes of the pumping complex, where the minimal dynamic loads in the hydronetwork and the quickest control of the pipeline valves are provided. It has been shown that in the operation mode of the pumping complex at the optimal law of control by the 5.2 kW variable-frequency electric drive the head increase in the pipeline decreases by 2.6 time. In addition, the time of the valve control is half as big as in the case when a non-controlled electric drive of the same power is used. In the emergency mode at the abrupt switch-off of the pump unit power supply and the optimal control of the stopcock electric drive the increase of the head in the hydronetwork and the time of the valves shutting reduce by 1.36 and 3.4 times, respectively.
12. We have substantiated the structure of the criterion of the quality of the system of automatic control of the pumping complex electric drive taking into account the cavitation processes, including three components: the deviation by the head at the check point of the network; the power losses caused by the presence of the cavitation processes; the power of the recuperation of the steam-gas mixture energy. We have developed the system of the control of the pumping complex electric drive. This complex makes it possible to improve the energy efficiency of the pumping plant via the decrease of the power losses caused by the cavitation processes in the pipeline network, at the stabilization of the head at the consumer and the simultaneous recuperation of the steam-gas mixture energy.
13. We have confirmed by the physical model the adequacy of the proposed mathematical models of the pumping complex, taking into account the transient processes occurring at the control of the stopcock electric drive and at the development of the cavitation in the hydrosystem.

References

[1] Leznov B. S. *Energy saving and adjustable drive in pumping and blowing units*. – M.: Energoatomizdat, 2006. – 360 p.

[2] Onishchenko G. B. *The electric drive of turbomechanisms*. – M.: Energiia, 1972. – 240 p.

[3] Popov V. M. *Drainage systems*. – M.: Nedra, 1990. – 110 p.

[4] Grishko A. P. *Stationary machines. Volume 2: Mine drainage, fan and pneumatic installations*: a textbook for universities. – M.: MGGU, 2007. – 586 p.

[5] Bizenkov A. P., Abramov V. N. *Stationary machines. Calculation of drainage installations of mining enterprises: a tutorial*. – Kemerovo: SU of Kuzbass State University, 2003. – 143 p.

[6] Karelin V. Ya., Minaev A. V. *Pumps and pumping stations: a textbook for universities*. – M.: ID BASTET, 2010. – 448 p.

[7] Voronov Yu. V., Alekseev E. V., Salomeev V. P., Pugachev E. A. *Wastewater*. – M.: INFRA-M, 2007. – 415 p.

[8] Soshnikov E. V., Tchaikovsky G. P. *Water pumping stations*: Tutorial. – Khabarovsk: Publishing House DVGUPS, 2005. – 115 p.

[9] Zhurba M. G., Sokolov L. I., Govorova Zh. M. *Water supply. Designing systems and structures:* a textbook – M.: ASV, 2003. – 228 p.

[10] Yuriev A. S. *Handbook on the calculation of hydraulic and ventilation systems*. – St. Petersburg: Peace and Family, 2001. – 1154 p.

[11] Bolshakov V. A., Popov V. N. *Hydraulics: general course: a textbook for engineer-techn. spec. universities*. – K.: Higher school, 1989. – 214 p.

[12] Idelchik I. E. *Handbook of hydraulic resistance*. – M.: Mechanical Engineering, 1992. – 672 p.

[13] Kurganov A. M., Fedorov N. F. *Hydraulic calculations of water supply and drainage systems*. Reference book. – L.: Stroyizdat, 1986. – 440 p.

[14] Lyamayev B. F., Nebolsin G. P., Nelyubov V. A. *Stationary and transient processes in complex hydrosystems. Methods of computer calculation.* – L.: Mechanical Engineering, 1978. – 192 p.

[15] Vishnevsky K. P. *Transient processes in pressure water supply systems.* – M.: Agropromizdat, 1986. – 135 p.

[16] Gurevich D. F. *Pipe fittings: reference manual.* – M.: Stereotype, 2014. – 368 p.

[17] Goshko A. I. *Pipeline valves of special-purpose designation: Choice. Exploitation. Repair.* – M.: Mechanical Engineering, 2003, book 1. – 432 p.

[18] Gurevich D. F. *Calculation and design of pipeline valves: Calculation of pipeline fittings.* – M.: Publishing house LCI, 2008. – 480 p.

[19] Inyakin V. N., Shevchenko V. V. Minimization of the consequences of man-made accidents in the housing and communal services // *Industrial Economy.* – 2010. – No. 2. – P. 41–48.

[20] Karelin V. Ya., Novoderezhkin R. A. *Pump stations with centrifugal pumps.* – M.: Stroiizdat, 1983. – 221 p.

[21] Horst-Joachim L., Bernd K. *KSB Know-How. Gidroudar*, no. 1. – St. Petersburg.: KSB, 2012. – 32 p.

[22] Arsenievsky N. N., Pospelov B. B. *Transients of large pumping stations.* – M.: Energia, 1980. – 111 p.

[23] Arsenievsky N. N. *Transitional hydromechanical processes in pressure water conduits and aggregates of hydroelectric power plants, hydroelectric power stations and large pumping stations*: Abstract of thesis. dis. for the scientific degree of Doct. tech. Sciences: spec. 05.14.10 Hydroelectric power stations and hydropower plants, Moscow, 1992. – 40 p.

[24] Vishnevsky K. P. *Modeling of transient processes in complex pressure systems with pumping stations*: the author's abstract. dis. for the scientific degree of Doct. tech. Sciences: spec. 05.23.16 Hydraulics and Engineering Hydrology. – Moscow, 1988. – 37 p.

[25] Goncharenko D. F., Aleinikova A. I., Pabolkov V. V., Starkova O. V. Sanitation of water mains using large-diameter glass-fiber pipes // *Collection of scientific works UkrDaST.* – 2014. – No. 143. – P. 136–145.

[26] *Plastic pipes: a comparative characteristic of polymeric materials:* [Electronic resource]. Access mode: http://plastinfo.ru/information/articles/310/.

[27] Aliekseeva Yu. O., Korenkova T. V., Borisov D. I. Analysis of operational characteristics of technological equipment of pumping stations // *Electromechanical and Energy Systems. Modeling and Optimization Methods. Conference proceedings of the 8th International conference of students and young researchers in Kremenchuk on April 8–9, 2010.* – Kremenchuk, 2010. – P. 292 – 293.

[28] Kuzenkov E. V. Piping systems made of high-strength cast iron for water, heat supply and wastewater networks in Russian regions // *Housing and communal services: Journal of the head and chief accountant*. – 2003. – No. 4. – Part 1. – P. 71–73.

[29] Mambretti S. *Water Hammer Simulations*. – WIT Press, 2014. – 200 p.

[30] Záruba J. *Water Hammer in Pipe-Line Systems* // Elsevier. – 1993. – Iss. 43. – 359 p.

[31] Round G. F. *Incompressible Flow Turbomachines: Design, Selection, Applications, and Theory*. – Butterworth-Heinemann, 2004. – 341 p.

[32] Kravets A. M., Korenkova T. V. Analysis of methods for determining the hydro impact parameters in pumping stations // *Transactions of Kremenchuk Mykhaylo Ostrogradskiy State Polytechnic University*. – Kremenchuk: KSPU, 2008. – Vol. 3/2008 (50), part 1. – P. 103–107.

[33] Kravets A. M., Korenkova T. V. On the task of control dynamic loads in hydrotransport systems under emergency conditions // *Transactions of Kremenchuk State Polytechnic University*, Kremenchuk: KSPU, 2006. – Vol. 3 (39). – P. 23–27.

[34] Beglyarov D. S. *Scientific substantiation of methods for calculating transient processes in pressure water supply systems with pumping stations*: Abstract of thesis. dis. for the scientific degree of Doct. tech. Sciences: spec. 05.23.16 Hydraulics and Engineering Hydrology. – Moscow, 2007. – 59 p.

[35] Koshkin N. M. *Substantiation and development of a set of technical means that increase the efficiency of closed irrigation networks and sprinklers*: the author's abstract. dis. for the scientific degree of Doct. tech. Sciences: spec. 06.01.02 Agricultural Reclamation. – Novocherkassk, 2000. – 49 p.

[36] Ganiev R. F., Nizamov Kh. N. *Stabilization of pressure fluctuations in pipeline systems of power plants*. – M.: Publish house: MSTU them. N.E. Bauman, 1993. – 183 p.

[37] Sulimenko V. V. *Development of methods for preventing emergencies in urban infrastructure systems*: Abstract of Cand. dis. for the scientific degree of Cand. tech. Sciences: spec. 05.26.02 Safety in emergency situations (in power engineering). – Moscow: Rus. University of Friendship of Peoples, 2007. – 20 p.

[38] Mamedov G. A., Zeynalova N. S. Control of large-sized electric gate valves in case of emergencies // *Problems of power engineering*. – Azerbaijan, 2007. – No. 4. – P. 45–49.

[39] Shatalov E. S. *Prevention of hydraulic impact in pressure pipelines of water supply systems by means of controlling the electric drive of a pump unit*: author's abstract. dis for obtaining a scientific degree of candidate. tech Sciences: special 05.23.04 Water supply, sewerage. – Kharkiv: Kharkiv State Technical University of Construction and Architecture, 2000. – 20 p.

[40] Tigist T. T. *Development and research of control systems for electric drives of main pipeline pumps*: Abstract of thesis. dis. for the scientific degree of Cand. tech. Sciences: spec. 05.09.03 Electrotechnical complexes and systems. – St. Petersburg: St. Petersburg state electrotechnical Univ., 2003. – 15 p.

[41] Adoevsky A. V. *Modeling the operation of oil pipelines equipped with systems for smoothing pressure waves*: author's abstract. dis. for the scientific degree of Cand. tech. Sciences: spec. 25.00.19 Construction and operation of oil and gas pipelines, bases and storage facilities. – Moscow: Rus. state. Univ. oil and gas them. I.M. Gubkina, 2011. – 23 p.

[42] Prokofiev A. B. *Development of a method for the complex analysis of the dynamics and strength of pipeline systems with absorbers of vibrations of a working fluid*: the author's abstract. dis. for the scientific degree of Doct. tech. Sciences: spec. 01.02.06 Dynamics, Strength of Machines, Devices and Apparatus. – Samara: Samara State Aerospace University. acad. S.P. Koroleva, 2008. – 39 p.

[43] Primenko V. N. *Development of means for preventing accidents in pipeline transport and researching the effectiveness of their work*: Abstract of thesis. dis. for the scientific degree of Cand. tech. Sciences: spec. 11.00.11 Protection of the environment and rational use of natural resources. – Moscow: Rus. univ. friendship of peoples, 1996. – 16 p.

[44] Nizamova G. Kh. *Development of technical means of protection of pipelines with piston pumps and compressors from emergency situations:* The author's abstract. dis. for the scientific degree of Cand. tech. Sciences: spec. 05.14.16 Technical means and methods of environmental protection (in engineering and power engineering). – Moscow: Rus. University of Friendship of Peoples, 2000. – 16 p.

[45] Manlin Zhu, Xiaohong Zhang, Yanhe Zhang, Tao Wang. Study on Water Hammer Prevention in Pumping Water Supply Systems by Multi-valves // *Proceedings of ICHIT Cheju Island.* – 2006. – P. 342–346.

[46] Bazargan-Lari M. R., Kerachian R., Afshar H., Bashi-Azghadi N. Developing optimal valve closure rule curve for real-time pressure control in pipes // *Journal of Mechanical Science and Technology*. – 2013. – Iss. 27 (1). – P. 215–225.

[47] Jing-Yang Yu, Zheng-Yi Wu, Yi-Xing Yuan, Ming Zhao, Chen-Guang Wu. Optimal valve closure for long-distance water transmission // *Sustain. Environ. Res.* – 2010. – Iss. 20 (5). – P. 287–291.

[48] Wu Juan, Ziming Kou. Study on the dynamic characteristic of the controllable gate valve in the mine automatic draining system // *Proceedings of ICAEE Beijing.* – 2010. – P. 35–39.

[49] Roy J. K., Roy P. K., Basak P. Water hammer protection in water supply system: A new approach with practical implementation // *Proceedings of ICCIA Kolkata.* – 2011. – P. 1–6.

[50] Dudlik A., Schlüter S., Hoyer N., Prasser H. M. Pressure surges – experimental investigations and calculations with software codes // *Proceedings of IAHR Stuttgart*. – 2003. – P. 12–23.

[51] Bergant A., Simpson A. R. Water Hammer Analysis of Pumping Systems for Control of Water in Underground Mines // *Proceedings of Mine Water Congress Ljubljana*. – 1991. – P. 9–20.

[52] Nicolet A C., Kaelbel T., Alligne S., Ruchonnet N., Allenbach P., Bergant A., Avellan F. Simulation of Water Hammer Induced Column Separation through Electrical Analogy // *Proceedings of Cavitation and Dynamic Problems in Hydraulic Machinery and Systems*. – 2011. – P. 1–9.

[53] Karakulov A. S. *Microprocessor control of an asynchronous electric drive of a gate valve:* the author's abstract. dis. for the scientific degree of Cand. tech. Sciences: spec. 05.09.03 Electrotechnical complexes and systems. – Tomsk: TPU, 2005. – 16 p.

[54] Langraf S. V. *Asynchronous torque electric drive with vector control for simulating the efforts of the stop valves of main oil pipelines*: the author's abstract. dis. for the scientific degree of Cand. tech. Sciences: spec. 05.09.03 Electrotechnical complexes and systems. – Tomsk: TPU, 2007. – 20 p.

[55] Fox D. A. *Hydraulic analysis of unsteady flow in pipelines*. – M.: Energoizdat, 1981. – 325 p.

[56] Rozhdestvensky V. V. *Cavitation*. – L.: Sudostroyeniye, 1977. – 247 p.

[57] Neppiras E. Acoustic cavitation // *Phys. Repts*. – 1980. – Vol. 61, no 3. – P. 159–251.

[58] Flannigan D., Suslick K. Molecular and atomic emission during single- bubble cavitation in concentrated sulfuric acid acoustics // *Research Letters Online*. – 2005. –Vol. 6, no 3. – P. 157–161.

[59] Storey B. D., Szeri A. J. A reduced model of cavitation physics for use in sonochemistry // *Proc. R. Soc. London, Ser. A*. – 2001. – Vol. 457. – P. 1685–1700.

[60] Brennen C. *Cavitation and bubble dynamics*. – Oxford University Press, 1995. – 254 p.

[61] Pirsol I. *Cavitation*. – Moscow: Mir, 1975. – 94 p.

[62] Pilipenko V. V., Zadontsev V. A., Natanzon M. S. *Cavitation self-oscillations and dynamics of hydrosystems*. – M.: Mashinostroieniie, 1997. – 352 p.

[63] Prandtl L. *Hydroaeromechanics*. – Izhevsk: Scientific and Publishing Center Regular and chaotic dynamics. – 2000. – 320 p.

[64] Karelin V. Ya. *Cavitation phenomena in centrifugal and axial pumps*. – M.: Mashinostroieniie, 1975. – 248 p.

[65] Karelin V. Ya. Wearing of vane pumps. – M.: *Mashinostroieniie*, 1983. – 168 p.

[66] Xavier E., Egusquiza E., Farthat M. et al. Detection of cavitation in hydraulic turbines // *Mechanical Systems and Signal Processing*. – 2006. – Vol. 20. – P. 983–1007.

[67] Predin A., Gregorc B., Biluš I. Cavitation swirl at the entrance of centrifugal pump // *Journal of Energy Technology*. – 2009. – Vol. 2, Iss. 3. – P. 67–82.

[68] Gregorc B., Predin A. Particle influence on cavitaiton development at blade profile // *Journal of Energy Technology*. – 2010. – Vol. 3, Iss. 2. – P. 67–82.

[69] Taleyarkhan R. P., West C. D., Cho J. S. et al. Evidence for Nuclear Emissions During Acoustic Cavitation // *Science*. – 2002. – Vol. 295. – P. 1868–1872.

[70] Altshul A. D., Zhivotovsky L. S., Ivanov L. P. *Hydraulics and aerodynamics*. – Moscow: Stroiizdat, 1987. – 414 p.

[71] Arzumanov Z. S. *Cavitation in local hydraulic resistances*. – M.: Energia, 1978. – 303 p.

[72] *Physics and technology of powerful ultrasound. Powerful ultrasonic fields* / [Nugolnykh KA. etc.]; Ed. L. D. Rosenberg. – Moscow: Nauka, 1968. – 268 p.

[73] Biluš I., Predin A. Experimental analysis of thermo-dynamical surge at water pump inlet // *Journal of energy technology*. – 2010. – Vol. 3, Iss. 3. – P. 67–74.

[74] Gregorc B., Hriberšek M., Predin A. Experimental and numerical analysis of the impact of particles on the development of cavitation flow around a hydrofoil // *Journal of energy technology*. – 2010. – Vol. 4. – PP. 47–64.

[75] Leonov G. V., Savina E. I. Modeling of cavitation, initiated by ultrasonic oscillators // *International Workshop and Tutorials on Electron Devices and Materials, EDM – Proceedings*. – *2006*. – № 1694105. – P. 259–262.

[76] Volokh K. Y. Softening hyperelasticity for modeling material failure: analysis of cavitation in hydrostatic tension // *Int. J. Solids Struct*. – 2007. – Vol. 44. – P. 5043–5055.

[77] Kiesbauer J. Predicting cavitation damage in control // *Hydrocarbon Processing Magazine*. – 2001. – Art. 80, № 6. – P. 89–100.

[78] Kozelkov V. P., Efimochkin A. F. Experimental study of cavitation self-oscillations in the hydrotransport system // *Proceedings of the meeting of the Academy of Sciences of the Ukrainian SSR Cavitation self-oscillations in pumping systems*. – 1976. –Vol. 1. – P. 71–80.

[79] Zadontsev V. A. Determination of the dependence of the volume of cavitation caverns on pressure and flow at the pump inlet from experimental data // *Proceedings of the meeting of the Academy of Sciences of the Ukrainian SSR Cavitation self-oscillations in pumping systems*. – 1976. – Vol. 1. – P. 126–130.

[80] Pilipenko V. V., Zadontsev V. A. About one mechanism of self-oscillations in a hydraulic system with Venturi tube // *Materials of the meeting of the Academy of Sciences of the Ukrainian SSR Cavitation self-oscillations in pumping systems*. – 1976. – Vol. 2. – P. 93–103.

[81] *Studies on advanced cavitation:* [collection. sci. works] – Novosibirsk: Institute of Thermal Physics, Siberian Branch of the USSR Academy of Sciences, 1976. – 144 p.

[82] Didenkulov I. N. Investigation of cavitation in a plane acoustic // *Proceedings of the Nizhny Novgorod State Technical University. R.E. Alekseeva*. – 2011. – No. 4 (91). – P. 38–46.

[83] Kravets A. M., Korenkova T. V. Features of dynamic modes in hydrotransport systems with controlled pipeline fittings // *Transactions of Kremenchuk State Polytechnic University,* Kremenchuk: KSPU, 2007. – Vol. 3 (44). – part 1. – P. 162–167.

[84] Kravets A. M., Korenkova T. V. Mathematical model of hydrotransport complex with controlled pipeline valves // Electromechanical and saving up systems. *Quarterly scientific production journal.* – Kremenchuk: KSU, 2009. – Iss. 2/2009 (6). – P. 29–32.

[85] Zagirnyak M., Korenkov E., Kravets O., Korenkova T. Virtual constructor used for research of hydrotransport system operation conditions // *Proceedings of International IEEE Conference EUROCON 2013.* – Zagreb, Croatia, 2013. – P. 1193–1200. IEEE Catalog Number CFP13EUR-USB, ISBN 978-1-4673-2231-7, Library of Congress Number 2009900519.

[86] Kravets A. M., Korenkova T. V. Dynamic processes in the hydrotransport complex with controlled pipeline fittings // *Electromechanical and Energy Systems. Modeling and Optimization Methods. Conference proceedings of the 8th International conference of students and young researchers in Kremenchuk on April 2–4, 2009.* – Kremenchuk: KrNU, 2009. – P. 87–90.

[87] Kravets A. M. Determination of the moment of resistance of the adjustable stop valve // *Transactions of Kremenchuk Mykhaylo Ostrogradskiy State Polytechnic University,* Kremenchuk: KSPU, 2009. – Vol. 3/2009 (56). – Part 1. – P. 20–22.

[88] Gurevich D. F., Kosykh S. I. *Pipe fittings with automatic control.* Reference book. – L.: Mechanical Engineering, 1982. – 320 p.

[89] Kravets A. M., Korenkova T. V. Features of frequency control of the electric drive of pipeline valves in the hydrotransport complex // *Transactions of Kremenchuk Mykhaylo Ostrogradskiy state university,* Kremenchuk: KSU, 2010. – Vol. 3/2010 (62). – part 1. – P. 18–21.

[90] Zagirnyak M. V., Kravets A. M., Korenkova T. V. Determination of the power of the regulated electric drive of pipeline valves // *Proceedings of high schools. Electromechanics.* – 2012. – no. 3. – P. 24–28.

[91] Kravets A. M., Korenkova T. V. Recommendations for determining the power and selection of the electric actuator of a regulated pipeline armature // *Problems of energy saving in electrical engineering systems. Science, education and practice: a scientific publication.* – Kremenchuk: KNU, 2011. – Iss. 1/2011 (1). – P. 88–89.

[92] Kravets A. M., Korenkova T. V. Requirements for a closed electromechanical system for reducing dynamic loads in a hydrotransport complex // *Transactions of Kremenchuk Mykhaylo Ostrogradskiy State University,* Kremenchuk: KSU, 2010. – Vol. 3/2010 (62). – part 2. – P. 26–28.

[93] Zagirnyak M. V., Kravets A. M., Korenkova T. V. Control of dynamic loads in the hydrotransport complex // *Proceedings of high schools. Electromechanics.* – 2011. – no. 4. – P. 124–128.

[94] Kravets A. M., Korenkova T. V. Method for reducing dynamic loads in a pumping complex with a controlled hydraulic valve // *Electromechanical and saving up systems. Quarterly scientific production journal.* – Kremenchuk: KSU, 2010. – Iss. 3/2010 (11). – P. 41–45.

[95] Korenkova T. V., Mykhailychenko D. A., Kravets A. M. Technological and power capabilities of a pump station with active parameter control // *Engineering in agricultural production, branch engineering, automation: scientific works of Kirovograd Technical University*. – 2004. – Iss. 15. – P. 78–85.

[96] Korenkova T. V., Perekrest A. L., Mykhailychenko D. A., Kravets A. M. Characteristics of the hydro turbine in the system of active control of the performance of pumping stations // *Transactions of Kremenchuk Mykhaylo Ostrogradskiy State Polytechnic University.* – Kremenchuk: KSPU, 2004. – Iss. 2 (25). – P. 21–27.

[97] Korenkova T. V., Kravets A. M. Increase of controllability of electromechanical systems of hydrotransport systems // *Mining electromechanics and automatics: a collection of sciences works of NMU*. – Dnipropetrovsk: NMU, 2005. – Iss. 75. – P. 115–121.

[98] Zagirnyak M. V., Kravets O. M., Korenkova T. V. Dynamic loads control in a pump complex with adjustable pipeline valves // *ISTET 2013: International Symposiumon Theoretical Electrical Engineering*: 24th – 26th June 2013: Pilsen, Czech Republic. – P. IV-21 – IV-22.

[99] Popovich M. G., Gavrilyuk V. A. *Theory of the Electric Drive* – K.: Higher School, 1993. – 494 p.

[100] Usoltsev A. A. *Frequency control of asynchronous motors:* [textbook]. – St. Petersburg: St. Petersburg State University ITMO, 2006. – 94 p.

[101] Korenkova T. V., Serdiuk O. O. Model of a hydrotransport system with a gas generation loop in the flowing part // *Transactions of Kremenchuk State Polytechnic University.* – Kremenchuk: KSPU, 2005. – Iss. 4/2005 (33). – P. 153–157.

[102] Serdiuk O. O., Korenkova T. V. Features of dynamic processes in hydrotransport systems with allowance for cavitation phenomena // *Transactions of Kremenchuk State Polytechnic University.* – Kremenchuk: KSPU, 2006. – Iss. 4/2006 (39). – P. 35–41.

[103] Korenkova T. V., Serdiuk O. O. Features of models of water pumping complexes // *Transactions of Kremenchuk State Polytechnic University.* – Kremenchuk: KSPU, 2007. – Iss. 3/2007 (44). – part 1. – P. 143–147.

[104] Serdiuk O. O., Korenkova T. V. Dynamic processes in hydrotransport complex taking into account cavitational events, *Bulletin of National Technical University "Kharkov Polytechnic Institute"*. – Kharkov: NTU «KPI», 2010. – no. 28. – P. 375–382.

[105] Serdiuk O. O., Korenkova T. V. Analysis of the time constants of the pipeline system with allowance for cavitation processes // *Electromechanical and Energy Systems. Modeling and Optimization Methods. Conference proceedings of the 7th International Conference of Students and Young Researchers in Kremenchuk on April 8–9, 2010*. – Kremenchuk: KSU, 2010. – P. 305–306.

[106] Serdiuk O. O., Korenkova T. V. Substantiation of ways to increase controllability of cavitation processes in the hydrotransport complex // *Transactions of Kremenchuk State Polytechnic University*. – Kremenchuk: KSPU, 2008. – Iss. 4/2008 (51). – part 1. – P. 163–168.

[107] Serdiuk O. O. Influence of the parameters of the pipeline system on the boundaries of the occurrence of cavitation phenomena // *Electromechanical and saving up systems. Quarterly scientific production journal*. – Kremenchuk: KSPU, 2009. – Iss. 1/2009 (5). – P. 43–46.

[108] Serdiuk O. O., Kozlenko I. O., Korenkova T. V. Influence of regulating regimes of pumping complex on parameters of cavitation processes // *Transactions of Kremenchuk Mykhaylo Ostrogradskiy state university,* Kremenchuk: KSU, 2010. – Vol. 4/2010 (63). – part 2. – P. 55–58.

[109] Karassik I. J., Messina J. P. Cooper P., Heald C. C. *Pump handbook*. – McGRAW-HILL, 2001. – 600 p.

[110] Firago B. I. *Calculations on the drive industrial machinery: study guide*. – Minsk: Tekhnoperspektiva, 2012 – 639 p.

[111] Ferreira T. E., Fong J. C. and De Almeida A. T. Ecoanalysis of variablespeed drives for flow regulation in pumping systems // *Industrial Electronics, IEEE Transactions*, 2011. – Vol. 58, no. 6. – P. 2117–2125.

[112] Marcelo Martins Stopa, Cardoso Filho, Carlos B. Martinez Incipient Detection of Cavitation Phenomenon in Centrifugal Pumps // *IEEE Transactions on Industry Applications*, 2014. – Iss. 50 (1). – P. 120–126.

[113] Kostishin V. S. *Simulation operation modes of centrifugal pumps based on electrohydraulic analogy: monograph*. – Ivano-Frankivsk, 2000. – 163 p.

[114] Karassik J. J., McGuire T. *Centrifugal Pumps*. – Springer Science & Business Media, 2012. – 780 p.

[115] Adam Adamkowski, Mariusz Lewandowski Improved numerical modeling of hydraulic transients in pipelines with column separarion // *3-rd International Meeting of the Workgroup on Cavitation and Dynamic Problems in Hydraulic Machinery and Systems, October, 14-16, 2009, Czech Republic*. – P. 419–432.

[116] Bergant A., Simpson A. R., Tijsseling A. S. Water Hammer with column separation: A historical review // *Journal of Fluids and Structures*, 2006. – Vol. 22 (2006). – P. 135–171.

[117] Dudik A., Schoenfeld S. B., Schluter S., Fahlenkamp H. Prevention of water hammer and cavitational hammer in pipeline system // *Chemical Engineering and Technology*. – Vol. 25 (2002). – No. 9. – P. 888–890.

[118] Kazuaki Inaba, Joseph E. Shepherd Dynamics of cavitating and flexural waves in fluid-filled tubes subject to axial impact // *Proceedings of the ASME 2010 Pressure Vessels & Piping Division,* K-PVP Conference PVP 2010 July 18-22, 2010. – Bellevue, Washington, USA, 2010. – P. 1–10.

[119] Keramat A., Ahmadi A., Majd A. Transient cavitating pipe flow due to a pump failure // *3rd International Meeting of the Workgroup on Cavitation and Dynamic Problems in Hydraulic Machinery and Systems, October, 14-16, 2009, Czech Republic.* – P. 475–485.

[120] Mohamed S., Ghidaoui Ming Zhao, Duncan A. Mclnnis, David H. Axworthy A Review of Water Hammer Theory and Practice //*Transaction of the ASME,* January 2005. – Vol. 58. – P. 49–76.

[121] Zagirnyak M., Kovalchuk V., Korenkova T., Berdai A. Estimation of dynamic loads in an electrohydraulic complex at different laws of supply voltage frequency variation // *Przeglad Elektrotechniczny*, 2017. – R. 93. – № 1. – P. 237–240. ISSN 0033-2097.

[122] Bronshteyn I. N. and Semendiaev K. A. *Handbook of mathematics for engineers and students of technical colleges.* – Moscow: Nauka, 1986. – 545 p.

[123] Zagirnyak M., Korenkova T., Kovalchuk V. Estimation of electromechanical systems power controllability according to instantaneous power components // *Proceedings of 2014 IEEE International Conference on Intelligent energy and power systems (IEPS).* – Kyiv, Ukraine, 2014. – P. 266–272. IEEE Catalog Number: CFP1405X–PRT. ISBN: 978-1-4799-2265-9.

[124] Zagirnyak M., Kovalchuk V., Korenkova T. Power Model of an Electrohydraulic Complex with Periodic Nonlinear Processes in the Pipeline Network. – *Proceedings 2015 International Conference on Electrical Drives and Power Electronics (EDPE), – Tatranska Lomnica, The High Tatras, Slovakia, 2015.* – P. 345–352. IEEE Catalog Number CFP15EDQ-USB. ISBN 978-1-4673-9661-5.

[125] Zagirnyak M., Rod'kin D., Korenkova T. Enhancement of instantaneous power method in the problems of estimation of electromechanical complexes power controllability // *Przeglad Elektrotechniczny*, 2011. – no 12b. – P. 208–212.

[126] Zagirnyak M., Korenkova T., Kovalchuk V. Harmonic Analysis of Power in an Electrohydraulic Complex with Nonlinear Processes in the Pipeline System // *Proceedings of 2016 IEEE 2nd International Conference on Intelligent energy and power systems (IEPS).* – Kyiv, Ukraine, 2016. – P. 143–148. IEEE Catalog Number: CFP1605X–PRT ISBN: 978-1-5090-1767-6.

[127] Zagirnyak M., Serdiuk A., Korenkova T. Limits of noncavitation operation of an electrohydraulic complex with a variable-frequency electric drive // *Przeglad Elektrotechniczny*, 2015. – R. 91. – № 1. –P. 212–216. ISSN 0033-2097.

[128] Zagirnyak M., Serdiuk O., Korenkova T. Pump complex electric drive control system taking into account cavitation processes in the pipeline // *Przeglad Elektrotechniczny*, 2015. – R. 91. – № 12. P. 224–227. ISSN 0033-2097.

[129] Zagirnyak M., Kovalchuk V., Korenkova T. Identification of electrohydraulic complex parameters using instantaneous power component // *Przeglad Elektrotechniczny* (Electrical review), 2013. – № 2b. – P. 286–289.

[130] Zagirnyak M., Kovalchuk V., Korenkova T. Power method of the tasks of determining electrohydraulic complex parameters // *Technical Electrodynamics*, 2016. – № 3. – P. 76–78.

[131] Wentzel E. S. *Elements of dynamic programming*. – M.: Science, 1964. – 176 p.

[132] Okulov S. M., Pestov O. A. *Dynamic programming* – M.: BINOM. Laboratory of Knowledge, 2012. – 296 p.

[133] Kravets O. M., Korenkova T. V. Optimum control law for pipeline fittings of the pump complex // *Electromechanical and Energy Systems. Modeling and Optimization Methods. Conference proceedings of the 10th International conference of students and young researchers in Kremenchuk on Mart 28–29, 2012.* – Kremenchuk: KrNU, 2012. – P. 130–131.

[134] Zagirnyak M., Kravets O., Korenkova T. The optimal control of dynamic loads in a pump complex with adjustable pipeline valves // *Naukovyi Visnyk Natsionalnoho Hirnychoho Universytetu*, 2016. – № 3 (153), C. 78-86 ISSN 2071-2227.

[135] Zagirnyak M., Alieksieieva I., Korenkova T. Energy and resource saving control system for pumping station. – *Przeglad Elektrotechniczny*, 2013. – № 2b. – P. 76–79. ISSN 0033-2097.

[136] Zagirnyak M., Manko R., Kovalchuk V., Korenkova T. Assessment of Pipeline Network Leakage Influence on Dynamic Characteristics of the Pump Complex, *Proceedings of The IEEE First Ukraine Conference on Electrical and Computer Engineering (UKRCON-2017).* – Kyiv, Ukraine, 2017. – P. 473–478. DOI: 10.1109/UKRCON.2017.8100534 Electronic ISBN: 978-1-5090-3006-4 USB ISBN: 978-1-5090-3005-7 Print on Demand(PoD) ISBN: 978-1-5090-3007-1.

[137] Pechenik M., Kiselychnyk O., Buryan S. Experimental research of interactive energy saving controller of water supply pump based on flow rate measurement // *Bulletin of the National Technical University Kharkiv Polytechnic Institute*. – Kharkiv: NTU KhPI, 2010. – No. 28. – P. 272–274.

[138] Sokolovsky O. F. Determination of the boundary of energy-saving properties of interactive control in pumping plants // *Electromechanical and energy saving systems. Thematic issue Problems of the automated electric drive. Theory and practice*. – Kremenchuk: KrNU, 2012. – Vol. 3/2012 (19). – P. 239–241.

[139] Serdiuk O. O., Korenkova T. V. Control system of cavitation processes in a hydrotransport complex // *Transactions of Kremenchuk Mykhaylo Ostrogradskiy state university,* Kremenchuk: KSU, 2010. – Vol. 4/2010 (63). – part 1. – P. 54–57.

[140] Popovich M. G., Pechenik M. V., Kovalchuk O. V., Kiselychnik A. I. Features of Extreme Electromechanical Systems of Automatic Control and the Problem of Energy Saving // *Electrical Engineering. The monthly scientific and technical journal is a collective member of the Academy of Electrical Engineering of the Russian Federation.* – 2003. – No. 3. – P. 12–17.

[141] Serdiuk O. O., Korenkova T. V. The optimal control system for the pumping complex, taking into account the cavitation processes in the pipeline // *Scientific works of the Donetsk National Technical University Electromechanics and Power Engineering.* – Donetsk: DonNTU, 2011. – Vol. 11 (186). – P. 336–343.

[142] Serdiuk O. O., Konokh I. S., Korenkova T. V., Degunkov D. V. Requirements for the construction of a control system for cavitation processes in the pumping complex // Electromechanical and Energy Systems. Modeling and Optimization Methods. *Conference proceedings of the 8th International Conference of Students and Young Researchers in Kremenchuk on April 7–8, 2011.* – Kremenchuk: KSU, 2011. – P. 139–140.

[143] Korenkova T. V., Serdiuk A. A., Perekrest A. L., Shokolov V. N. Experimental complex-simulator of a hydrotransport unit // *Transactions of Kremenchuk State University.* – Kremenchuk: KSU, 2010. – Iss. 4 (63). – P. 64–67.

[144] Kravets O. M., Korenkova T. V. Experimental system for reducing dynamic loads in a hydrotransport complex // Electromechanical and Energy Systems. Modeling and Optimization Methods. *Conference proceedings of the 9th International Conference of Students and Young Researchers in Kremenchuk on April 7–8, 2011.* – Kremenchuk: KrNU, 2011. – P. 201–202.

[145] Kravets O. M. Transient processes in the pumping complex in the control of pipeline valves // *Electromechanical and saving up systems. Quarterly scientific production journal.* – Kremenchuk: KSU, 2013. – Iss. 2/2013 (22). – P. 44–51.

[146] Kravets O. M., Biluk A. V., Neborak A. I., Korenkova T. V. Experimental characteristics of a hydrotransport unit in the control of shut-off and control valves // Electromechanical and Energy Systems. Modeling and Optimization Methods. *Conference proceedings of the 11th International Conference of Students and Young Researchers in Kremenchuk on April 9–11, 2013.* – Kremenchuk: KrNU, 2013. – P. 110–111.

[147] Korenkova T. V., Serdiuk O. O., Perekrest A. L., Shokolov V. N. Experimental complex-simulator of hydrotransport unit // *Transactions of Kremenchuk Mykhaylo Ostrogradskiy State University,* Kremenchuk: KSU, 2010. – Vol. 4/2010 (63). – part 2. – P. 64–67.

[148] Serdiuk O. O., Korenkova T. V. Algorithms for studying cavitation processes in a pumping complex // Electromechanical and saving up systems. *Quarterly scientific production journal*. – Kremenchuk: KSU, 2010. – Iss. 1/2010 (9). – P. 58–62.

[149] Serdiuk O. O., Korenkova T. V. Characteristics of the physical model of the pumping complex taking into account cavitation processes // Electromechanical and saving up systems. *Quarterly scientific production journal*. – Kremenchuk: KSU, 2012. – Iss. 4/2012 (20). – P. 57–62.

[150] Zagirnyak M. V., Korenkova, T. V. Power estimation of electromechanical systems controllability // *Proceedings of XIX International Conference on Electrical Machines, ICEM 2010*. – Rome, Italy, 2010. – Paper RF-009458. IEEE Catalog Number CFP1090B-CDR, ISBN 978-1-4244-4175-4, Library of Congress Number 2009901651.

[151] Korenkova T. Indices of the processes of energy conversion in an electric hydraulic complex // *Technical Electrodynamics*, 2014. – no 5. – P. 128–130.

[152] Zagirnyak M., Kovalchuk V., Korenkova T. The automation of the procedure of the electrohydraulic complex power harmonic analysis // *Przeglad Elektrotechniczny*, 2018. – №1. – P. 1–4. ISSN 0033-2097 R.94 NR 1/2018. DOI: 10.15199/48.2018.01.01

[153] Zagirnyak M., Rodkin D., Korenkova T. Estimation of energy conversion processes in an electromechanical complex with the use of instantaneous power method // *16th International Power Electronics and Motion Control Conference and Exposition, PEMC*, Antalya, Turkey, 2014. – P. 319–326. ISBN 978-1-4799-2062-4, 2014

[154] Zagirnyak M. V., Rod'kin D. I., Korenkova T. V. Estimation of carrying capacity of power channel according to instantaneous power indexes in of electromechanics problems // *Technical Electrodynamics,* 2012. – no. 3. – P. 81–82.

APPENDICES

APPENDIX A

The Characteristic of Water- and Heat-Supply Pumping Complexes

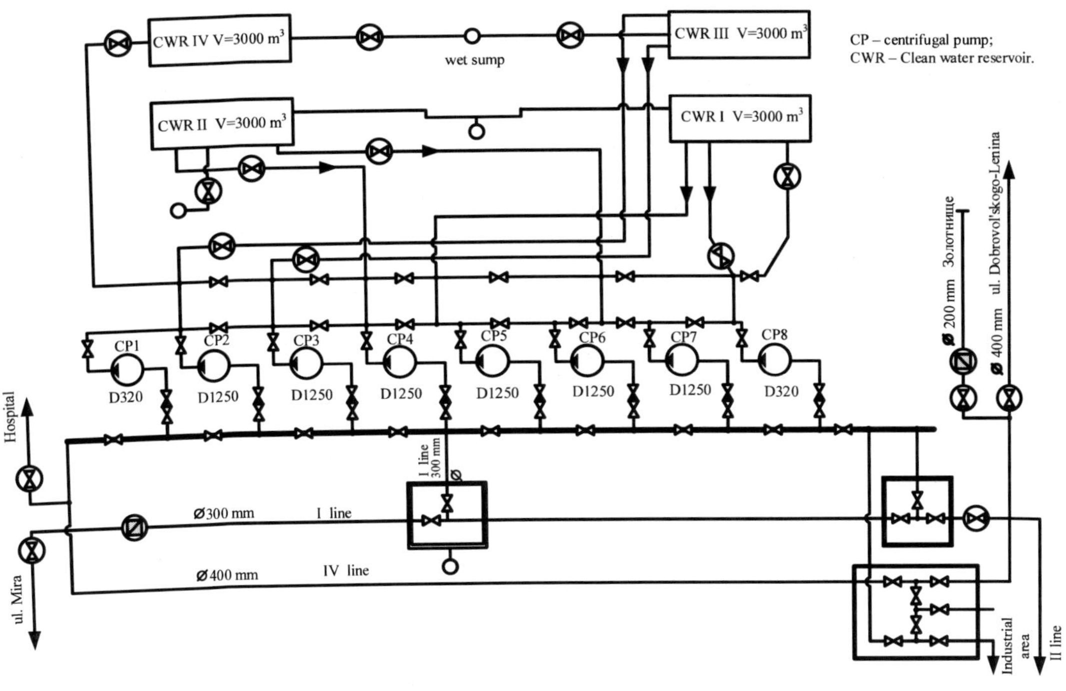

Figure A1. The flowsheet of the pumping stationof the water supply system.

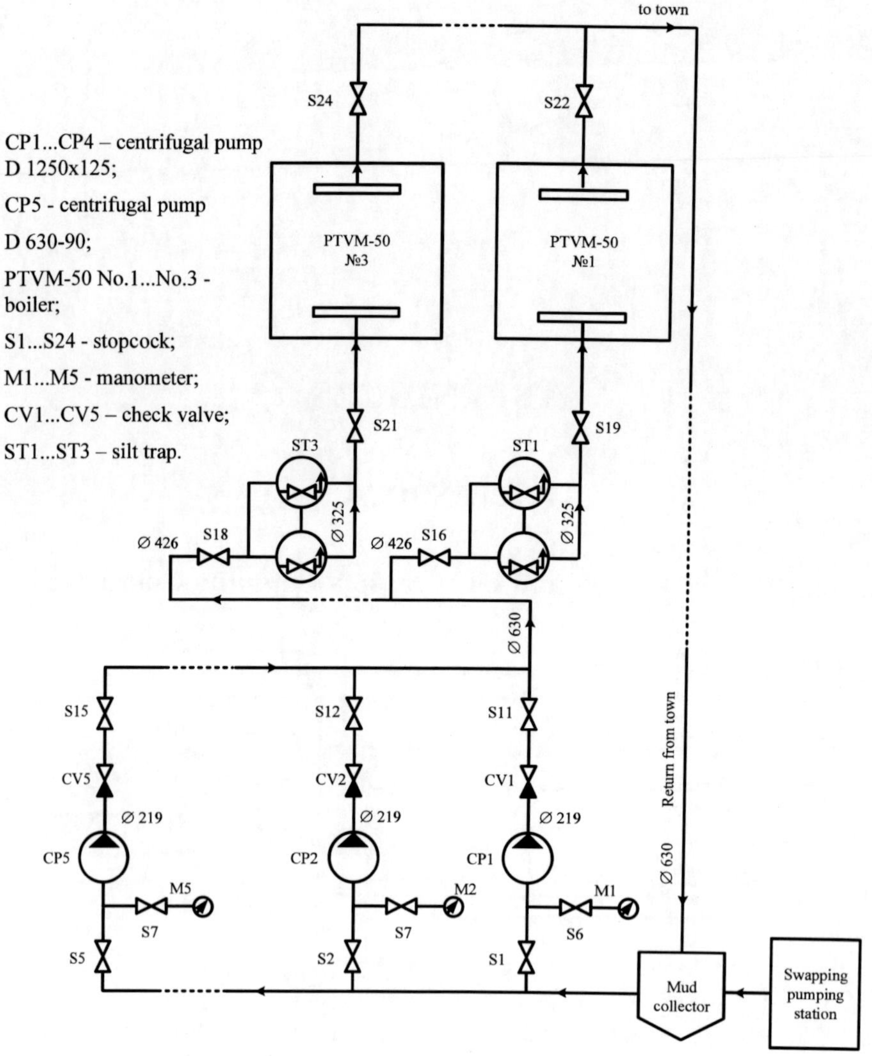

Figure A2. The flowsheet of the pumping station of the heat supply.

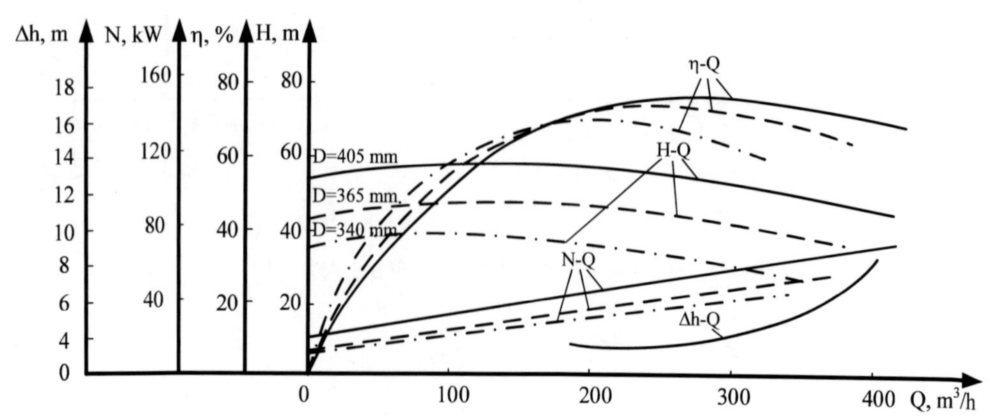

Figure A3. The published characteristics of D320-50 pump.

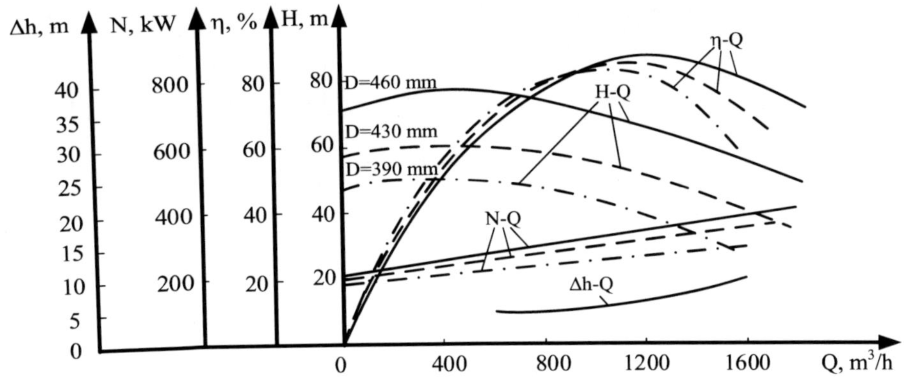

Figure A4. The published characteristics of D1250-62.5 pump.

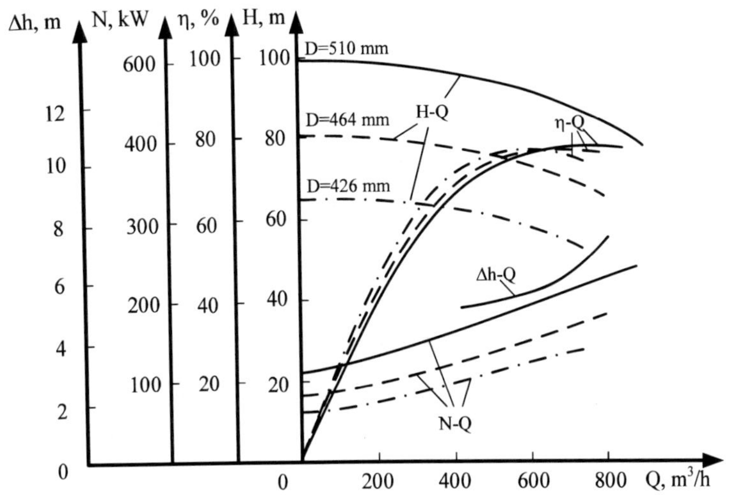

Figure A5. The published characteristics of D630-90 pump.

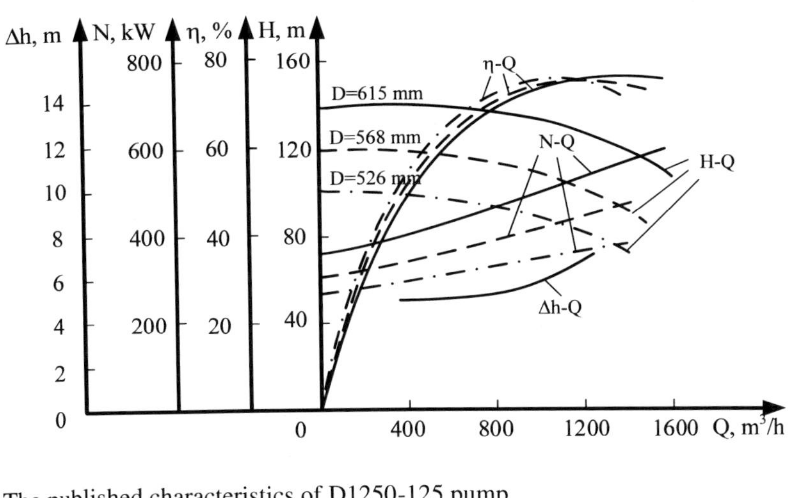

Figure A6. The published characteristics of D1250-125 pump.

Table A1. The published data of the centrifugal pumps

Parameters	Pump name			
	D320-50	D1250-62.5	D630-90	D1250-125
Feed, m^3/h	320	1250	630	1250
Head, m	50	62.5	90	125
Rotation frequency, rev/min	1450	1450	1450	1450
Allowable cavitation reserve, m	4	6.5	3.5	5.0
Pump efficiency, %	75	86	75	76
Motor power, kW	75	250	250	630

Table A2. The coefficients of the approximation of the head-discharge and energy characteristics of the pumps of the systems of water and heat supply of the town of Komsomolsk

Pump type	H-Q characteristic			N-Q characteristic		
	A_2	B_2	C_2	A_3	B_3	D_3
D320-50	53	210.5	$-2.748 \cdot 10^3$	416.1	381.176	19
D1250-62.5	70	52.615	-199.385	436.731	-103.091	110
D630-90	87	191.914	-999.035	-355.442	8071.91	65
D1250-125	127	124.214	-374.325	871.481	807.896	200

Figure A7. The cavitation destruction of a body in the liquid flow.

Table A3. The basic types of local resistances in the pipeline

Head losses	Diagrammatic representation
Sudden expansion: $h_m = \zeta_{vr}\dfrac{V_2^2}{2g}$; $\zeta_{vr} = \alpha_1\left(\dfrac{D_2^2}{D_1^2}-1\right)^2$.	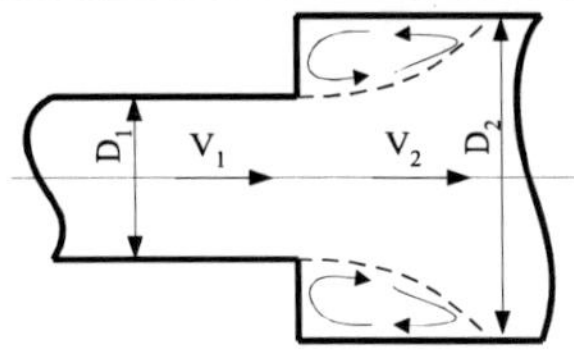
Sudden narrowing: $h_m = \zeta_{vs}\dfrac{V_2^2}{2g}$; $\zeta_{vs} = 0{,}5\left(1-\dfrac{D_2^2}{D_1^2}\right)$.	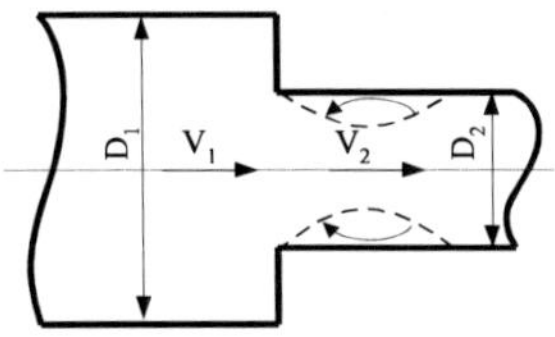
Diffusors: $h_m = \zeta_{dif}\dfrac{V_2^2}{2g}$; $\zeta_{dif} = k_{sm}\left(\dfrac{D_2^2}{D_1^2}-1\right)^2 + \zeta_{tr}$.	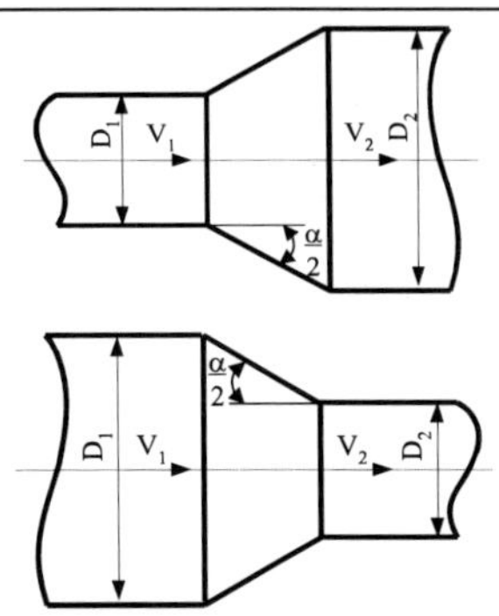
Stopcock, valve, butterfly gate: $h_m = \zeta_z\dfrac{V^2}{2g}$ $\zeta_z = \left(\dfrac{\omega}{\omega_0\varepsilon}-1\right)^2$.	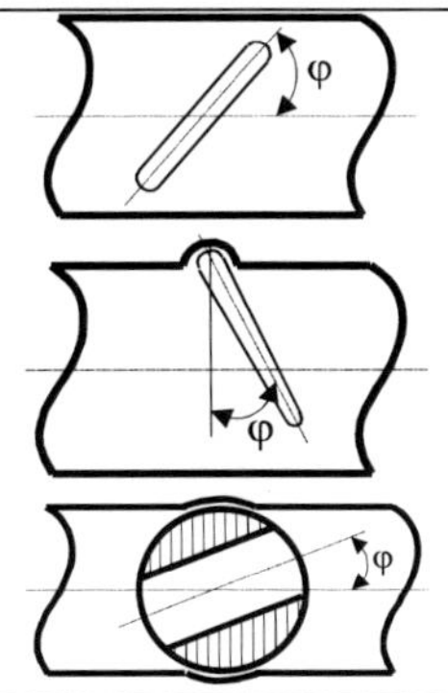
Sharp turn: $h_m = \zeta_k\dfrac{V^2}{2g}$; $\zeta_k = a + b\alpha + c\alpha^2 + d\alpha^3 + e\alpha^4$.	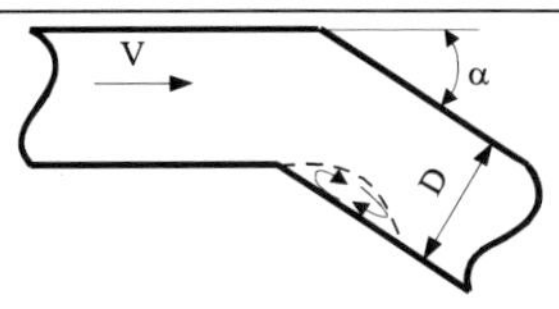
Smooth turn: $h_m = \zeta_\alpha\dfrac{V^2}{2g}$; $\zeta_\alpha = \zeta_{90^\circ}a$.	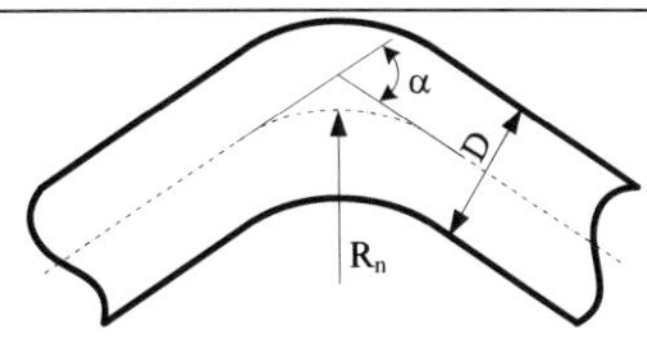

Table A3. (Continued)

Head losses	Diagrammatic representation
Flows separation: $$h_{p.p.} = \frac{V_0^2 + V_p^2 - 2\frac{\omega_p}{\omega_0}V_p^2 - 2\frac{\omega_b}{\omega_0}V_b^2 \cos\alpha}{2g};$$ $$h_{b.p.} = \frac{V_0^2 + V_b^2 - 2\frac{\omega_p}{\omega_0}V_p^2 - 2\frac{\omega_b}{\omega_0}V_b^2 \cos\alpha}{2g}.$$	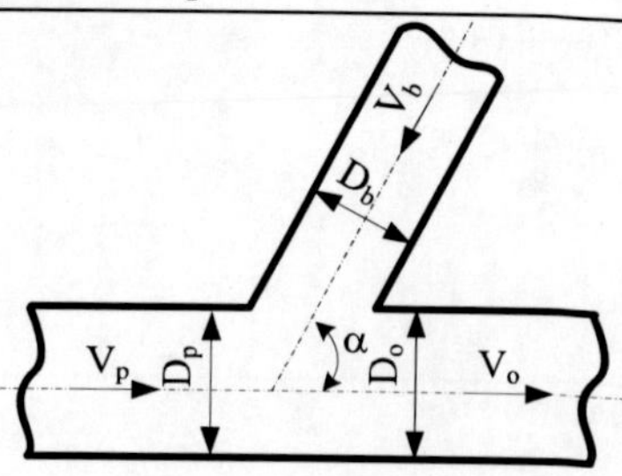

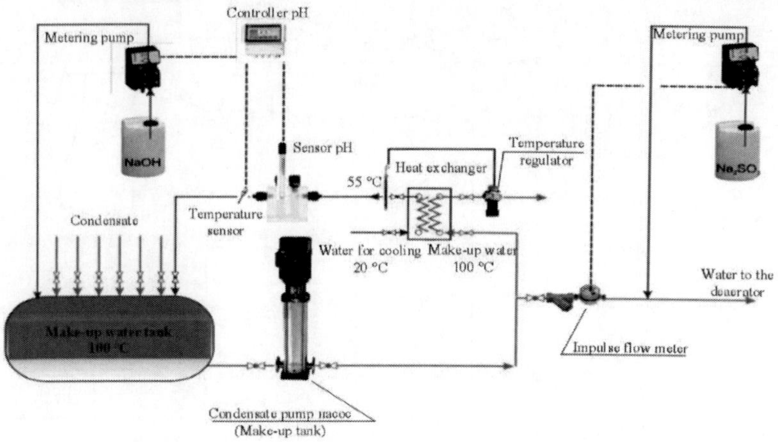

Figure A8. The flowsheet of chemical water deaeration.

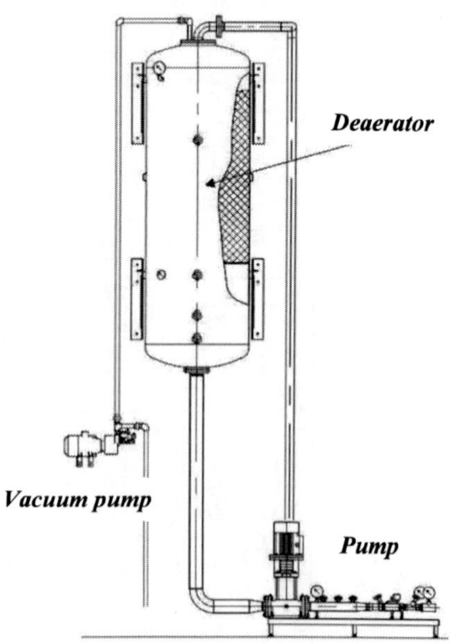

Figure A9. The flowsheet of the installation of the vacuum deaerator.

APPENDIX B

The Block Diagram of a Pumping Complex Model

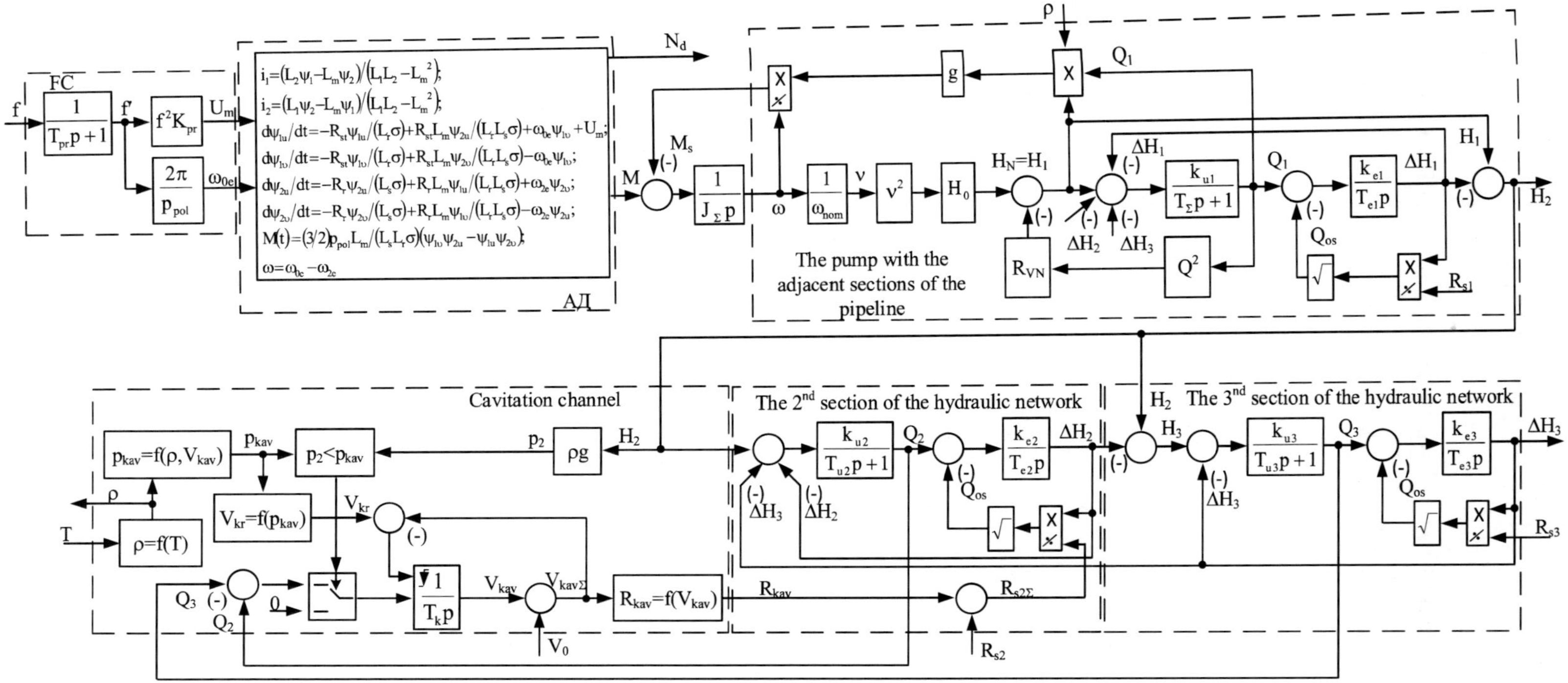

Figure B1. The flowsheet of a model of PC with a cavitation channel.

APPENDIX C

The Algorithm of the Operation of the Electromechanical Device of Cavitation Processes Control

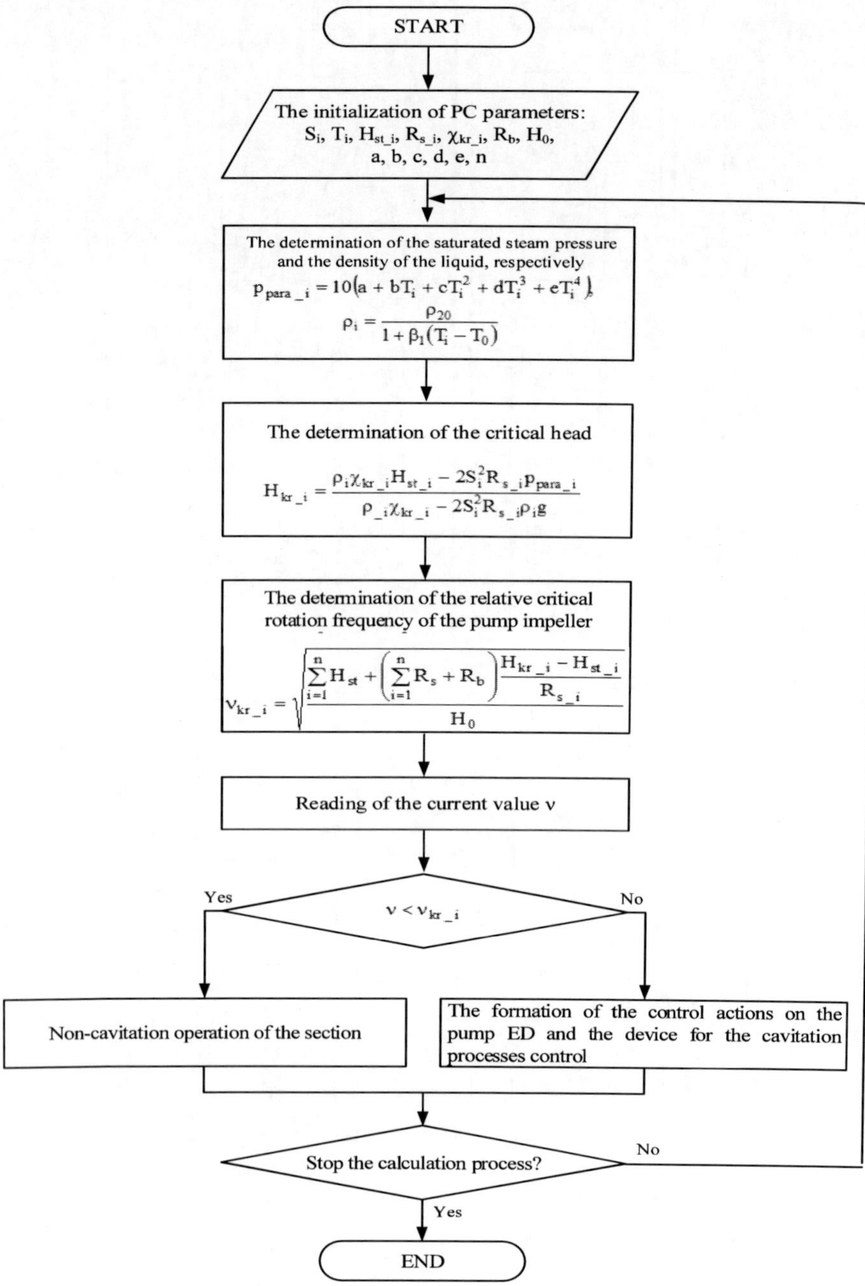

Figure C1. The algorithm of the determination of the relative critical rotation frequency of the pump ED for the *i*-th section of the hydrodynamic network.

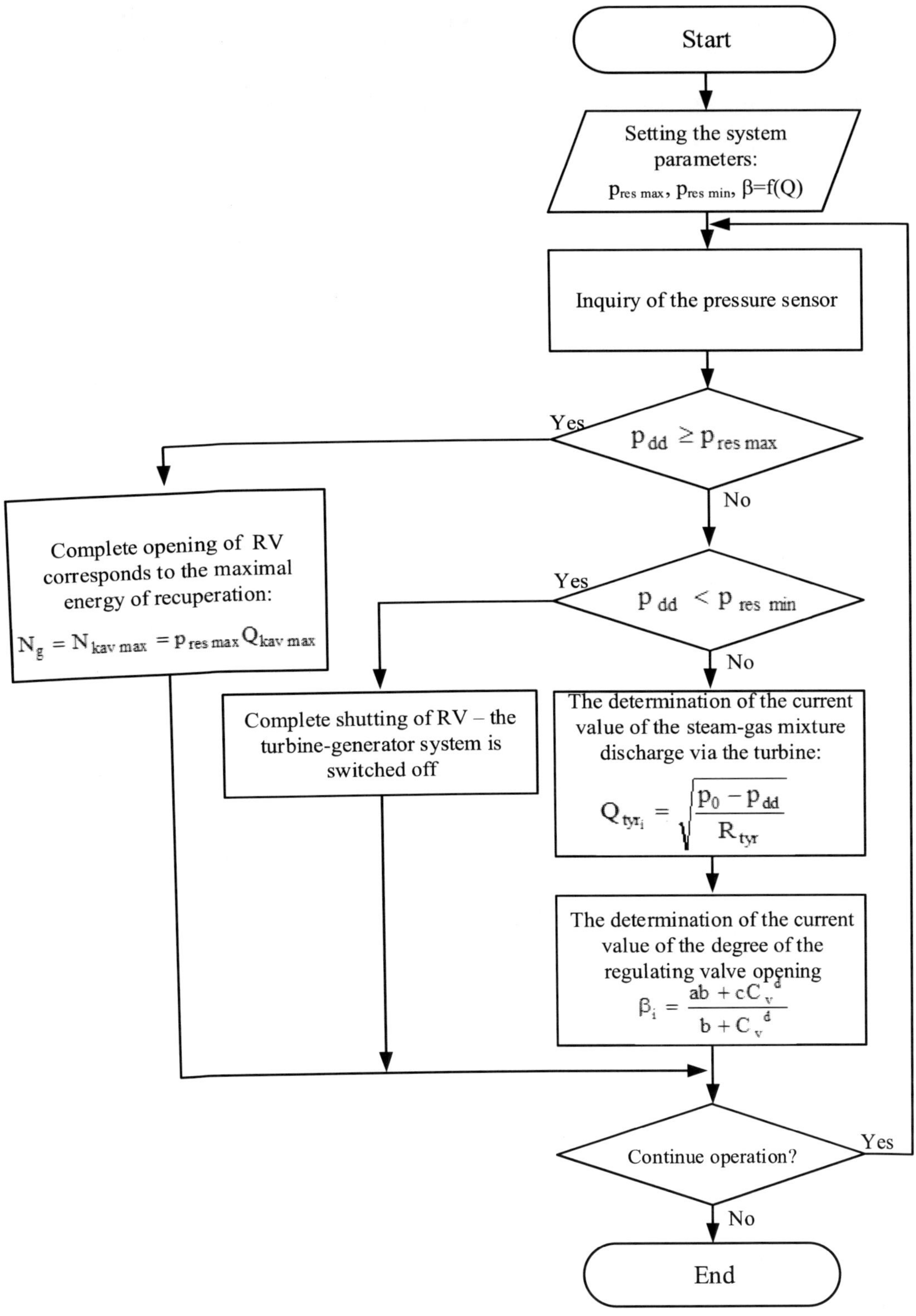

Figure C2. The algorithm of EDCCP operation.

APPENDIX D

The System of the Automatic Control of the Electric Drives of the Pumping Complex

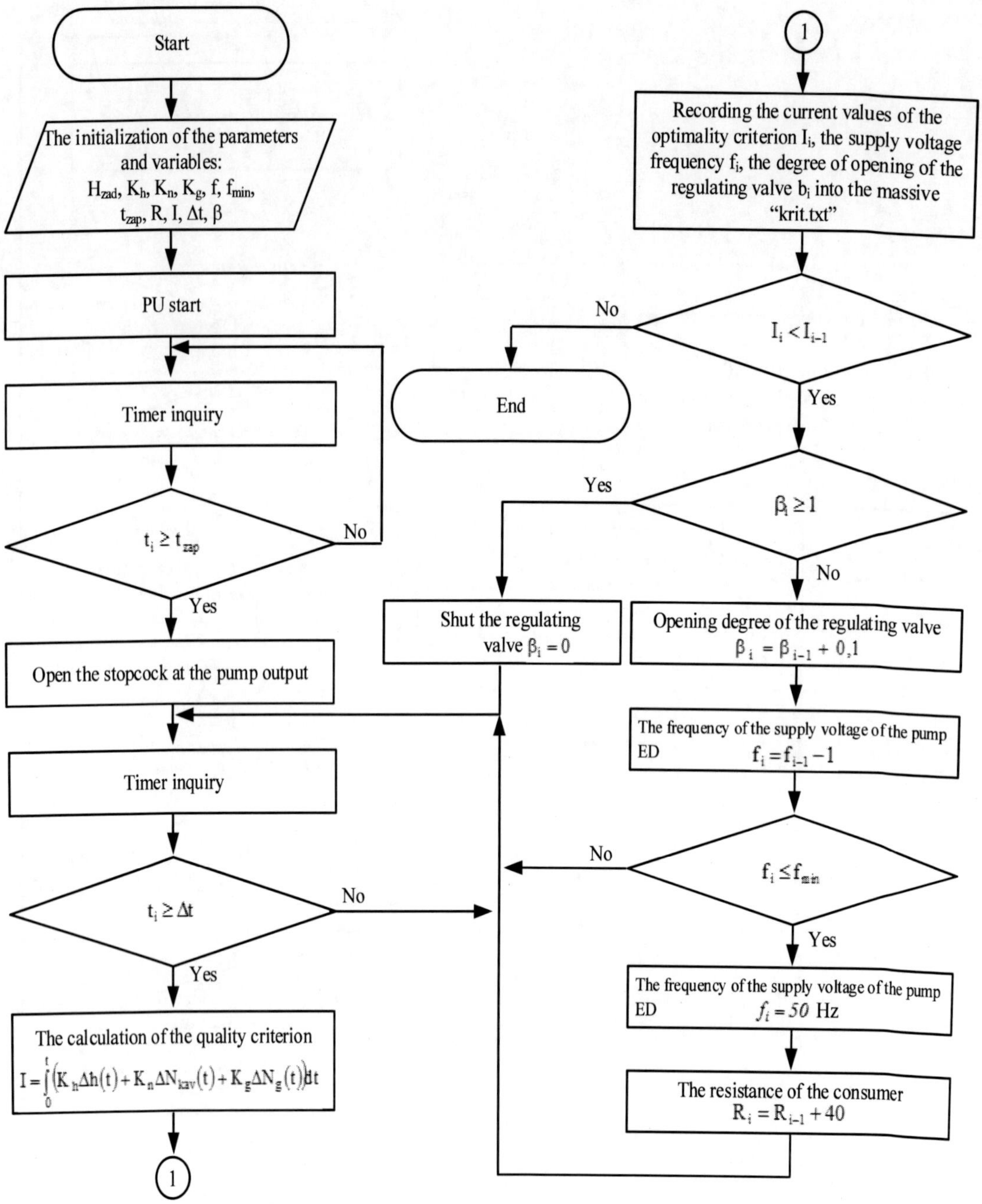

Figure D1. The algorithm of the determination of the minimum of the quality criterion.

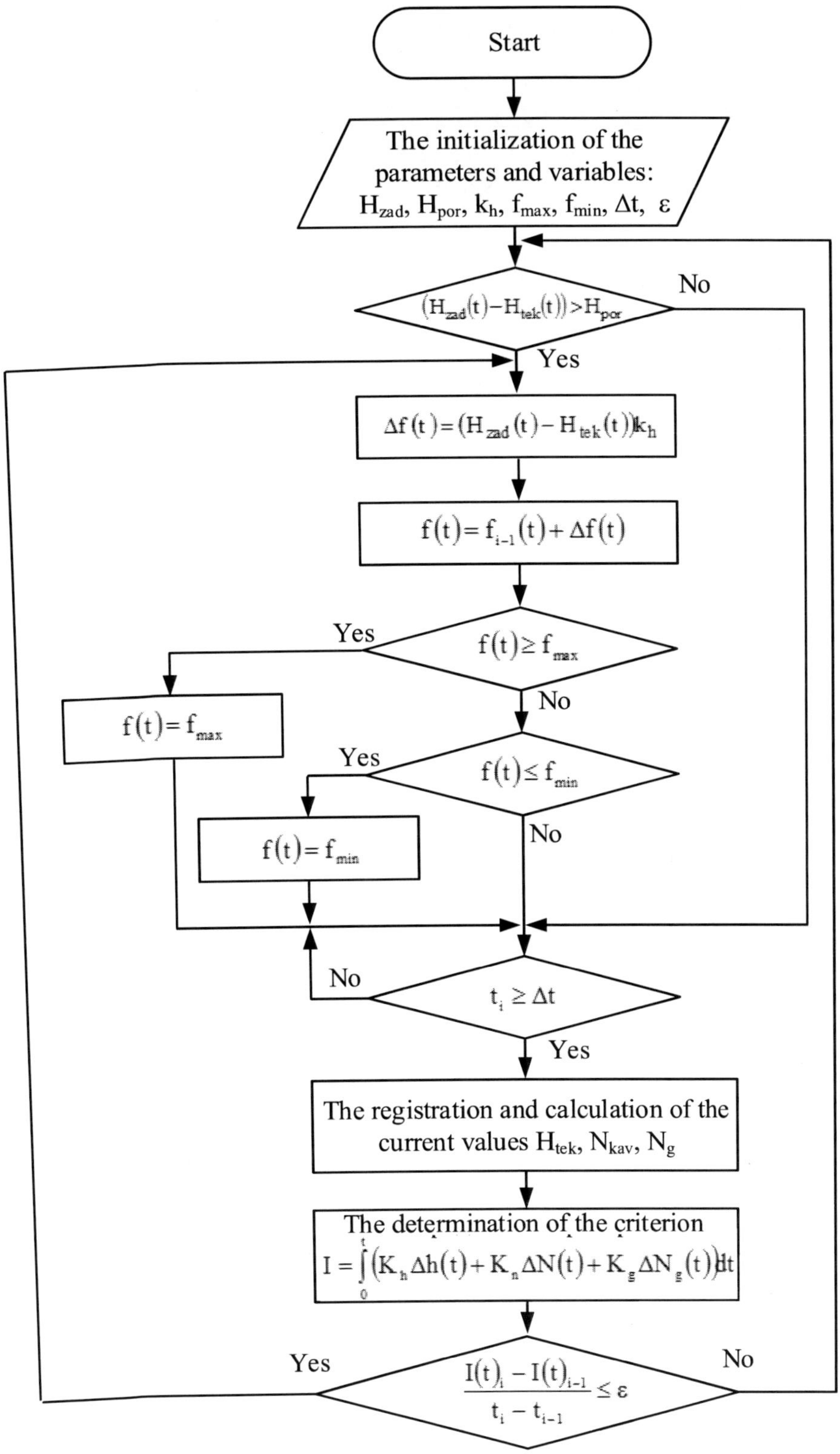

Figure D2. The algorithm of the formation of the frequency of the voltage supplied to PU in ACS of PC ED taking into account the cavitation processes.

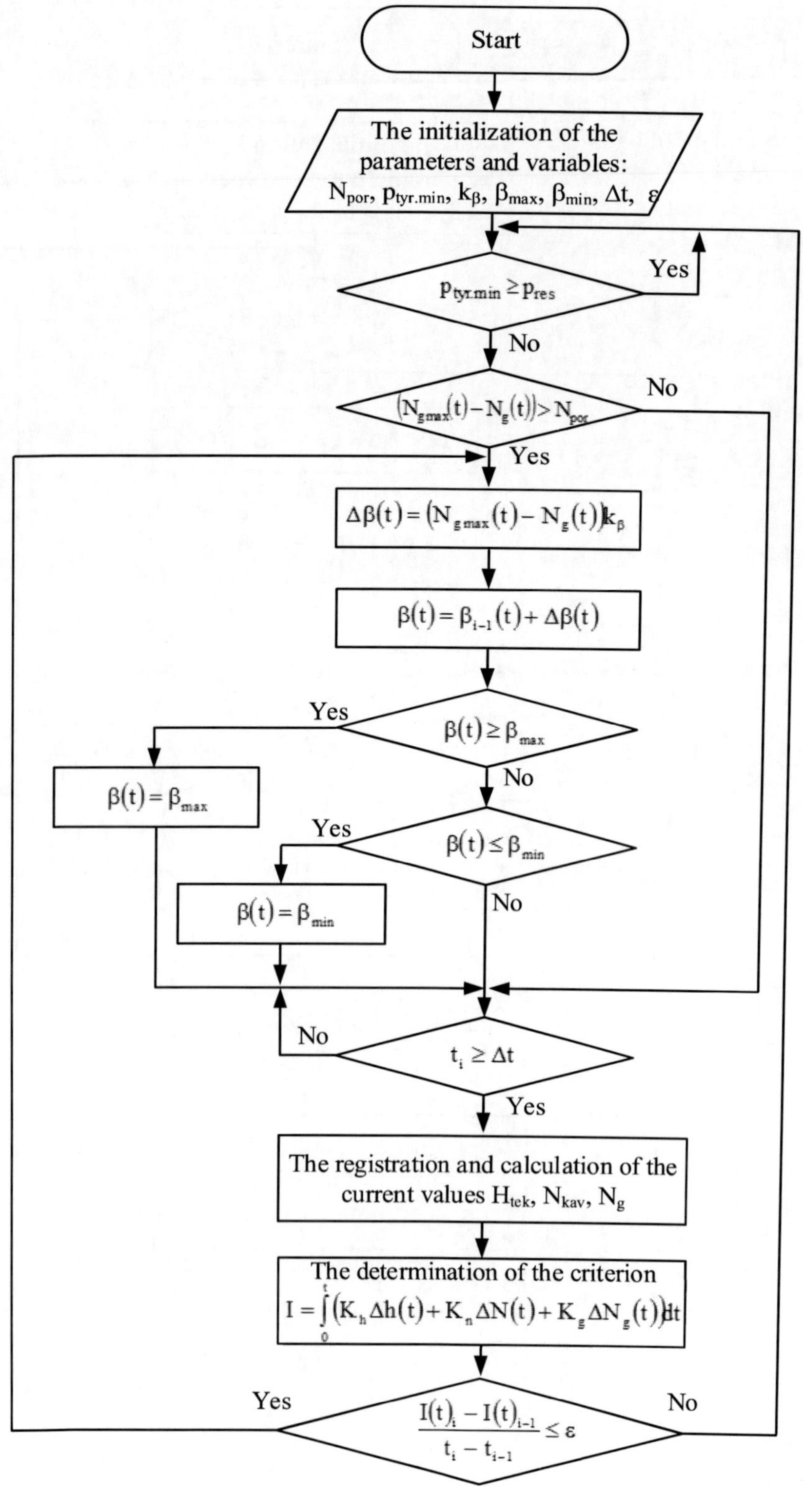

Figure D3. The algorithm of the formation of the degree of opening of the regulating valve in ACS of PC ED taking into account the cavitation processes.

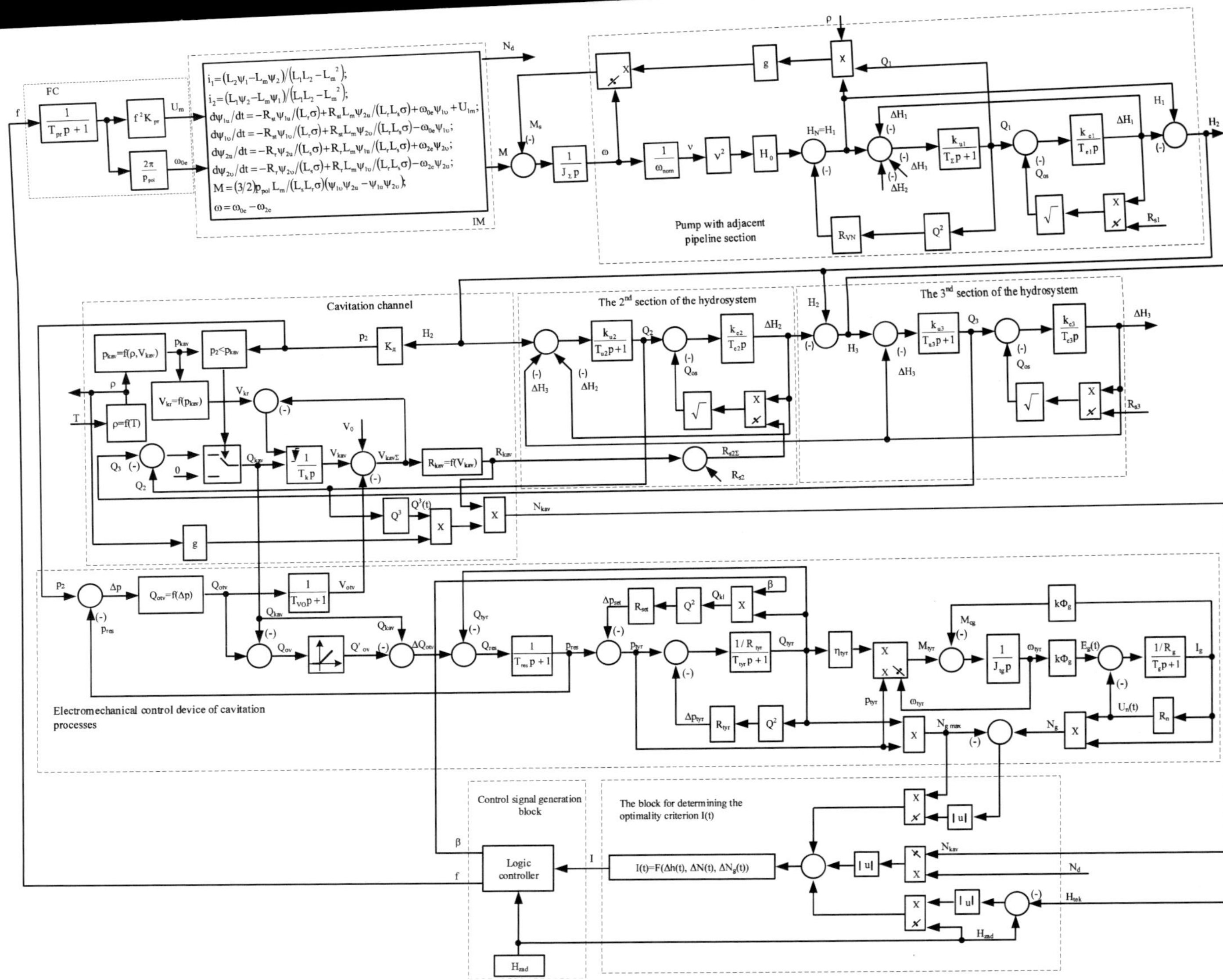

Figure D4. The block diagram of the model of the system of automatic control of PC ED.

APPENDIX E

The Characteristic of the Experimental Laboratory Plant

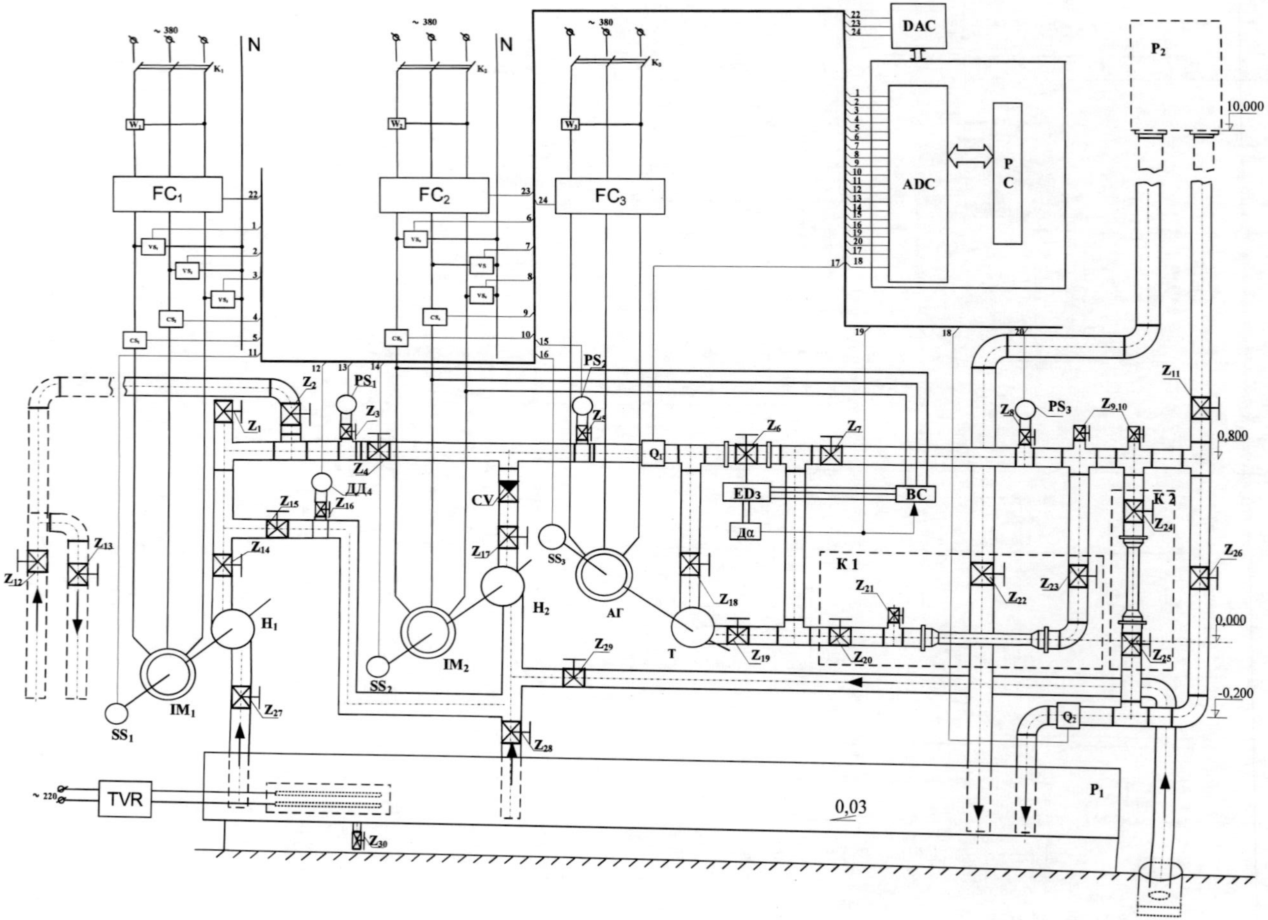

Figure E1. The functional diagram of PC physical model.

Table E1. The technical data of the turbomechanisms

Designation in the diagram	Pump type	Feed m^3/h	Head, m	Rotation frequency, rev/min
H_1	Willo MHI 202.1	5	22	3000
H_2	Willo MHI 402.1	8	18	3000

Table E2. The technical data of the electric motors

Designation in the diagram	Electric motor type	N, kW	U_n, V	cos φ	η, %	n_0, rev/min
IM_1	MHI 202-1/E/3	0.83	380	0.72	66	3000
IM_2	MHI 402-1/E/3	0.83	380	0.72	66	3000

Table E3. The technical parameters of regulating stopcock Z_6

Parameter name	Value
The diameter of the nominal bore	50 mm
Operating pressure	16 bar
The temperature of the working medium	0..+80° C
Connection	flange
The material of the body	cast iron GGG-40

Table E4. The technical characteristics of the motor reducer

Parameter name	Value
Type	C03218.WMHB1A.75E
Rated power	0.75 kW
The number of revolutions	2871 rev/min
Rated voltage	380 V (AC)
Rated current	1.81 A
The moment of inertia	0.002 kgm^2
Reduction rate	8.59
The moment at the output	36 Nм
cos φ	0.79

Table E5. The technical parameters of the control device of regulating stopcock Z_6

Parameter name	Value
Type	PKP1I-SHCH1.RS
The supply voltage of the alternating current	90-264 V (AC) (47-63 Hz)
The supply voltage of the direct current	20-34 V (DC)
The number of the control inputs	3
The minimal duration of the control signals	0.1 s
The rated supply voltage of the sensors	24 V (DC)
The number of digits of the digital indicator	4
The maximal current commutated by the relay contacts	10 A (~220 V)
Computer interface	Double-wire RS-485
Data transfer rate	2.4-115.2 kbit/s
Communication line length	up to 100 m

Table E6. The technical parameters of position sensor PS

Parameter name	Value
Type	IA18DSF05PO
Supply voltage	10-30 V (DC)
Protection type	short circuit, reverse polarity
Output current	maximum 200 mA

Table E7. The technical parameters of interface converters IC

Parameter name	Value
Type	AC4
Supply voltage	5 V (DC) from USB bus
Consumed power	0.5 VA
The admissible voltage of the galvanic insulation of the inputs	minimum 1500 V
Computer interface	USB 2.0
Computer interface	Double-wire RS-485
The length of the line of communication with the computer	maximum 3 m
The length of the line of communication with the external device	maximum 1200 m
Data transfer rate	up to 115.2 kbit/s

APPENDIX F

The Typical Flowsheets of Pumping Complexes

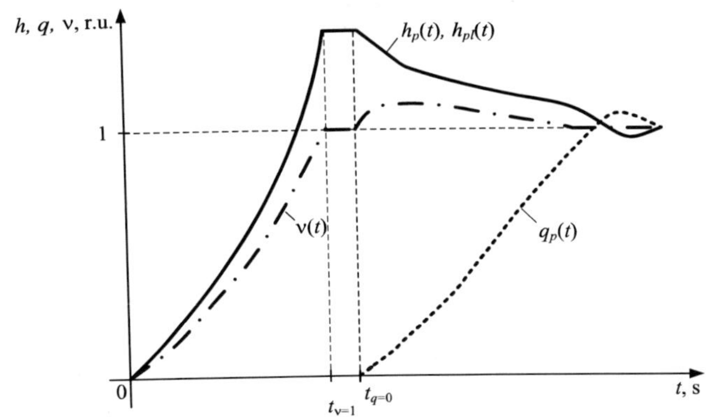

Figure F1. The transient processes at the pump start with the shut stopcock: $h_p(t)$, $h_{pl}(t)$ – the relative head at the pump output and in the pipeline network, respectively; $q_p(t)$ – relative feed of the pump; $\nu(t)$ – relative rotation frequency of the pump impeller.

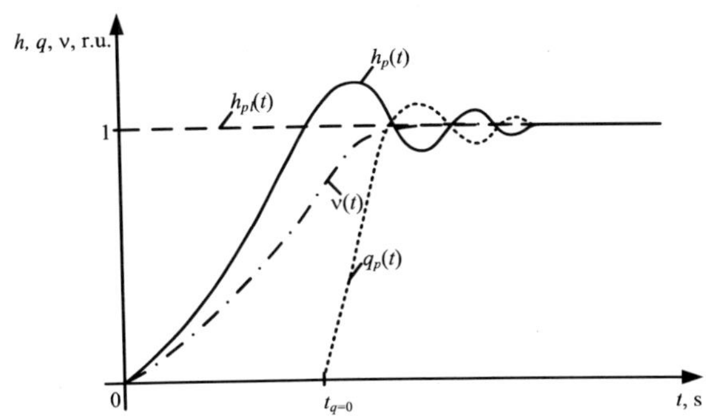

Figure F2. The transient processes at the pump start with the open stopcock in case when the pipeline section from the pump to the check valve is filled with liquid.

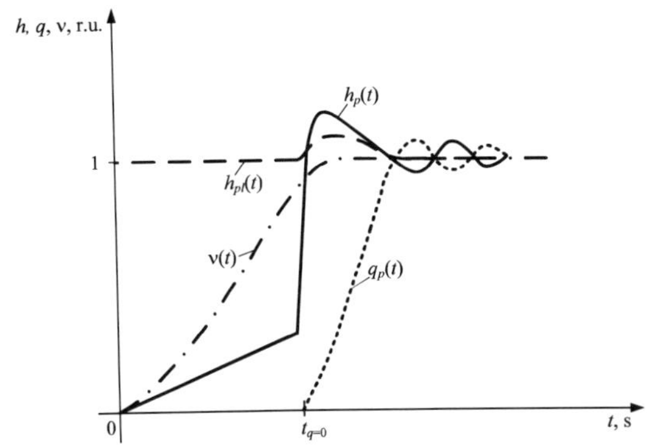

Figure F3. The transient processes at the pump start with the open stopcock in case when the pipeline section from the pump is not filled with liquid.

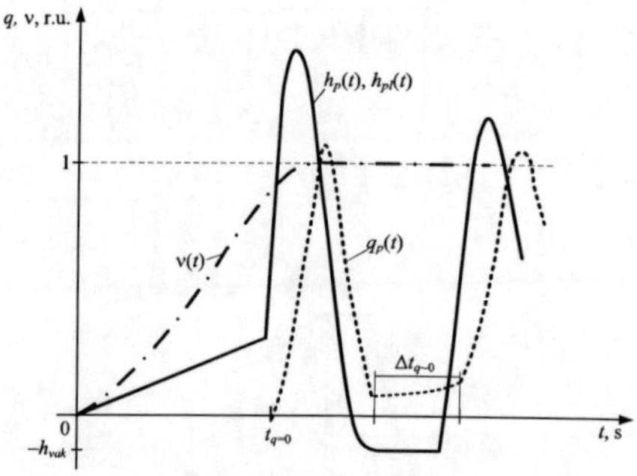

Figure F4. The transient processes at the pump start with the open stopcock and emptied pipeline.

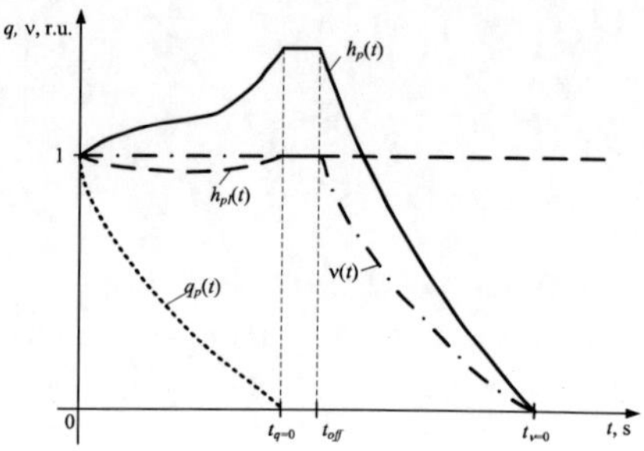

Figure F5. The transient processes at the pump stop.

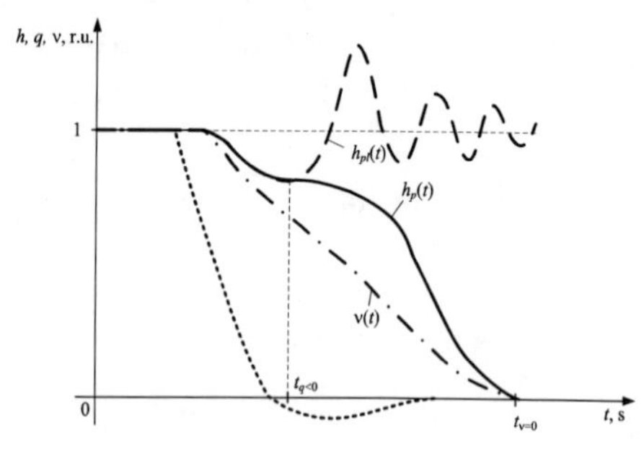

Figure F6. The transient processes at abrupt disappearance of voltage in PU equipped with a check valve.

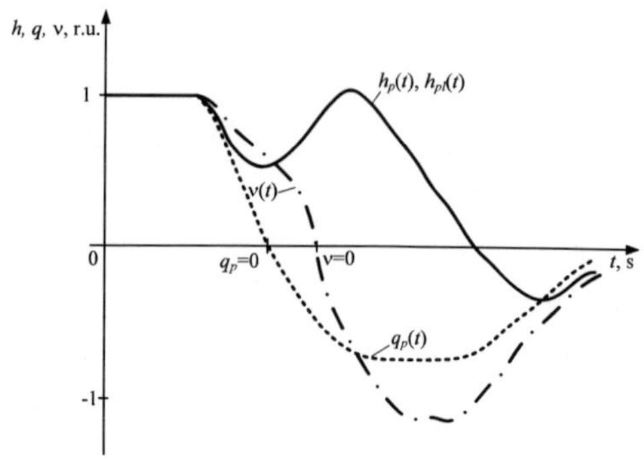

Figure F7. The transient processes at abrupt disappearance of voltage in PU in the absence of a check valve.

APPENDIX G

The Comparative Characteristic of the Pipeline Valves as the Means of Hydrodynamic Protection of Modern PC

Table G1. The comparative characteristic of the pipeline valves as the means of hydrodynamic protection of modern PC

Index name	Stopcock	Tap	Damper	Hydrovalve	
	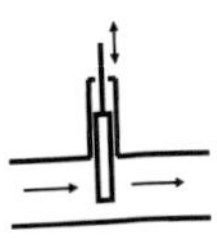	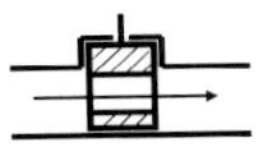	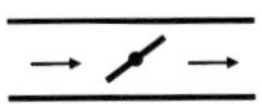	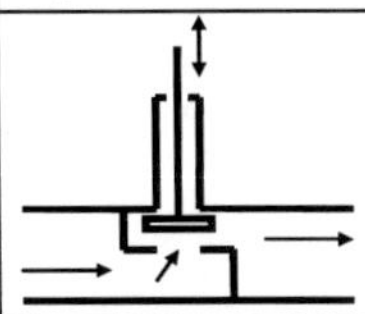	
				controlled	uncontrolled
Location	pump head collectors; the pipelines of the systems of the municipal and industrial water supply, sewage, mining and quarry water pumping				
Performed functions	PP parameters regulation; hydraulic protection of the system				hydraulic protection of the system
Used types of drives	electric drive; electromagnetic drive; hydraulic drive; pneumatic drive				-
Electric drive power, kW	0.01-15				-
Diameter of the nominal bore, mm	50-2000	up to 1400	up to 2200	up to 1200	up to 2800
Material	bronze, cast iron, steel, stainless steel				
Speed of operation, s	300-480			20-300	0.02-10
Standard lifespan, year	5-7	10-14	5-7	3-10	10-15
Factual lifespan, year	1.5-2.5			1.5-3	1-1.5
Average cost, (mm)	up to 6000 (up to 1200)	up to 9000 (up to1200)	up to 1500 (up to 1200)	up to 2500 (up to1200)	up to 500 (up to 700)

APPENDIX H

The Hydraulic Characteristics of the Pipeline Valves of the Pumping Complexes

Table H1. The basic hydraulic characteristics for different types of the pipeline valves

Shut-off and control valves
Stopcock
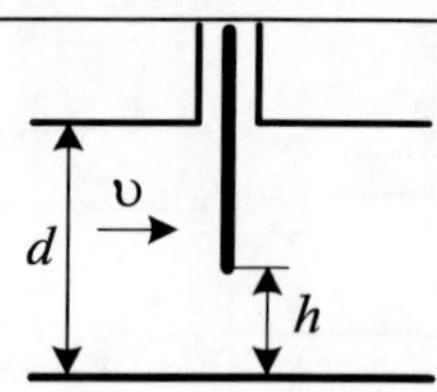
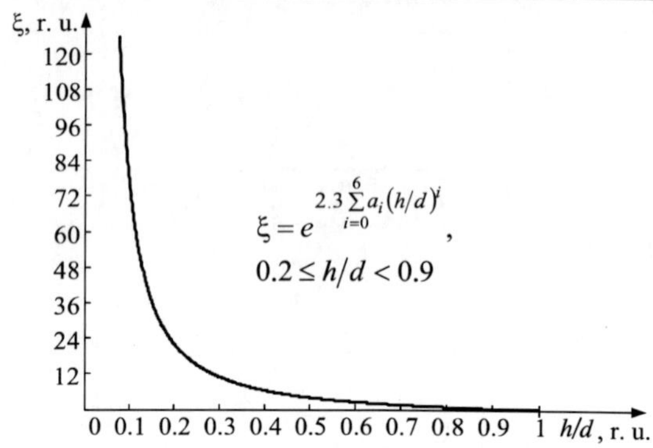 $\xi = e^{2.3\sum_{i=0}^{6} a_i (h/d)^i}$, $0.2 \le h/d < 0.9$
Disc shutter
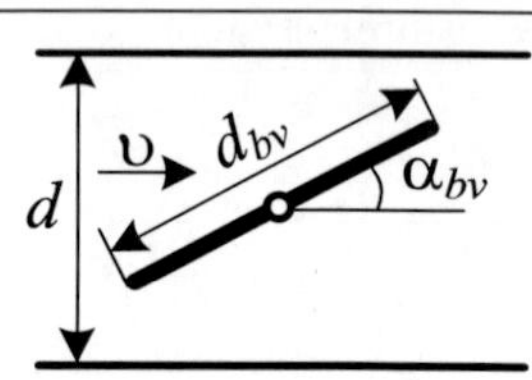
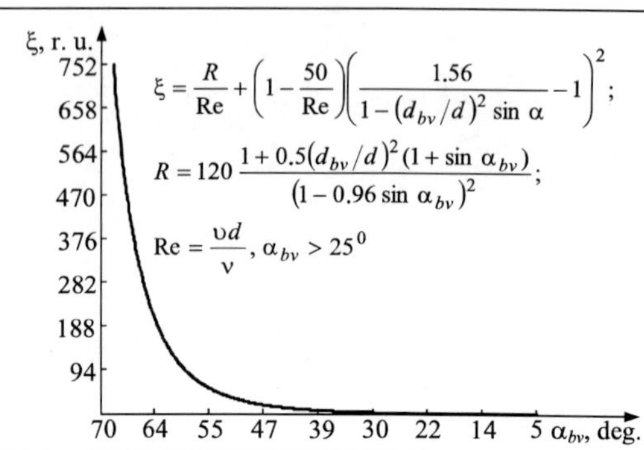 $\xi = \frac{R}{\text{Re}} + \left(1 - \frac{50}{\text{Re}}\right)\left(\frac{1.56}{1 - (d_{bv}/d)^2 \sin\alpha} - 1\right)^2$; $R = 120\,\frac{1 + 0.5(d_{bv}/d)^2(1 + \sin\alpha_{bv})}{(1 - 0.96\sin\alpha_{bv})^2}$; $\text{Re} = \frac{\upsilon d}{\nu},\ \alpha_{bv} > 25^0$

Control valves
Control valve
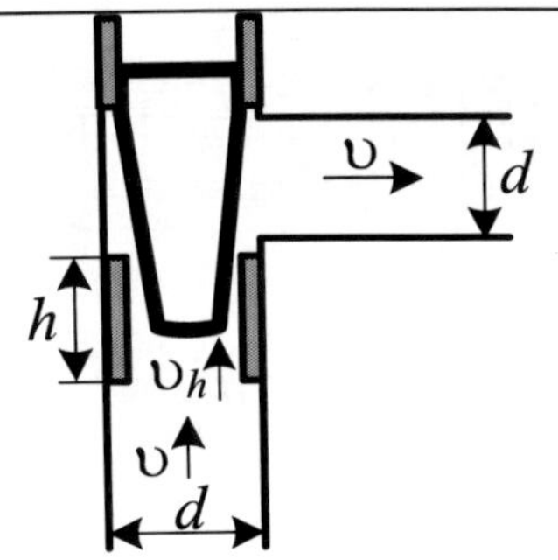
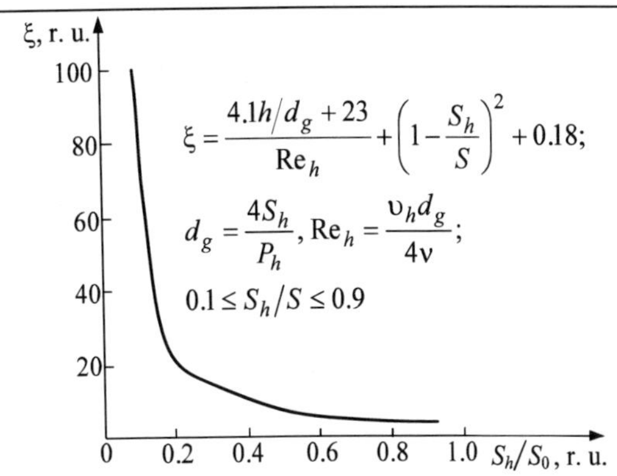
Protective valves
Check valve
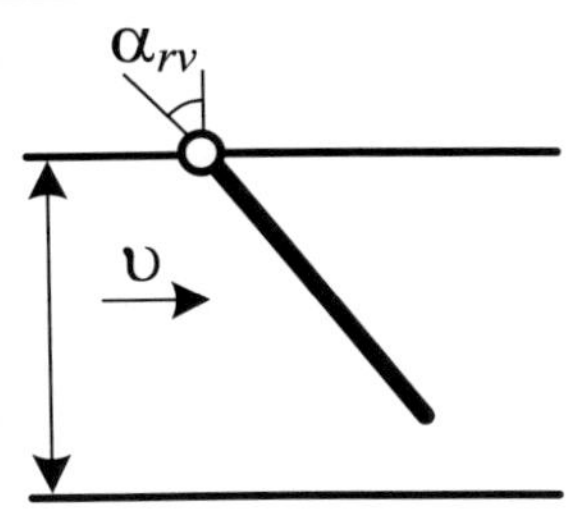
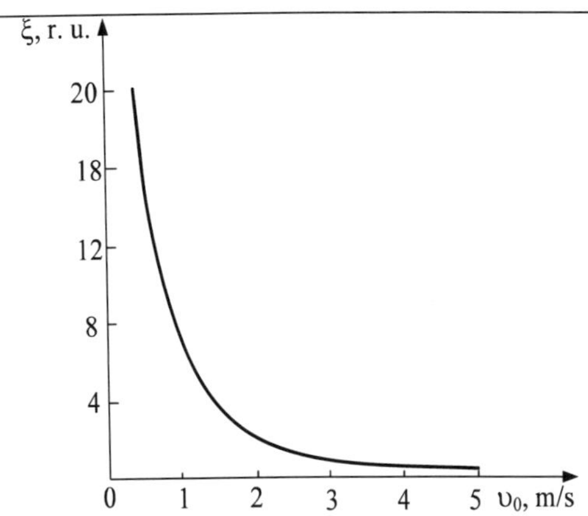

APPENDIX I

The Closed-Loop Electromechanical System of the Reduction of the Dynamic Loads in the Pumping Complex

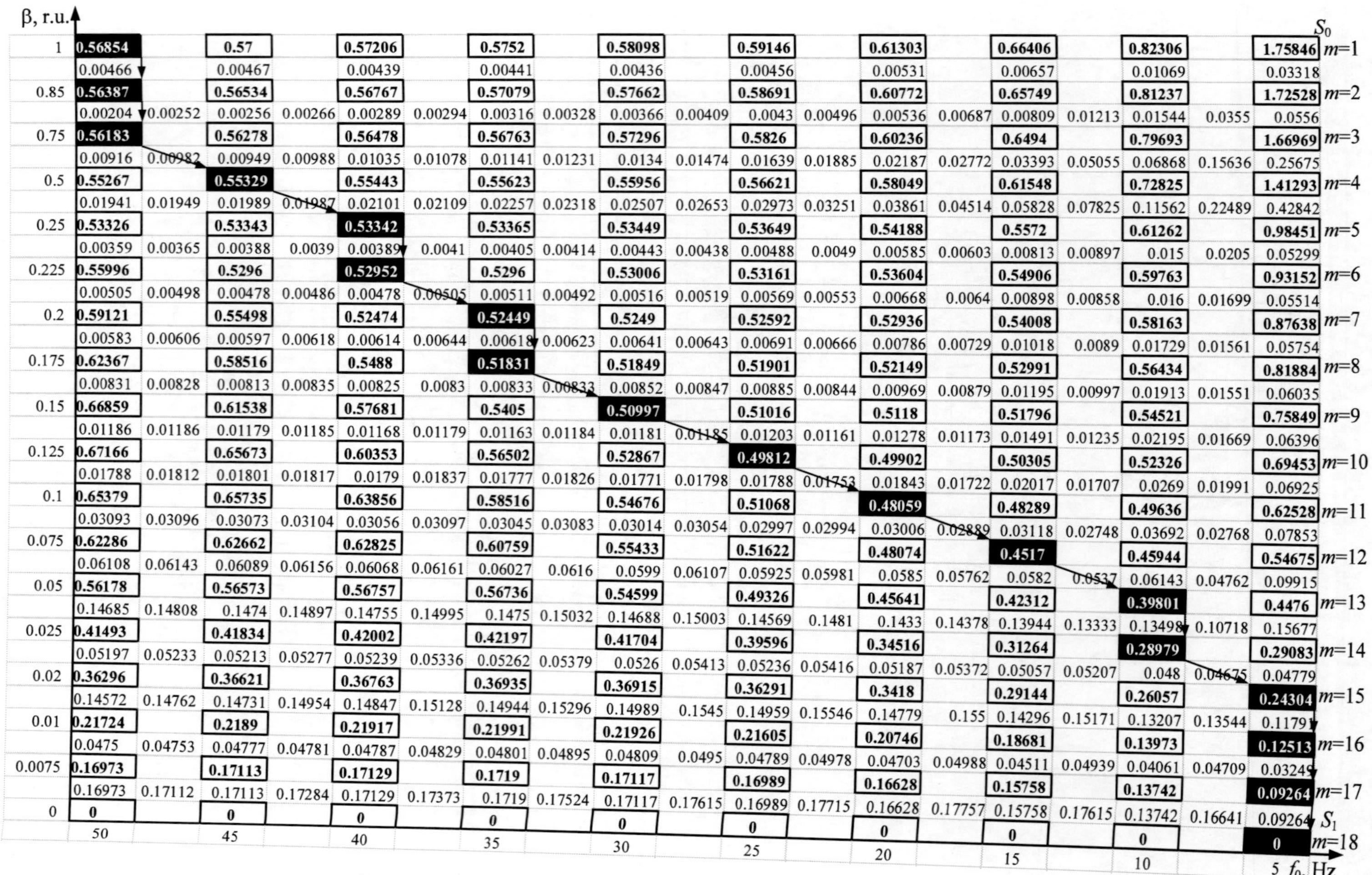

Figure I1. The procedure of the search for the optimal law of the control of the stopcock ED in PC operation mode at the static head h_s=40 m.

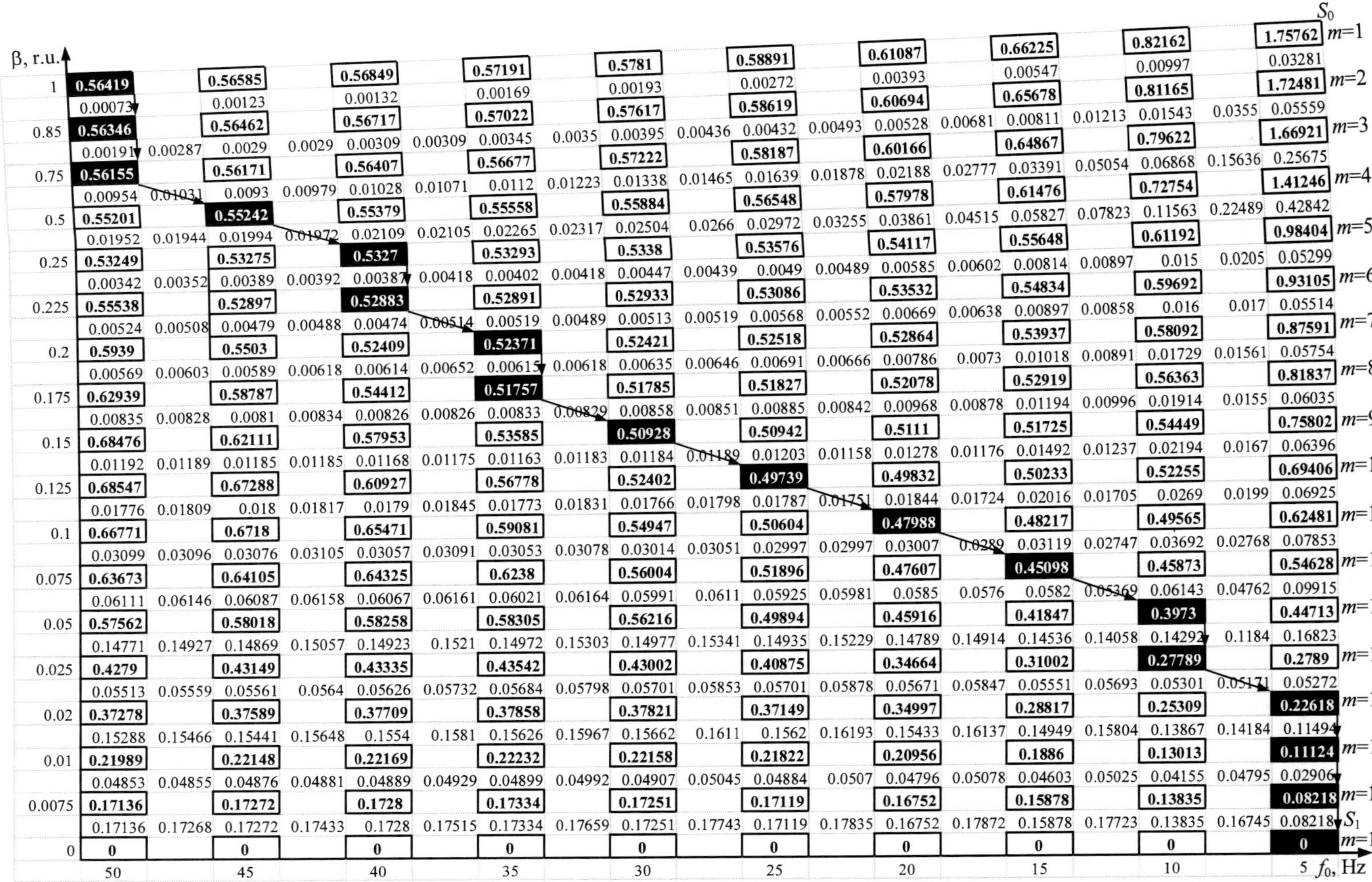

Figure I2. The procedure of the search for the optimal law of the control of the stopcock ED in PC operation mode at the static head h_s=60 m.

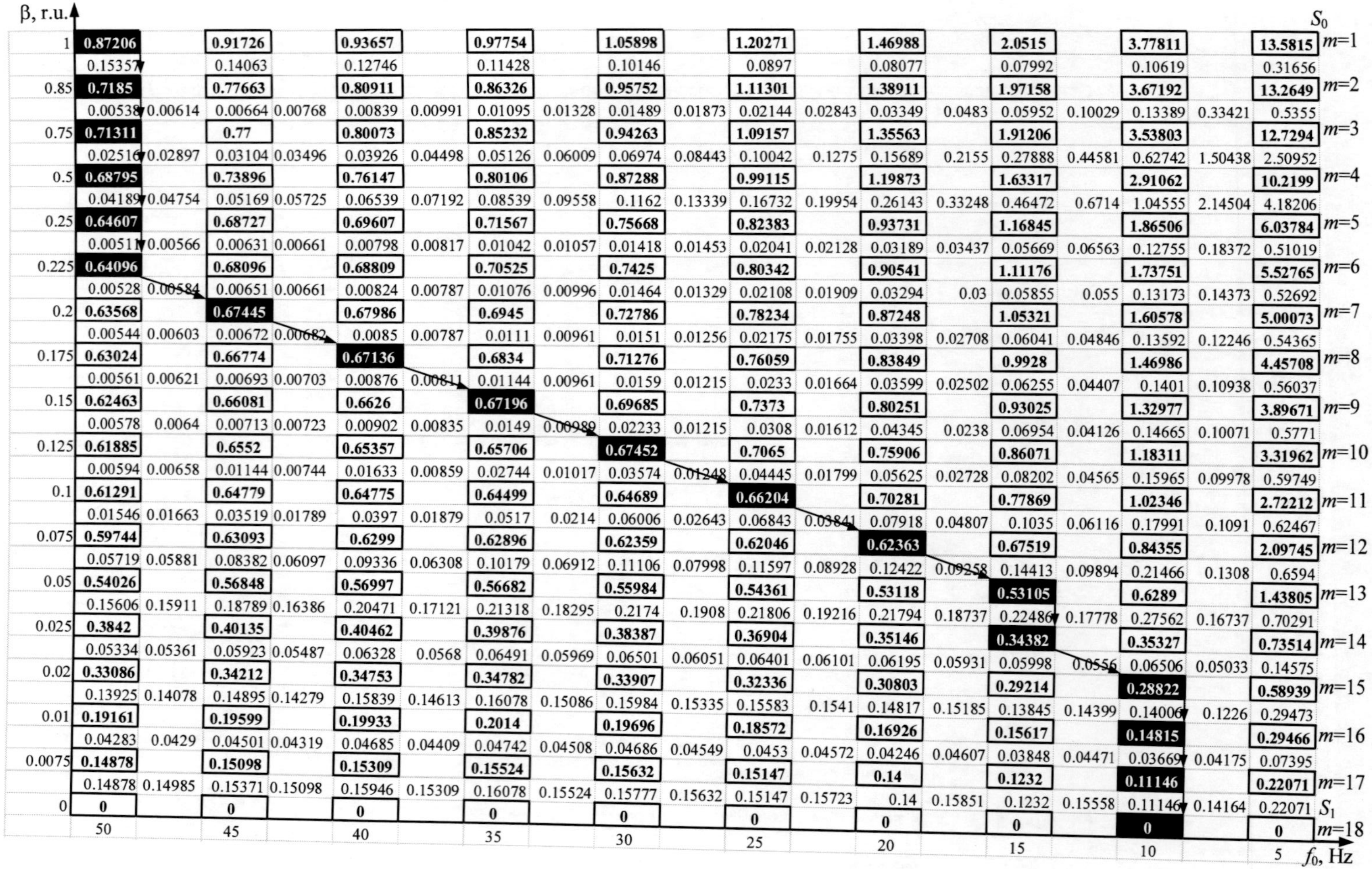

Figure I3. The procedure of the search for the optimal law of the control of the stopcock ED in PC emergency mode at the static head h_s=40 м.

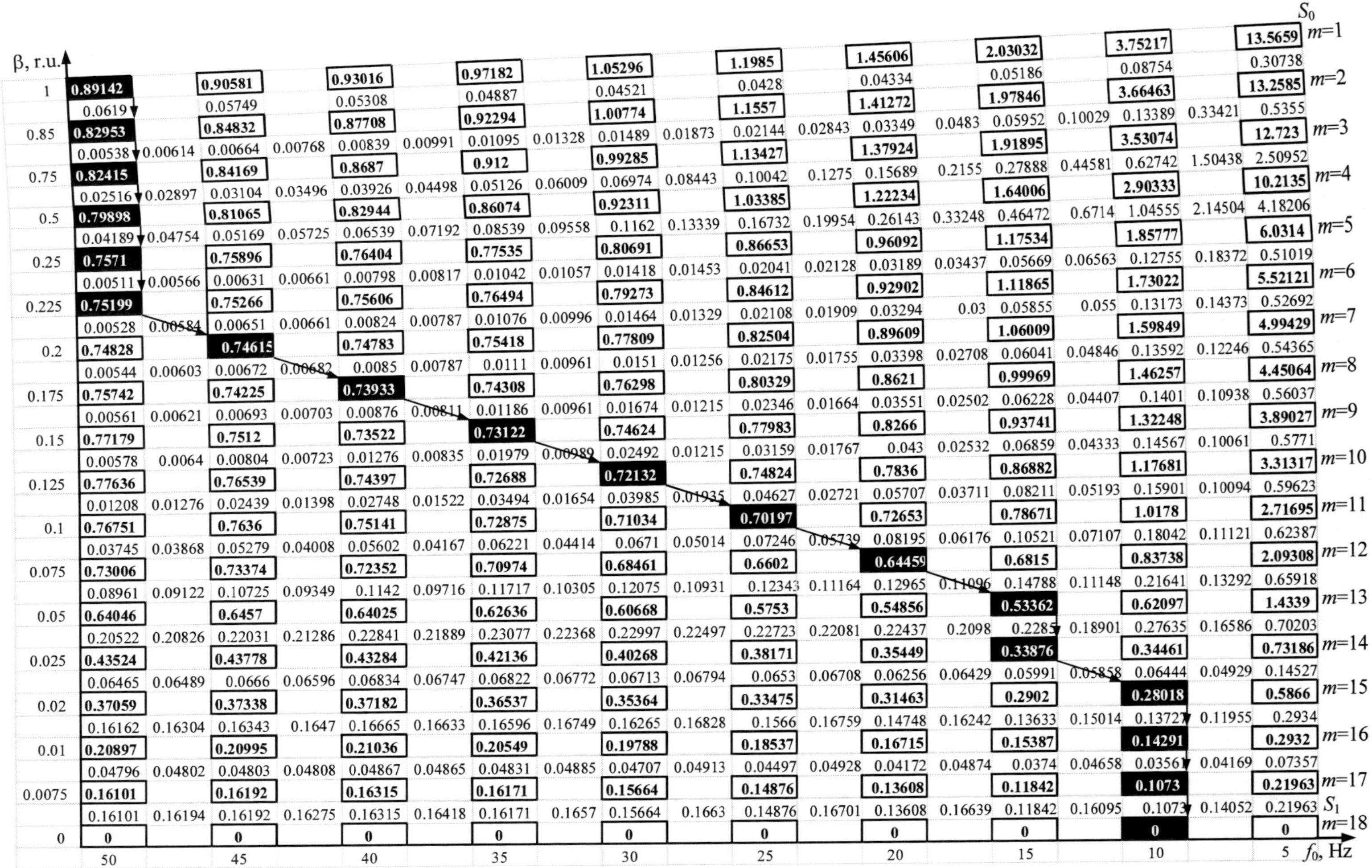

Figure I4. The procedure of the search for the optimal law of the control of the stopcock ED in PC emergency mode at the static head h_s=60 м.

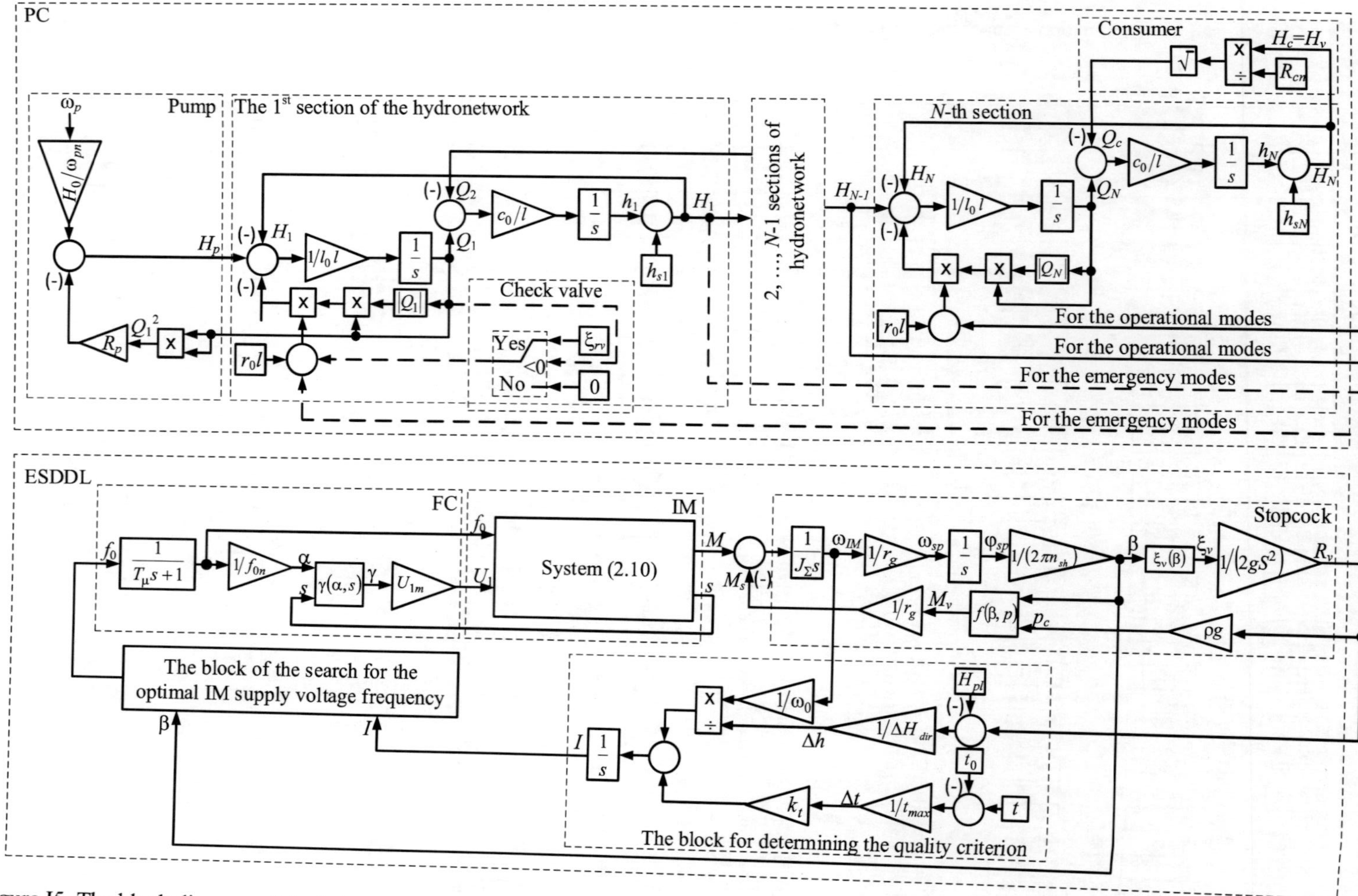

Figure I5. The block diagram of the imitation model of PC with closed-loop ESDDL.

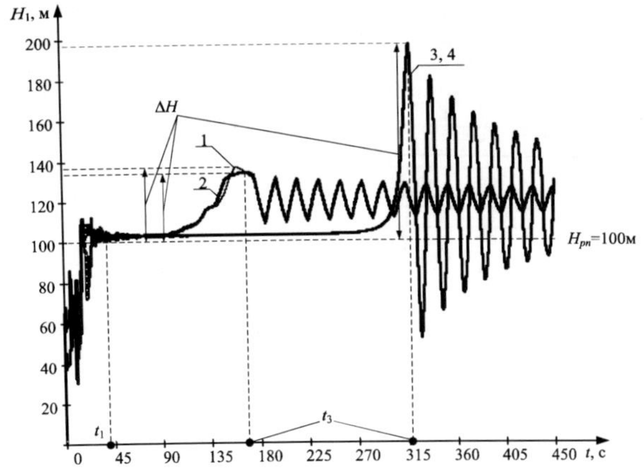

Figure I6. The curves representing head variation $H_1(t)$ before stopcock SA2 in the operation mode of PC with ESDDL and without it.

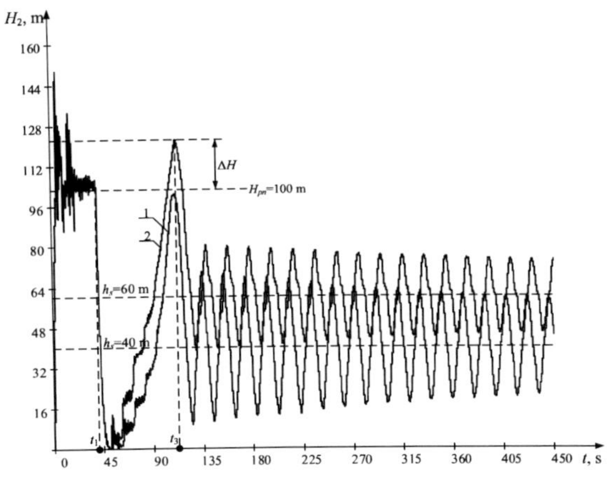

Figure I7. The curves representing head variation $H_2(t)$ after stopcock SA1 in the emergency mode of PC with ESDDL.

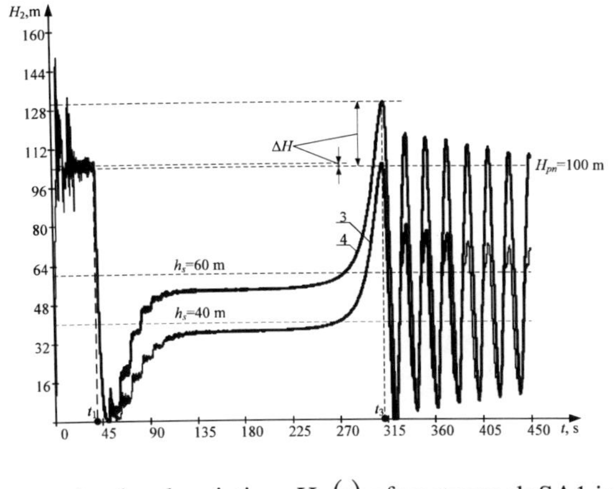

Figure I8. The curves representing head variation $H_2(t)$ after stopcock SA1 in the emergency mode of PC without ESDDL.

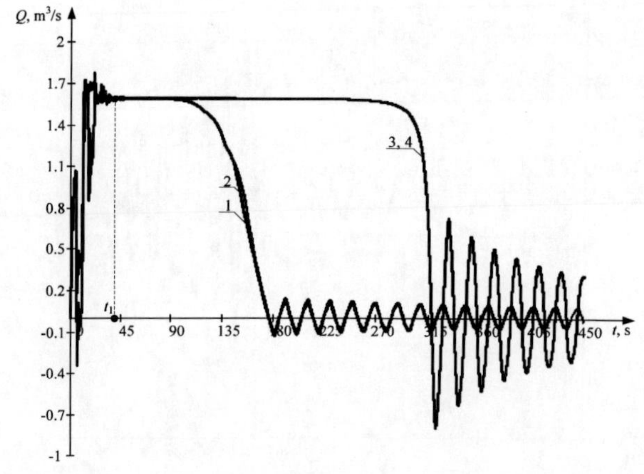

Figure I9. The curves representing discharge variation $Q(t)$ in the operation mode of PC with ESDDL and without it.

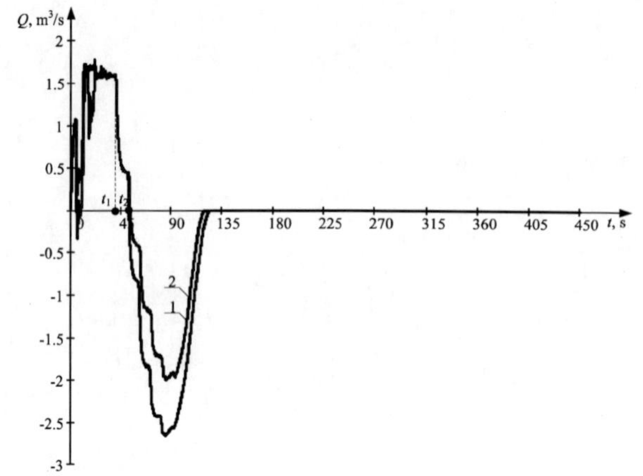

Figure I10. The curves representing discharge variation $Q(t)$ in the emergency mode of PC with ESDDL.

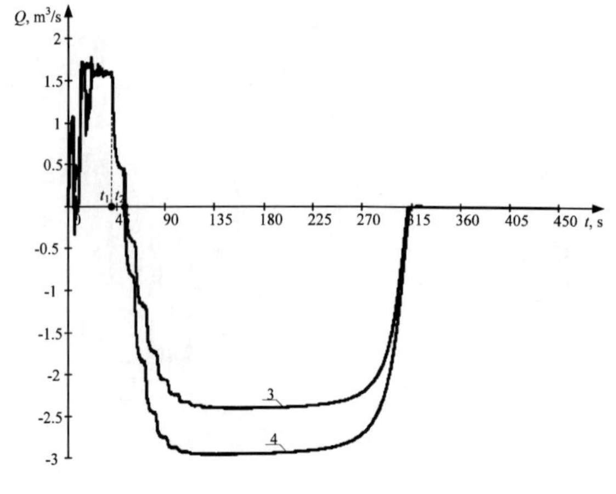

Figure I11. The curves representing discharge variation $Q(t)$ in the emergency mode of PC without ESDDL.

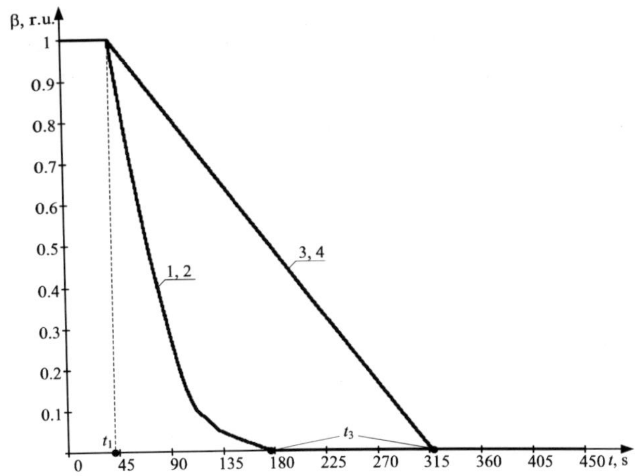

Figure I12. The curves representing the time variations of the relative degree β of opening of stopcock SA2 in PC operation mode.

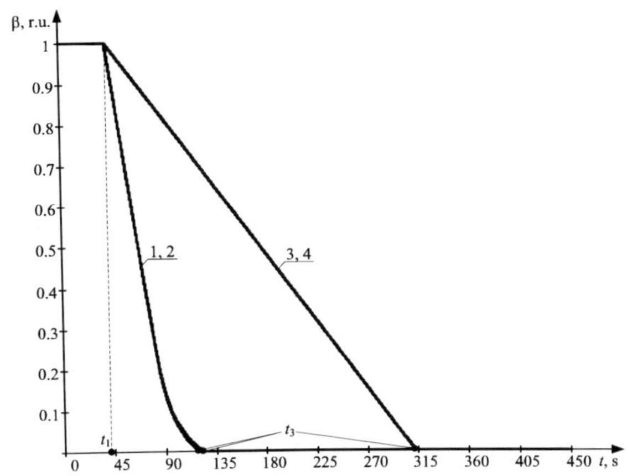

Figure I13. The curves representing the time variations of the relative degree β of opening of stopcock SA1 in PC emergency mode.

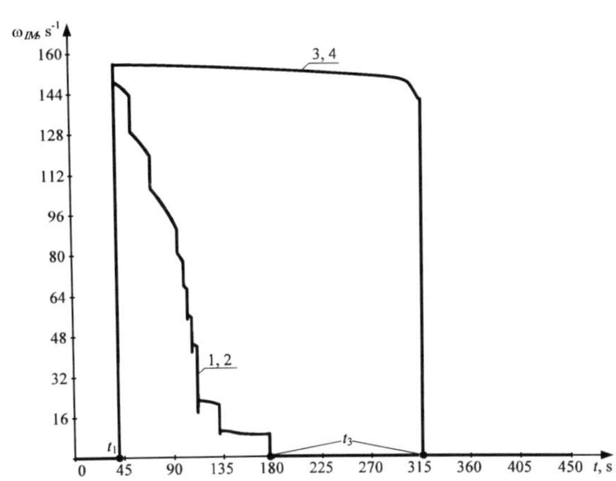

Figure I14. The curves representing the variations of the rotation frequency $\omega_{IM}(t)$ of stopcock SA2 ED in PC operation mode.

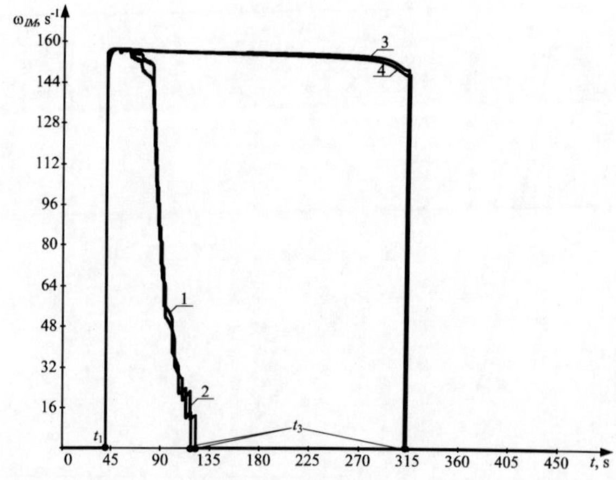

Figure I15. The curves representing the variations of the rotation frequency $\omega_{IM}(t)$ of stopcock SA1 ED in PC emergency mode.

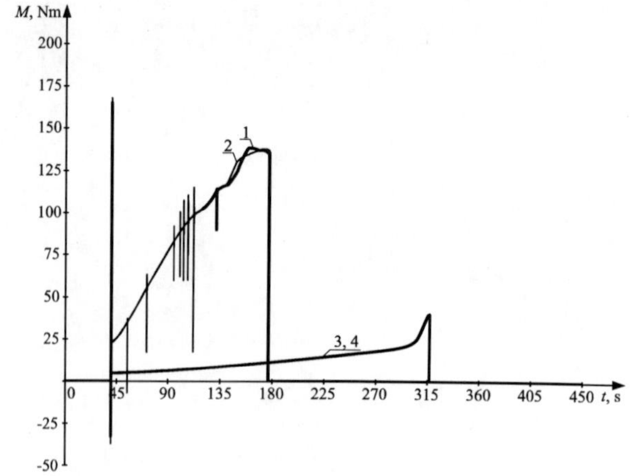

Figure I16. The curves representing the variations of moment $M(t)$ of stopcock SA2 IM in PC operation mode.

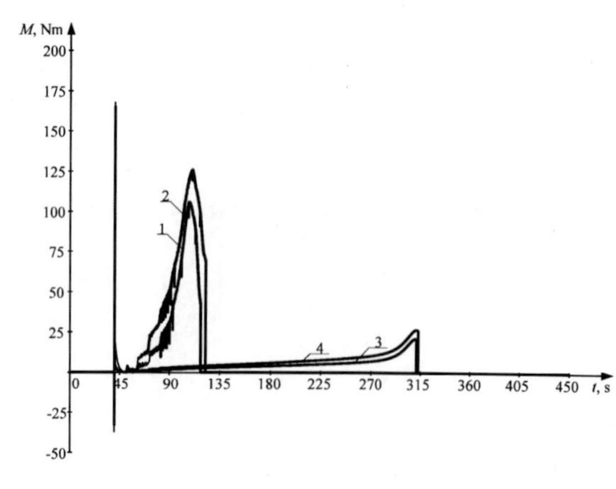

Figure I17. The curves representing the variations of moment $M(t)$ of stopcock SA1 IM in PC emergency mode.

The Script for the Determination of the Values of the Criterion of ESDDL Operation Quality

```
// The declaration of the used variables
global frequencies;
global grid;
global opening;
global n;
global m;
global k;
global trajectories;
global fs;
global cost;
global criterion_k;
global cost_moving;
global temp;
// The determination of the massive of the stopcock ED supply voltage frequencies f0
frequencies =
[50 45 40 35 30 25 20 15 10 5];
// The determination of the massive of the relative degrees β of stopcock opening
opening =
[1 0.85 0.75 0.5 0.25 0.225 0.2 0.175 0.15 0.125 0.1 0.075 0.05 0.025 0.02 0.01 0.0075
0];
grid = zeros(0,0);
k = 1;

// partitioning the coordinate grid βf0 0 into sections
for n = 1: length(frequencies)
   for m = 1: length(opening) – 1
      grid(m, n) = frequencies(n);
   end
end

// the formation of the stopcock shutting trajectories required for the method of
dynamic programming
trajectories = zeros(length(opening) – 1, (2 * length(frequencies) – 1) +
((length(opening) – 1) – 2));

for n = 1: length(frequencies)
   for m = 1: length(opening) – 1
```

```
        trajectories(m, k) = frequencies(n);
      end
      k = k + 1;
    end
    for n = 1: length(frequencies) – 1
      for m = 1: length(opening) – 1
        if m + (n – 1) < = length(frequencies)
          trajectories(m, k) = grid(m, m + (n – 1));
        else
          trajectories(m, k) = – 1;
        end
      end
      k = k + 1;
    end

    for n = 2:(length(opening) – 1) – 1
      for m = 1: n
        trajectories(m, k) = frequencies(1);
      end
      for m = 1: (length(opening) – 1) – n
        if m + 1 < = length(frequencies)
          trajectories(m + n, k) = grid(m + n, m + 1);
        else
          trajectories(m + n, k) = – 1;
        end
      end
      k = k + 1;
    end

    // the determination of the values of the quality criterion I in the crosses of coordinate grid
βf00

    cost = zeros(0, 0);
    for n = 1: length(trajectories)
      fs = trajectories(:, n);
      criterion_k = zeros(1, 0);
      open('muz.mdl');
      sim('muz.mdl');
      for m = 1: length(opening) – 1
        if m < = length(criterion_k)
          cost(m + 1, n) = criterion_k(m);
```

```
        else
            cost(m + 1, n) = – 1;
        end
    end
end
```

// the determination of the values of increments ΔI of the quality criterion at the transition from one cross of the coordinate grid $\beta f_0 0$ to another for the formed trajectories of the stopcock shutting

```
cost_moving = zeros(0, 0);
for n = 1: length(trajectories)
    for m = 1: length(opening)-1
        if (cost(m + 1, n) > 0)
            cost_moving(m, n) = cost(m + 1, n)-cost(m, n);
        else
            cost_moving(m, n) = – 1;
        end
    end
end

temp = zeros(2 * (length(opening) – 1) – 1 + 1, 2 * length(frequencies – 1);
k = 1;
for n = 1: length(frequencies)
    for m = 1: length(opening)-1
        temp(2 * m – 1, 2 * n – 1) = cost_moving(m, n);
    end
    k = k + 1;
end

for n = 1: length(frequencies) – 1
    m = 2;
    while m < = length(opening) – 1
        if cost_moving(m, k) > 0
            temp(2 * m – 1, 2 * m – 2 + 2 * (n – 1)) = cost_moving(m, k);
        end
        m = m + 1;
    end
    k = k + 1;
end
```

```
for n = 1: (length(opening) – 1) – 2
   m = 3 + (n – 1);
   while m <= length(opening) – 1
      if cost_moving(m,k) > 0
         temp(2 * m – 1, 2 * (m – 1) – 2 – 2 * (n – 1)) = cost_moving(m, k);
      end
      m = m + 1;
   end
   k = k + 1;
end

// the function of the record of the values of quality criterion I in the massive
function out = poisk(u)
global opening
global fs
global criterion_k
for i = 1: length(opening)-1
   if (u(1) < = opening(i)) && (u(1) > = opening(i + 1))
     out = fs(i);
   if (u(1) > = opening(i + 1) – 0.001) && (u(1) < = opening(i + 1) + 0.001)
     criterion_k(i) = u(2);
   end
   end
end;
```

Table I1. The technical parameters of uninterruptible power supplies

Input parameters									
Rated voltage				230 V alt. current 1 (phase) or 230/400 V alt. current (3 phases)					
Output parameters									
Rated voltage				230 V alt. current 1 (phase) or 230/400 V alt. current (3 phases)					
Rated frequency				50/60 Hz (automatic choice), synchronized with the network					
Output power	VA	1000	2000	3000	5000	6000	8000	10 000	15 000
	kW	700	1400	2100	3500	4200	6400	8000	12 000

Table I2. The time of the autonomous operation of uninterruptible power supplies at different load

Load		Time of autonomous operation							
kW	VA								
500	700	16 min	22 min	1 hour 16 min	1 hour 16 min	1 hour 20 min	2 hours 19 min	2 hours 19 min	4 hours 59 min
700	980	10 min	14 min	54 min	54 min	54 min	1 hour 43 min	1 hour 43 min	3 hours 43 min
1000	1400		8 min	36 min	36 min	39 min	1 hour 13 min	1 hour 13 min	2 hours 40 min
1400	1960		4 min	24 min	24 min	26 min	52 min	52 min	1 hour 55 min
2000	2800			15 min	15 min	17 min	35 min	35 min	1 hour 19 min
3400	4760				7 min	8 min	18 min	18 min	44 min
4000	5000					6 min	15 min	15 min	37 min
6000	7500						8 min	8 min	22 min
8000	10000							5 min	15 min
12000	15000								8 min
Full load	24 min (0.35 kW)	10 min (0.7 kW)	4 min (1.4 kW)	14 min (2.1 W)	6 min (3.5 kW)	5 min (4.2 kW)	7 min (6.4 kW)	5 min (0.8 W)	8 min (12 kW)

AUTHORS' CONTACT INFORMATION

Mykhaylo Zagirnyak
Rector, D. Sc. (Eng.), professor
Kremenchuk Mykhailo Ostrohradskyi National University
20, Pershotravneva ul, Kremenchuk, Ukraine
Email: mzagirn@kdu.edu.ua; mzagirn@gmail.com

Tetyana Korenkova
Associate Professor, Ph. D. (Eng.).
Kremenchuk Mykhailo Ostrohradskyi National University
20, Pershotravneva ul, Kremenchuk, Ukraine
Email: tanya74kor@gmail.com

Oleksandr Serdiuk
Associate Professor, Ph. D. (Eng.).
Kremenchuk Mykhailo Ostrohradskyi National University
20, Pershotravneva ul, Kremenchuk, Ukraine
Email: serduk@link.pl.ua

Oleksii Kravets
Senior Lecturer, Ph. D. (Eng.)
Kremenchuk Mykhailo Ostrohradskyi National University
20, Pershotravneva ul, Kremenchuk, Ukraine
Email: kdu7008@ukr.net

Viktoriya Kovalchuk
Assistant, (Eng.).
Kremenchuk Mykhailo Ostrohradskyi National University
20, Pershotravneva ul, Kremenchuk, Ukraine
Email: ikysia123@gmail.com

INDEX

A

B

C

D

E

F

H

Glycolysis: Regulation, Processes and Diseases

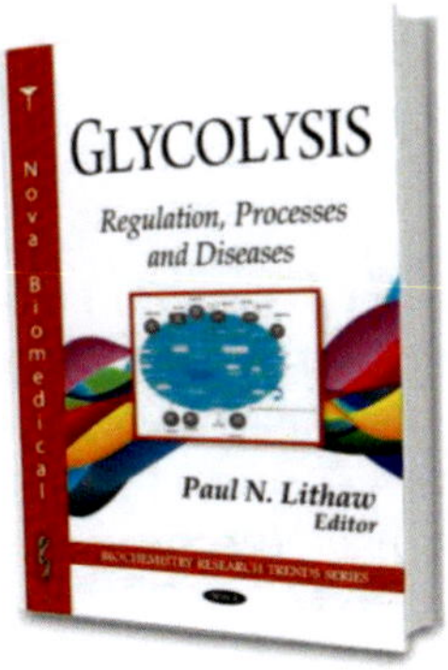

Editor: Paul N. Lithaw

Series: Biochemistry Research Trends

Book Description: Glycolysis literally means "splitting sugars." In glycolysis, glucose (a six carbon sugar) is split into two molecules of a three-carbon sugar.

Hardcover ISBN: 978-1-60741-103-1
Retail Price: $195

Pheromones: Theories, Types and Uses

Editor: Isabelle M. Gregory

Series: Biochemistry Research Trends

Book Description: This book examines trail pheromones and sex pheromones in termites and their potential use in pest management and insect repellents. Also discussed is splendipherin, the aquatic male sex pheromone of the tree frog as well as the study of "human pheromones."

Hardcover ISBN: 978-1-61668-283-5
Retail Price: $265